BARBARA MAE CHRISTY
121 W. ARRELLAGA APT. 3
SANTA BARBARA, CALIF. 93104

T.V. = Turner + Verhoogen

INTRODUCTION TO PETROLOGY

INTRODUCTION

Prentice-Hall, Inc., Englewood Cliffs, New Jersey

O PETROLOGY

BRIAN BAYLY

Rensselaer Polytechnic Institute

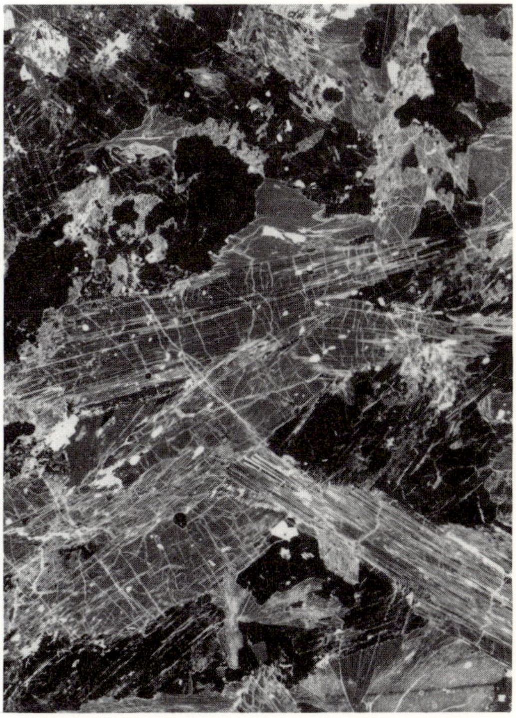

Cover photo: The Surtsey volcanic eruption, Iceland, which began November 14, 1963. In the effusive activity shown, towering vapor columns rise as lava enters the sea on the Southwest side of Surtsey. Courtesy of Sunfilm, Reykjavik, Iceland.

INTRODUCTION TO PETROLOGY
BRIAN BAYLY

Current printing (last digit):
10 9 8 7 6 5 4 3 2

Library of Congress catalog card number
68-11273

Printed in the United States of America

Prentice-Hall International, Inc. **London**
Prentice-Hall of Australia, Pty. Ltd. **Sydney**
Prentice-Hall of Canada, Ltd. **Toronto**
Prentice-Hall of India Private Ltd. **New Delhi**
Prentice-Hall of Japan, Inc. **Tokyo**

To H. R.

PREFACE

If we ask why science is regarded as a desirable ingredient of education, part of the answer must be that it helps us to direct our lives. If we have conceived an intention, scientific training should enable us to distinguish good ways from bad ways of putting it into effect. Whether the matter be banking or burglary, training should enable us to base judgment on observation, experiment and theory; we should be able to conceive a hypothesis, test it and predict from it. These activities are common to all purposive action and we study science partly in order to become more skillful at them. Because petrology is midway between the experimental and the observational sciences, it brings these activities clearly to light. The purpose of this book, then, is to make petrology effective as part of a general education.

To be consistent with this purpose, the text aims to be self-contained. Those familiar with elementary geology, physics, chemistry and mineralogy will be better placed to use the text than unprepared readers, but the subject of petrology is developed from the fundamentals; topics from classical chemistry and physics are introduced where needed; and at selected points, the trend of development is indicated beyond the introductory stage. As regards "difficulty," no attempt has been made to preserve a consistent level. Rather, the attempt is to trace a path from the most familiar of facts toward some less familiar theories; and not only to trace it, but to let the reader participate in the process; not to feed the reader knowledge but to let him discover how knowledge grows.

A second feature consistent with the purpose of the book is the attention paid to the procedures of science. The segregation of fact from hypothesis on one hand and methodology on the other is attempted; and the general topic of classification is repeatedly discussed, accurate ideas about classes being increasingly recognized as among the most fundamental for any logical thought about human affairs. Those who wish to be informed about current beliefs without concern for how the beliefs arose are asked to have patience with these discussions of procedure.

A feature that has been excluded for the sake of scientific procedure is the description of actual field examples. A text can present only a selection of the facts observable at any location and will almost automatically present that selection which fits a particular hypothesis; the presentation can only be authoritarian and make no provision for the reader's curiosity or initiative. It is hoped that omission of "textbook examples" will lead readers to consult their own experience or, at worst, their colleagues'. I would prefer that a reader learn from an imperfect example that he can investigate at first hand rather than accept what he is told about a more perfect example he cannot visit.

Those whose studies in petrology extend beyond this text will realize how heavily it relies on established works, both for information and for the assumption that topics treated scantily here can be followed up in greater detail elsewhere. Acknowledgment for isolated details is made in the text but I wish to record here my general debt to those who, by writing textbooks more erudite and compendious than this one, have codified past researches; by bringing the results into accessible form and illuminating them with their own insight, these authors have influenced to a great, though invisible, extent the development of the present work.

It has been a privilege to discuss parts of the text in detail with H. S. Yoder, Jr. and G. M. Friedman. Without these readers' patience and goodwill, the undertaking would have been immeasurably more difficult; as it is, the text inadequately reflects the thought and time generously given it by these critics. I have also had the benefit of comments and suggestions from G. V. Middleton, J. H. Maxwell, R. M. Garrels, R. J. Silva, J. McLelland and Y. W. Isachsen; these all advised me with the courtesy and acumen of scholars. I should like finally to acknowledge my debt to my many mentors and colleagues; I should be happy if I could transmit even a small fraction of the wisdom these men have shared with me.

BRIAN BAYLY

CONTENTS

PART ONE
INTRODUCTION

ONE
THE CRUST OF THE EARTH
3

Features of the Earth's Crust
3

Pressure Distribution in the Crust
7

Temperature Distribution in the Crust
8

The Occurrence of Magma in the Crust
8

TWO
PROCESSES IN THE CRUST
11

Cyclic Processes in the Crust
11

Noncyclic Processes in the Crust
13

THREE
ENERGY AS A BASIS FOR CRUSTAL EVENTS
19

Total Energy and Free Energy
21

FOUR
CLASSIFICATION, DEFINITION AND DESCRIPTION
23

Classification
24

Definition and Description
26

PART TWO
MAGMATIC ROCKS

FIVE
FORM AND TEXTURE OF COMMON MAGMATIC ROCKS
31

Volcanoes and Magmas
31

Rock Bodies Formed by Volcanic Activity
35

Textures of Volcanic Rocks
38

Recognition of Volcanic Rocks
42

Rock Bodies Formed by Subterranean Magmatic Activity
45

Textures and Recognition of Subterranean Magmatic Rocks
48

SIX
CLASSIFICATION AND NOMENCLATURE OF MAGMATIC ROCKS
52

Classification of Magmatic Rocks
52

Naming of Magmatic Rocks
53

SEVEN
MINERAL VARIETIES, TRENDS AND ASSOCIATIONS IN MAGMATIC ROCKS
56

Major Trends in the Composition of Magmatic Rocks
56

Departures from the Main Trend
60

Forms and Abundance of the Less Common Magmatic Rocks
65

EIGHT
PETROGENESIS 1: BINARY SYSTEMS
67

Binary Systems
68

Binary Diagrams and the Balance Principle
69

The Quartz-Albite System
71

The Quartz-Water System
79

The Albite-Anorthite System
82

NINE
PETROGENESIS 2: APPLICATION TO ROCKS
86

Polycomponent Systems and the Fractional Crystallization of Basalt
86

Fractional Melting of Sediments to Granite
94

Genesis of Less Abundant Rocks
98

TEN
THE VOLATILE FRACTION
104

ELEVEN
THE PHYSICS OF MAGMATIC PROCESSES
109

The Cooling of Intrusions
109

The Melting of Rocks
114

The Effect of Gravity on Magmatic Bodies
116

Physics of Volcanic Activity
119

TWELVE
METHODS OF STUDY OF MAGMATIC ROCKS
122

PART THREE
SEDIMENTARY ROCKS
Except Pyroclastic Sediments

THIRTEEN
OBSERVABLE FEATURES IN SEDIMENT FORMATION
129

Mechanisms for Sedimentary Rocks
130

Forms, Textures and Compositions
134

Diagenesis
157

FOURTEEN
RECOGNITION, CLASSIFICATION AND NOMENCLATURE OF SEDIMENTARY ROCKS
162

Recognition
162

Classification and Nomenclature
164

FIFTEEN
CHEMICAL ASPECTS OF SEDIMENTARY ROCK GENESIS
171

Stability of Bulk Phases
172

Surface Effects
186

Effects of Temperature Change
189

Rates and Mechanisms
190

SIXTEEN
PHYSICAL ASPECTS OF SEDIMENTARY ROCK GENESIS
193

Uptake, Transport and Deposition
194

Diagenesis
211

SEVENTEEN
METHODS OF STUDY OF SEDIMENTARY ROCKS
215

PART FOUR
METAMORPHIC ROCKS
And Subjects Between Metamorphism And Magmatism

EIGHTEEN
VARIETIES OF METAMORPHISM
223

NINETEEN
CLASSIFICATION AND NOMENCLATURE OF METAMORPHIC ROCKS
227

Textural Terms
228

Compositional Terms
229

Genetic Terms
230

TWENTY
PRODUCTS OF METAMORPHISM
232

Dynamic Metamorphism
232

Metamorphic Minerals and the Phase Rule
233

Dynamothermal Metamorphism
239

Thermal Metamorphism
247

Recognition of Metamorphic Rocks
250

TWENTY-ONE
REACTION CONDITIONS IN METAMORPHISM
256

Dry Reactions
256

Reactions Involving Fluids
258

Reactions Involving Solid Solutions
261

Recognition of Metamorphic Intensity
265

TWENTY-TWO
MOVEMENT OF MATERIAL IN METAMORPHISM
272

Magnitude of the Movements
272

Mechanisms of the Movements
274

Development of Preferred Orientation 279

Metamorphic and Metasomatic Differentiation 282

TWENTY-THREE
VEINS AND VEINING 286

Classification and Nomenclature of Veins 286

The Nature of the Vein-forming Fluid 290

Minerals in Veins 294

TWENTY-FOUR
METAMORPHISM AND MAGMATISM IN GEOSYNCLINES 297

TWENTY-FIVE
METHODS OF STUDY OF METAMORPHIC ROCKS 302

ABBREVIATIONS AND UNITS 307

PART FIVE APPENDICES

ONE
DATA TABLES
311

TWO
IGNEOUS ROCK NOMENCLATURE
316

THREE
THE USE OF NORMS
319

Normative Classification
326

FOUR
PERITECTIC RELATIONS AND BOWEN'S DISCONTINUOUS REACTION SERIES
327

FIVE
THE USE OF TRIANGULAR DIAGRAMS
332

Triangular Coordinates
332

Crystallization and Melting in Three-componet Systems
335

Three-component Diagrams in Metamorphism
338

SIX
REFERENCES AND FURTHER READING
345

References
345

Further Reading
349

INDEX
353

INTRODUCTION TO PETROLOGY

PART ONE

INTRODUCTION

There are two senses in which petrology is part of a larger whole: on the one hand, the matter studied is only a part of the physical universe; on the other hand, the study of rocks is only a part of man's attempt to find order in the data provided by his senses. The four ensuing chapters draw attention to salient features of these backgrounds, against which a study of petrology is conducted. They are intended to give reader and author some reference points in common, in preparation for the systematic development of Parts Two through Four.

CHAPTER ONE

THE CRUST OF THE EARTH

Petrology, while partly concerned with the moon, the earth's deep interior and other presently inaccessible places, is concerned mainly with the earth's outermost layers, within about 50 km of the surface. The deepest mines reach no more than 4 km below the surface, and although holes have been drilled to 10 km, there are very few so deep. Consequently, knowledge, even of the outermost 50 km, is largely indirect. However, it is possible to compose a diagram (see Fig. 1-1) with which a majority of scientists would agree. Let us consider the features one by one and the evidence on which they are based.

FEATURES OF THE EARTH'S CRUST

The Mohorovicic Discontinuity

When an earthquake occurs, the first manifestation to reach a distant observer is a compression wave—the same kind of wave that travels down a steel rod when one end is struck with a hammer. The times taken for this

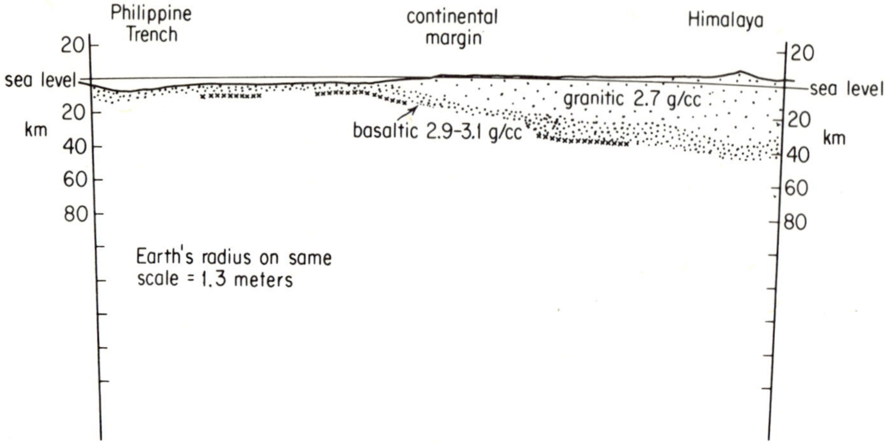

Fig. 1-1 Features of the crust. The diagram is approximately true to scale except for the distance from the Himalaya to the Philippine Trench; the diagram is drawn as if the Himalayan Range passed through the Philippine Islands. The terms *granitic* and *basaltic* are used only as approximate indicators of the crust's chemistry; in particular the basaltic layer may be quite different from basalt in mineralogy and texture. Unconsolidated surface sediments are not shown. The lines x x x indicate typical locations for the Mohorovicic discontinuity.

wave or shock to reach points at different distances from the center of disturbance tell us about velocities of wave propagation within the earth. In many parts of the earth, though not all, it appears that the velocity of propagation of the compression wave increases steadily with depth for several kilometers, increases rapidly over a short interval and then returns to steady increase (see Fig. 1-2). There is probably no actual discontinuity in the change of velocity with depth, but the term *Mohorovicic discontinuity* is nevertheless widely used to refer to the kink in the velocity profile. The change occurs at 30 km or more under the continents, but at only 5–10 km under the ocean floor. Under both oceans and continents,

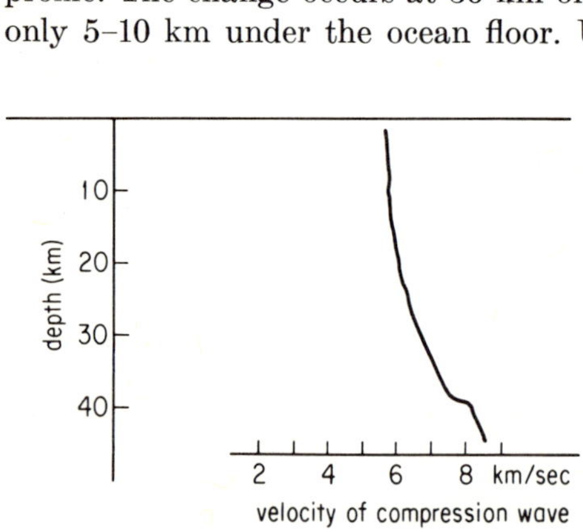

Fig. 1-2 Change of velocity with depth for seismic compression waves. The curve is a generalization based on measurements at many different places, and is intended to show merely the general trend of the change with depth.

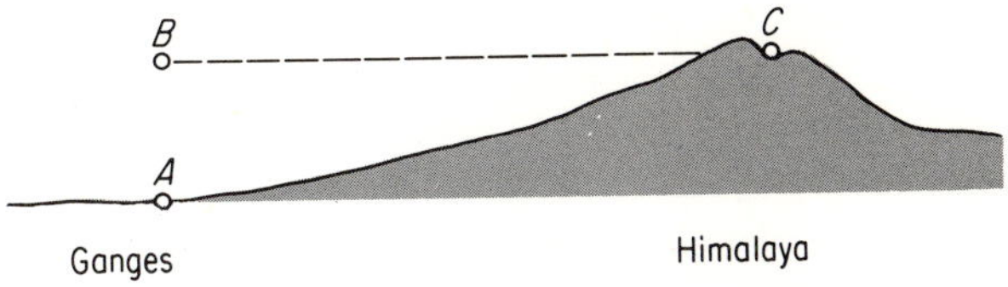

Fig. 1-3 Three points with different gravity fields (see text).

the velocity of propagation below the change is close to 8 km/sec. Although in many parts of the world such a kink is not well defined, the constancy of the value 8 km/sec wherever one is found has led to belief in a more-or-less worldwide layer of uniform composition below the discontinuity. We shall not speculate on the character of the layer at this point, but refer to it simply as the *mantle*, to distinguish it from the rocks above, which we shall call the *crust*.

The Density Distribution

The rocks on the surface of the continents are the only ones whose density is thoroughly known; the rocks a few kilometers down are rarely seen, and the rocks 1 km below the ocean floor are as unknown as the rocks on the moon, as far as direct sampling is concerned. The densities in Fig. 1-1 are therefore mainly speculations, based to a great extent on measurements of compression wave velocities, discussed in the previous paragraph, and of gravity.

If we measure the acceleration due to gravity on the plains of the Ganges (point A in Fig. 1-3), then by calculation we can obtain the acceleration to be expected at point B (somewhat less than at A, since B is further from the earth's center of gravity). An estimate can be made of the volume and weight of the rocks composing the Himalayas, and so the acceleration at C can be forecast; it should be greater than that at B by an amount due to the extra shaded mass. Measurement shows that accelerations due to gravity at points such as C all over the world are consistently less than forecast. The favored explanation of this is that every major mountain range is underlain by rocks of lower density than underlie the adjacent plains; or, more exactly, the plains are underlain by rocks of high density at comparatively shallow depth, while the high-density rocks under the mountain ranges lie rather deep. On the basis of measurements of this kind, the hypothesis can be formulated that the elevation of the crustal rocks is related to their density as it would be if they were floating on a liquid substratum. This is known as the theory of *isostasy* (see Fig. 1-4).

The theory of isostasy has important consequences in many branches of geology. Here it is applied as follows. Suppose we see one log floating almost completely submerged in water and another projecting above the surface perhaps one fifth of its diameter: because the former floats so low,

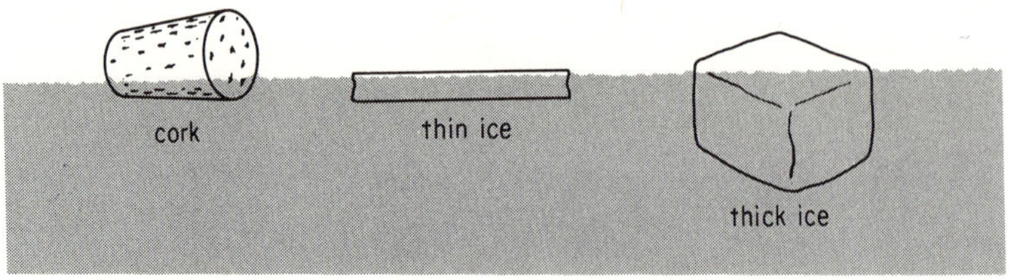

Fig. 1-4 The behavior of floating solids. The thickness of solid that projects above the surface of the liquid in which it floats is determined by the liquid's density, the solid's density and the solid's dimensions.

we can deduce that its density must be greater than that of the higher log by about one part in ten. Similarly, the floor of the Pacific Ocean "floats," over much of its area, 4 or 5 km lower than the land surface of the continents. Pursuing the analogy, we conclude that the rocks below the ocean floor are denser than the rocks of the continents. To decide how much denser, we must make a guess about the thickness of the lightweight continental rocks, and here the distance to the Mohorovicic discontinuity is a useful guide. By putting all available information together, geophysicists have concluded that the rock below the ocean floor probably has a density around 3.0 g/cc. An independent clue to the rock's density is provided by the velocity of seismic wave propagation, and evidence of this kind agrees with the value based on the isostatic argument.

The Composition Layering

The continents are largely covered by a thin veneer of sedimentary rocks, but enough drilling has been done to show that these are almost universally underlain by more compact crystalline rocks such as are exposed at the surface in Labrador, Sweden, Australia and elsewhere—rocks whose density of 2.7 is attributable to their being composed largely of quartz (2.65) and feldspars (2.5–2.7), with smaller amounts of the denser ferromagnesian minerals (see Table A1-2). Since granite is a rock of similar density and mineralogy, the term *granitic* is applicable to the crystalline continental subsurface; but it must be emphasized that most continental rocks are not similar to granite in texture or in mode of origin or in local composition—it is only when averaged that their composition is granitic.

The description of the deeper layers as *basaltic* is less easy to justify. Because their density is higher, they must contain less quartz and denser feldspar than the continental subsurface, and so may easily approach basalt in average composition [the main constituents of basalt are pyroxene, c. 3.3 g/cc, and plagioclase feldspar, c. 2.7 g/cc (see Chap. 7 for details)]. But again, it must be emphasized that there are no grounds for supposing that they have the precise texture, origin or composition of basalt.

The rocks of the Pacific islands are sometimes regarded as evidence that the oceanic crust is basaltic. The Hawaiian group, for instance, is almost pure basalt where seen and probably so throughout its submerged bulk—a button-shaped tumor rising 10 km above the general level of the ocean floor and measuring more than 150 km across its base. The islands, however, are not typical of the oceanic crust; they are excrescences upon it and may or may not resemble the submerged crust a thousand miles to east or west.

To sum up, when we say that the great oceans and the granitic continents probably both have basaltic rocks beneath them, we mean merely that the rocks are richer than granite in the denser minerals. Indirect measurements and theories set limits to their properties, but direct observation has supplied no details.

PRESSURE DISTRIBUTION IN THE CRUST

Having discussed the composition of the crust, let us turn to the pressure and temperature within it.

In parts of the crust where shearing forces are negligible, the pressure at a point within the rock will be equal to the weight of the column of rock above it; that is,

$$\text{pressure} = \rho g h$$

where ρ = density of overburden, g = acceleration due to gravity, h = height of column = thickness of overburden. Suppose $\rho = 3$ g/cc and $g = 1000$ cm/sec^2; then

$$\begin{aligned} &\text{pressure in dynes/cm}^2 &&= 3000\,h \text{ if } h \text{ is in centimeters} \\ \text{or } &\text{pressure in atmospheres} &&= 0.3\,h \text{ if } h \text{ is in meters} \\ \text{or } &\text{pressure in kilobars} &&= 0.3\,h \text{ if } h \text{ is kilometers} \end{aligned}$$

(1 bar = 0.99 atm = 1.02 kg/cm^2, so that within the precision of most geological calculations, we may take these units as interchangeable.)

To what extent are shearing forces negligible in comparison with hydrostatic pressures? Near the surface, where hydrostatic pressures are low, shearing forces are not negligible; on the contrary, they are of vital concern to builders of embankments and skyscrapers. But at depth they are *comparatively* less important, because there is an upper limit to the shearing stress a rock can withstand, whereas hydrostatic pressure can increase indefinitely. With these thoughts in mind, we shall frequently make the assumption that pressures in the crust are due merely to overburden, but the fact that the assumption is only approximate needs always to be kept in view.

TEMPERATURE DISTRIBUTION IN THE CRUST

Temperatures in the crust are estimated on the basis of two kinds of evidence. Firstly, they can be measured directly wherever man makes a hole. It has to be remembered that the rock temperature recorded at the bottom of a hole will not be the temperature of the rock before the hole was made; but the difference will often be small. Secondly, temperatures can be estimated by measuring the surface heat flow and the rocks' conductivity. Suppose the heat flow f is 1.5×10^{-6} cal/cm²-sec and the conductivity c is 5×10^{-3} cal/cm-sec-°C; then the temperature gradient at the point of observation must equal f/c, 3×10^{-4} °C per cm or 3°C per 100 m. On this basis, the temperature at 10 km depth would be 300°C higher than at the surface.

As a matter of fact, measurements in holes corroborate the value for the surface gradient: figures ranging from 0.6–15°C per 100 m have been recorded (Birch, 1942), but the most common value on the continents is close to 3°C per 100 m. However, the extrapolation to 10 km is subject to two important uncertainties: (1) the conductivity at that depth, and (2) the importance of heat sources within the crust itself. Radioactivity generates considerable heat within the top 10 km of the crust, which means that less heat rises across the 10 km level than reaches the surface. This would mean in turn that the temperature gradient at 10 km is less than that at the surface, if the conductivity were the same. In fact, the conductivity is probably not the same: increase in temperature lowers conductivity and increase of pressure probably raises it, within the range of circumstances likely to be encountered in the earth's crust, but the magnitudes of the effects are not known. Hence, we are uncertain of the temperature at 10 km and more uncertain at greater depths—temperature estimates have far less precision than those of pressure. It is certain, however, that the temperature will vary strongly from place to place. As with pressure, it must be remembered that what is postulated about temperature beneath one point on the surface is not necessarily applicable at other points.

THE OCCURRENCE OF MAGMA* IN THE CRUST

The relevant data are summarized in Fig. 1-5; in studying the figure, be mindful of the uncertainties mentioned in the two preceding sections. The intersection at Z indicates that *if* the rock at 20 km depth is granitic, *if* it is saturated with water and *if* the temperature rises along curve D, then the rock at this point will melt; but if with increasing depth, the change

**magma* = molten rock.

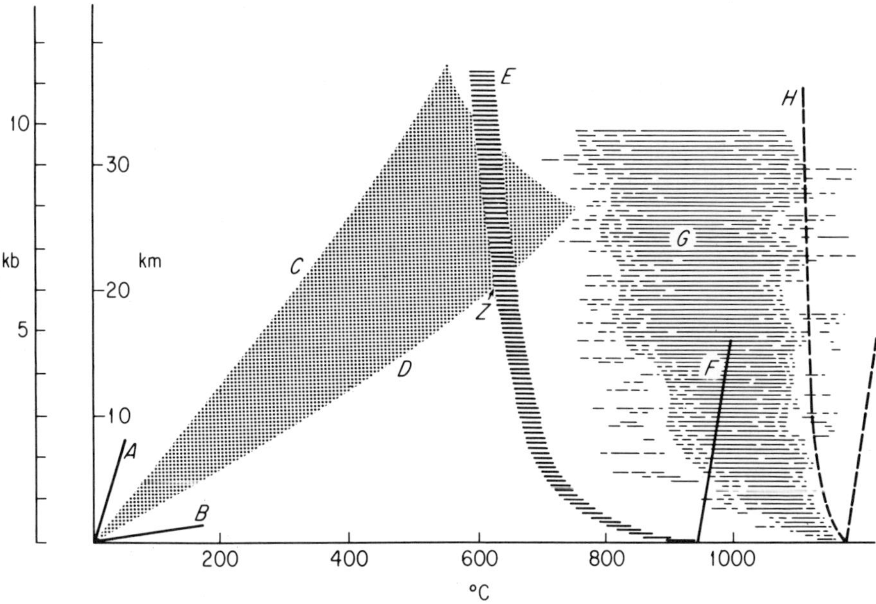

Fig. 1-5 Conditions for melting in the crust. (*A–B*) Range of measured geothermal gradients; (*C–D*) range of geothermal conditions believed common; (*E*) melting range of granite in equilibrium with water vapor.

For most rocks of granitic composition, the temperature range is quite small between the first signs of melting and melting being almost complete, as long as a certain amount of water is present (see Chaps. 8 and 9 for details). If less water than the required amount is present, melting will still begin along curve *E* but will not be complete until nearer curve *F*.

(*G*) Melting range of basalt. Unlike granite, basalt always requires a rise in temperature of 200°C or more between first signs of melting and completion. Melting begins close to range *E*; if water is sufficient, melting will be complete before curve *H* but if water is insufficient, melting may not be complete until nearer curve *J*.

(Data from Barth, 1962; Yoder and Tilley, 1962; Tuttle and Bowen, 1958.)

to more basaltic composition occurs before the melting temperature of granite is reached (as probably happens where conditions more closely follow curve *C*), melting will be deferred and may not become extensive until 50 km. (In fact, melting may not occur at all; the tendency for materials with high melting points to predominate at greater depth may be strong enough to defer melting for 2000 km or more; but such speculations are quite outside the scope of this chapter.) Here the purpose is to note:

1. That the probability of a granitic melt forming passes through a maximum at 20 to 30 km depth (at deeper levels, the crust's composition is unfavorable);

2. That a basaltic melt is unlikely to form in bulk until 40 km depth or more;

3. That within the crust at least, melting is more the exception than the rule.

The preceding remarks refer to continental regions. Under the oceans, where the crust is rarely over 10 km thick, melting within the crust is improbable. The abundant lava on Hawaii and similar volcanic islands is generated in the little-known regions below the Mohorovicic discontinuity.

CHAPTER TWO

PROCESSES IN THE CRUST

CYCLIC PROCESSES IN THE CRUST

The conclusion of the preceding chapter, that there are two distinct major sources of magma, is borne out by observation of rock bodies now exposed on the earth's surface. There are the basalt sheets whose origin as magma is a matter of historical observation, and there are granitic bodies whose relation to their surroundings make magmatic origin almost as certain: the surrounding rocks are split by fissures into which the granite has penetrated; fragments from the surroundings have been broken off and dislodged into the granitic fluid; and the heat of the invader is recorded in mineralogical changes at its margin.

It is equally certain that the materials introduced into the crust as magma are continually being eroded into sediments: the feldspars become clay, the quartz becomes sea sand, and the calcium, washed to the sea in solution, is "fixed" by organisms as carbonate and deposited after their death as beds of limestone. If we ask, however, what were the rocks whose melting at 20–30 km generates granite magma, the answer, provided by careful studies over many years and in many places, is "In large part,

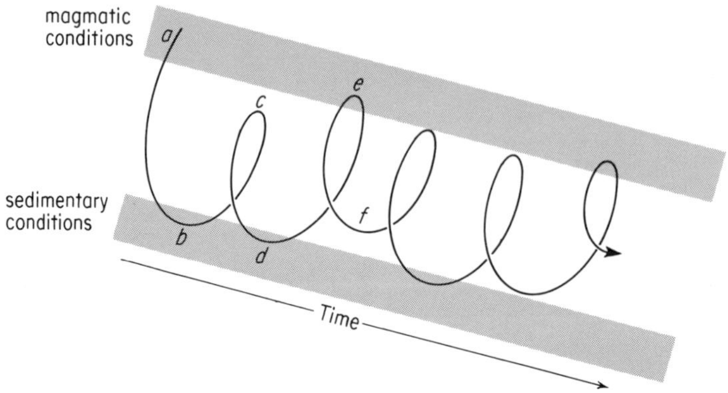

Fig. 2-1 Cycles of erosion, metamorphism and magmatism.

sediments." The history of the crust is, in fact, cyclic. Sediments are deposited, overlain, buried, perhaps even sucked down by convulsions in the mantle, and then remelted and, under favorable circumstances, forced or admitted back to upper crustal levels. The process involves a chemical separation or *fractionation:* the average sediment is not chemically equivalent to granite, so that the formation of a granitic magma usually leaves a less siliceous residue. The mechanism by which this residue makes its way into the basaltic family of magmatic rocks is not at all well understood, but the essential repetitiveness of crustal events has long been recognized—James Hutton wrote*, "In the economy of the world I can find no traces of a beginning, no prospect of an end."

Of course, there is a prospect of an end; every episode in the cycle transfers a quantity of heat from the mantle to the surface in a non-recoverable way; but as with a spring-driven clock, the machinery goes through innumerable cycles before the originally stored energy is all spent out.

Events in the crust are not limited, however, to this simple cycle. Figure 2-1 attempts to portray more realistically the variety of possibilities. The direct transition *a–b* represents total weathering from magmatic rock to sediment, and *d–e* represents fusion back to the magmatic state. But not every heating episode a sediment passes through is severe enough to melt it, and not every adjustment to lower temperature that a rock suffers disintegrates it completely to detritus. The term *metamorphism* covers the varied processes in the middle ground, by which the nature of a rock can be substantially changed while it retains its essentially solid crystalline state. The conversion of soft chalky limestone to crystalline marble is such a change and might occur along *b–c*, while the conversion of dunite (a rock

**Theory of the Earth* (Edinburgh, 1785).

consisting almost wholly of olivine crystals) to serpentine and talc goes forward with falling temperature, e.g., along *e–f*. Nor is temperature the only variable: changes in pressure can also provoke changes in the mineral composition of a rock.

There naturally arises a threefold classification of rocks into *igneous* (= magmatic), *metamorphic* and *sedimentary*, but it should be realized that not all rocks fall into one or the other class. If we tried to give a concise meaning to the terms we might say:

> Igneous rocks are those formed by solidification from a melt;
>
> Sedimentary rocks are those formed by the settling of particles out of water or air under gravity.*

On this basis, a rock can be both igneous and sedimentary—for example, a bed of ash thrown out by a volcano. There are also rocks that are neither igneous nor sedimentary, in the senses proposed—for example, veins of calcite; these do not form from a molten state but from aqueous solution, yet gravity plays no part in their formation.

Events in the crust, though confined to a basically cyclic routine, are thus quite varied. There are superimposed on them also the noncyclic changes, the "running-down" of the earth—changes that occur as the materials of the earth sort themselves out and descend from a high-energy arrangement in space to an arrangement with lower energy.

NONCYCLIC PROCESSES IN THE CRUST

The main noncyclic processes are of two kinds:

1. The escape of energy to the surface and thence out into space;
2. The escape of low-density material to the surface, with corresponding concentration of high-density materials in the center.

A link between these two exists in that process (2), the density rearrangement, generates some of the heat that escapes in process (1). It is possible, in fact, that density rearrangements are the main source of the earth's heat, in which case the two processes are merely separate aspects of the same degeneration. On account of this link, it is appropriate to describe density rearrangements first.

*This suggestion does not exactly match current usage; for example, it excludes coral reefs which are usually taken into the class of sediments. It is not the intention to recommend departure from current usage—the definition given here is for the present discussion only.

Natural Density Adjustments

When an egg slips from one's fingers and falls to the floor, potential energy is lost, in amount equal to (weight of egg × height of drop). What else happens? Kinetic energy is gained, converted to elastic strain energy at impact, and converted back to kinetic energy with some dissipation as sound. Eventually friction brings all particles to a standstill and the net effect is a slight rise in the temperature of the environment. There is also a net *upward* displacement of air equivalent to elevating 20 cc or so through 4 ft. The process is in fact a *density adjustment*, in which dense material sinks and light material of equal volume rises, with an overall loss of energy to the environment. Exactly the same kind of adjustment occurs continually within the earth.

According to current theories about planets, the earth at an earlier time was far less well organized in terms of density than it is now. A rough stratification was very soon established, however, with matter of density over 5 g/cc concentrating at the center, lighter solids above, and volatile materials mainly on the outside, where we now find the sea and air. We may ask: Is this fractionation now complete? In this respect, as in others, our knowledge of the mantle is very slight, but the crust is well enough known for examples of density inversion to be demonstrable. Drilling has penetrated salt beds (2.2 g/cc) below limestone (2.6 g/cc) and granite (2.6 g/cc) below metamorphosed rocks (2.7–2.9 g/cc). It is also known that water and oil exist at great depths, whereas in a well-organized earth the water would be above the rocks, but below the oil. Of course, the burial of salt and most hydrocarbons is due to sedimentation and is not a feature inherited from the primitive earth, so that it is therefore pertinent to ask whether all observed inversions are of this kind; or can we show that some of the observed inversions are primeval?

The fact is that we cannot. On the other hand, if we believe that anomalies once existed, we cannot believe that they have now wholly disappeared, because as perfect stratification is approached the driving force toward further improvement becomes more feeble. It is more sensible to ask, therefore, *at what rate* light materials are now escaping from the mantle into the crust. The question is critical, as will be seen in later chapters, regarding water and potassium ions, and is also relevant to the great basaltic outpourings of India, South Africa, Iceland and Hawaii: to what extent do these upwellings constitute material newly risen from the mantle, by differentiation from materials even more dense? At present we have very little idea, though there is some hope that the determination of isotope ratios will eventually give a basis for an answer. The most concrete fact bearing on the question is that methane rises in the Kimberley diamond pipes. Diamonds form only under great pressure; to be sure, one can see that great pressure has been at work, because the diamond-bearing rock consists largely of shattered fragments—stress in the rock has clearly risen to rupturing magnitude. But the formation of diamond requires more than the stress a rock can support by itself (its *breaking stress*)—a large hydrostatic pressure is needed as well. Consequently, the diamond-bearing rocks are presumed to have risen from below the crust, and the environment is such that they are unlikely to have picked up methane on the way. Hence methane is now separating from the mantle; hence other low-density constituents may be doing so too.

Less direct but more quantitative evidence has been adduced, for example by Runcorn (1964), to show that the separation of silicates from denser sulfides, oxides and native metals is not yet quite complete. Whether the process is complete or not, an estimate can be made of the heat generated *so far* by the sinking of the denser components to the earth's core. It is necessary to estimate the present distribution of weight in the earth; then, if it is assumed that the present materials were at an earlier time distributed with uniform density from surface to center, the heat derivable turns out to be sufficient to raise the temperature of the whole earth about 3000°C. Density adjustments thus pervade geology, and are important from the largest to the smallest scales.

Escape of Energy

Energy escapes primarily as heat. When an earthquake occurs, some energy travels to the surface in seismic waves and is expended in destroying buildings and so on; but these events, though occasionally spectacular, are totally insignificant in the earth's energy budget. Heat transfer occurs in the earth, as in the laboratory, by convection, conduction and radiation. At the surface, conduction is the major process—a quantitative estimate was made in Chap. 1. Within the earth, radiation is probably very vigorous; human experience is mainly of radiation through gases, but radiation of heat from one part of a solid body through the solid to another part is also possible, indeed inevitable, at temperatures over 1000°C. Between the deep, hot core and the cool, stiff crust is the field of convection—the transfer of heat that is involved in the bodily movement of hot materials. Suppose a body of lava is erupted from a source-area at 1000°C and cools on the surface to approximately 0°C. The specific heat of basalt is about 0.3 cal/gm or 1 cal/cc, so that every cubic centimeter of basalt delivers 1000 cal to the atmosphere. If the lava flow measures 10 km by 1 km by 10 m, the heat transfer is 10^{17} cal. Let us compare, for example, the convective output of Mount Etna with the conductive output for the rest of Sicily: these would be equal if Mount Etna delivered one such flow every ten years, assuming that conductive flow in Sicily is near the continental average, about 1.5×10^{-6} cal/cm^2-sec.

On this basis it appears that there are few places in the world where transfer by convection has approached that by conduction during historic time, but four comments on this generalization are appropriate:

1. Volcanic events on a far greater scale than the modest flow mentioned have occurred, but there have been too few during historic time for an "average frequency" to be estimated;

2. If magma rises to a shallow depth and there solidifies, heat escapes at the surface by conduction that traversed most of the crust by convection;

3. The world average heat loss by conduction may have been greater or less at other times than it is now;

4. In the distant past, total heat flow must have been greater than it is now on account of radioactivity, unless the extra heat generated was taken up in heating the earth. Barth (1962) calculates that in the early Precambrian (3 billion years ago) there might have been five times as much radioactive potassium in the crust as there is today, enough to double the rate of generation of radioactive heat. As a result, temperature gradients were probably higher, magmatic activity more vigorous, and heat-driven density adjustments, to which we now turn, were probably more frequent.

Heat-driven Density Adjustments

It was stated earlier that density adjustments generate heat, and it is equally true that heat can generate density adjustments. This is because most melts have lower density than the corresponding solid rock (see Table A1-5), so that if sufficient heat is supplied to the lower part of a rock mass to cause melting, a density inversion is created. Figure 2-2 shows the sequel. The final state is equivalent to the first in density distribution; all that the cycle has achieved in terms of energy is the transfer of a quantity of heat from a lower to a higher crustal level. Such an episode may be only part of one of the cycles in Fig. 2-1; in short, the rising of magma and burial of sediments is cyclic in terms of material but noncyclic in terms of heat transfer.

Fixation of Calcium as Carbonate

Although calcium, with the other elements, rises as magma and sinks as sediments, a noncyclic change in its behavior occurred with the evolution of skeletal organisms; first algae, and later the infinite variety of shelly animals, enormously increased the amount of calcium existing near the surface of the crust as carbonate. We should therefore distinguish an inor-

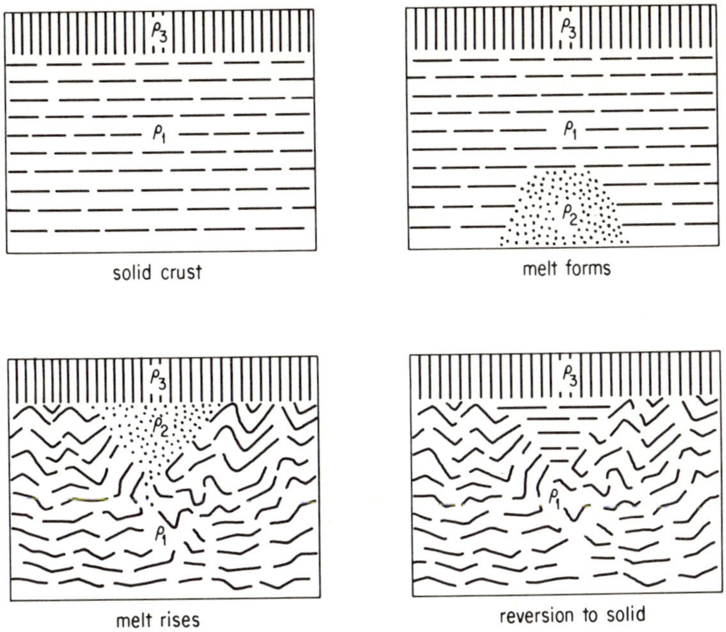

Fig. 2-2 Escape of heat by a convective cycle. In order of decreasing density, the materials involved are: solid crust, density ϱ_1; a melt formed from the crust, density ϱ_2; surface sediments, density ϱ_3.

ganic calcium cycle and an organic calcium cycle, and admit that there is a noncyclic tendency for more and more calcium to become involved in the latter. This drift only began in the second half of the earth's history and swelled enormously when floating foraminifera became abundant in Cretaceous time. The foraminifera acquire some of their calcium from the continents (via rivers and seawater) and deposit it in the deep oceans, so that the bulk composition of the continents is at present being changed. What effect this will have on the pattern of crustal events has yet to be seen.

SUMMARY

1. Except for local irregularities, the crust is graded in composition, the rocks becoming less siliceous and richer in high-density minerals with depth.

2. The arrangement is disturbed by cyclic activity, with the rising of magma and the burial of sediments.

3. The rise of magma may often be due to its density being lower than that of its crystalline environment.

4. The net effect of the cycle is mainly upward transfer of heat.

5. There may be a small imbalance in the cycle so that besides heat transfer there is also some preferential transfer of low-density constituents to the earth's surface layers.

CHAPTER THREE

ENERGY AS A BASIS FOR CRUSTAL EVENTS

Processes tend to occur in the natural world that result in a material's energy content being reduced. An example, the falling of an egg, has already been mentioned: an egg on the floor has less potential energy than an egg in one's hand, and in the absence of any restraints the tendency is for the latter to become the former. It is equally true that a body in a high-pressure environment has greater energy than the same body in a low-pressure environment and that in the absence of restraints bodies will run from high-pressure sites to low-pressure sites: when a balloon bursts, for example, the air that was contained does this. The two tendencies sometimes oppose each other, as they do in the atmosphere: the action of gravity is to pull molecules from the stratosphere down into the lower regions, but a molecule that follows this course gains as much energy through moving to a region of higher pressure as it loses through "falling." If many molecules sank downward, the space near the earth would be overcrowded and the pressure would be higher than it is now—in this situation, a molecule could *lose* energy by migrating upward again, and many molecules would do so. The change of pressure with height in the atmosphere is in fact such that the total energy due to both pressure and gravity is now as small as it can be: any deviation from the present state would involve an increase

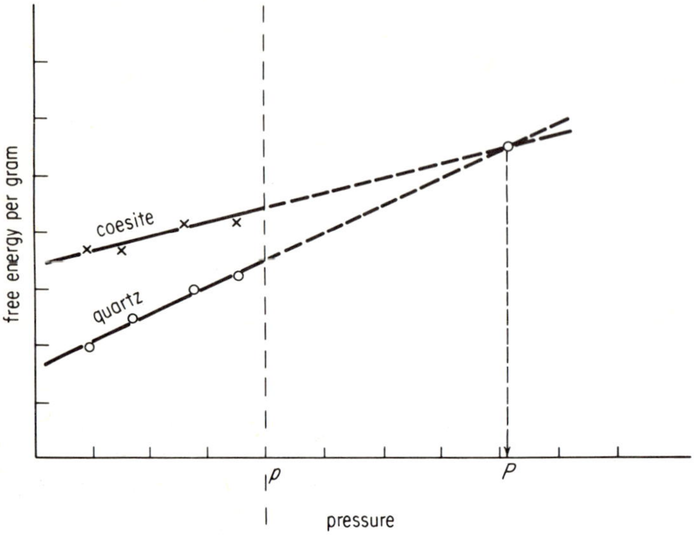

Fig. 3-1 Estimation of transition pressure. Suppose experiments could not be conducted above pressure *p*: a series of determinations at lower pressures would permit *P* to be found by extrapolation. (In fact, *P* is accessible to experiment and this procedure is not necessary.)

in the total energy and so the atmosphere tends to stay as it is. The tendency of rocks to arrange themselves with the heavier at the bottom has already been described and is another example of the tendency to take up the minimum-energy arrangement.

The same kind of consideration governs not only change of a material's position in space, but also change of its state. Solid SiO_2 has many forms, of which quartz and coesite are two that differ noticeably in density; the ions are packed together in the coesite structure more closely than in the quartz structure. Clearly, increasing the pressure will tend to make a quartz crystal convert to coesite, but to specify the exact amount of pressure required we need to remember that quartz will turn to coesite as soon as the energy per gram of the coesite form is less than the energy of the same gram in the form of quartz. A petrological prediction could be made on this basis: Fig. 3-1 shows the results that might be obtained if the energy per gram of coesite and quartz could be measured in a laboratory at several different pressures. Even if the transition pressure were too high to be reached in a laboratory, an idea of its value could be gained by extrapolating the two lines until the line for coesite came *below* the line for quartz; at this point coesite would be the stable form. (As a matter of fact, the transition pressure is within the range of experiment, and even if it were not, there are better ways to make an estimate than the one described; but the present concern is with principles, not techniques.)

TOTAL ENERGY AND FREE ENERGY

To make what has been said accurate, it is necessary to realize that the energy of a crystal can be divided into two portions, (1) the energy the ions would have if all were stationary and arranged without any kinks or displacements and (2) the extra energy the ions have because they are all vibrating, and some at least are displaced from their ideal sites. The total energy is the sum of (1) and (2), whereas the quantity that determines transition-points etc. is (1) alone—it is mainly electrostatic and is referred to as the *free energy*, F.

The distinction between free energy and total energy explains a paradox: at $+1°C$, a piece of ice turns to water spontaneously, yet absorbs heat in doing so. If spontaneous changes always reduced the energy-content of a material, we might expect ice to *give out* heat on melting at $+1°C$, as an egg does on falling. The difference comes from the fact that in water the ions vibrate much more vigorously than in ice: the change at $+1°C$ from ice-structure to water-structure liberates heat, but this heat, and more, is taken up by the ions when they begin to vibrate more vigorously with their new-found freedom. So, in general, spontaneous reactions can be exothermic or endothermic.

The other part of the total energy, item (2) above, may be referred to as *energy of disorder;* the term *bound energy* is sometimes used to emphasize the distinction from free energy, but has no other justification. It is given the symbol TS because S, the *entropy* of a material, is a measure of the scope it has for changing its internal arrangement and T, the temperature, controls the vigor with which it exploits this freedom.

Total energy can also be partitioned into two terms, U and PV (see Fig. 3-2). Here P is the pressure on the crystal in question and V is its volume, so that PV measures the work that would be done on the crystal if it escaped, without change of size, from its environment at pressure P into a vacuum. Because the PV term is energy that the crystal has solely by virtue of the pressure around it, it may be called *external energy* and the remainder, U, *internal energy*. Either pair adds up to the total energy of the crystal, so that

$$F + TS = U + PV$$

By rewriting this in the form $F = U + PV - TS$, we can estimate the effect on F of changes in temperature as well as changes in pressure—an

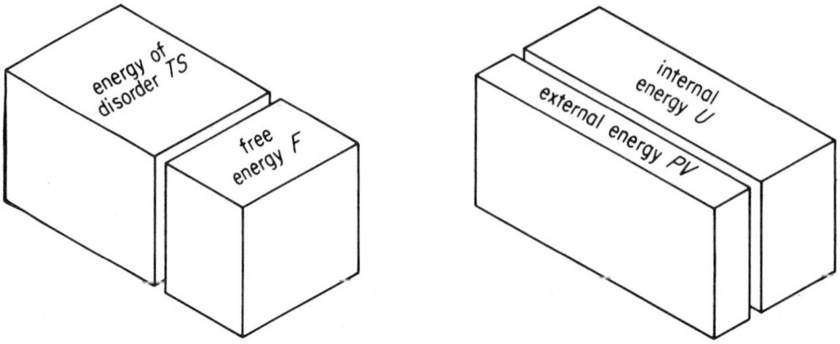

Fig. 3-2 Two ways of partitioning the total energy of a crystal.

essential prerequisite for predicting transition points between minerals in the crust.

Aside from this somewhat limited objective, almost every process in geology can be regarded as a special case of the general rule: *A system tends to change in ways that result in a lowering of the total free energy of the system.** All such considerations come under the heading of thermodynamics, whose principles must be thoroughly understood by serious students of petrology. At this point, it is sufficient to repeat that no process occurs without some energy transfer and that in subsequent chapters, we shall recurrently attempt to analyze the energy budget of the processes described.

*This rule is consistent with the definition given for free energy only when pressure is independent of the state of the system. In geology, this condition is never fulfilled exactly, but it is usually the most realistic of simple approximations.

CHAPTER FOUR

CLASSIFICATION, DEFINITION AND DESCRIPTION

In the ensuing chapters, several classifications will be used and many definitions offered. The reader is therefore entitled to some indication of the author's point of view in these matters. However, readers already conscious of the virtues and pitfalls in defining classes may pass directly to Chap. 5.

In the world of minerals, nature has provided divisions; a mineral may be an amphibole or a pyroxene* but it cannot be intermediate between the two. Between one rock and another, however, natural divisions are very few; for the most part, if one wishes to discuss intelligently a particular class of rocks, he has to define the class himself and also decide which rocks fall into it. Such divisions involve the *act of classifying*, which is not only a necessary part of petrology but one of man's most common mental acts.

Besides being common, the act of classifying is important. When we say "Jack Jones is trustworthy" or "Jones stock is a good investment," we make a classification act; and in general, the course of the world is affected more by such statements than by ones like, "473 g of water make

*All mineral names used are explained in Table A1-3.

a pint." Thus, not only does classifying things help us to master petrology, but also petrology helps us to master classifying; and such mastery is of sufficient value to call for a general discussion.

CLASSIFICATION

Given fifteen snails to be classified, one may begin in either of two ways: (1) one may decide upon a criterion such as weight and, further, decide that all snails below 5 g will form one class, and all those above another class; or, (2) one may pick two snails that are conspicuously different and decide to group all remaining snails with one or the other according to which they most resemble. If the fifteen actually contain just two species of snail, the second procedure will probably work well and pose no problems; but if the fifteen are all one species, or of several species, it is likely that a snail will be found which resembles one prototype in some properties and the other prototype in other properties. The operator is then obliged to decide which properties to use as diagnostic properties or criteria. At this point, procedure (2) becomes similar to procedure (1): both involve the establishment of a classification rule or rules, such as:

> "If a snail weighs over 5 g, put it in class one,"

or

> "If a snail's height is greater than its length, put it in class two,"

and so on. In devising such a set of rules, two precepts are important:

1. Every item should fall into some class;
2. No item should be eligible for two classes at once.

Either of the examples above serves as a suitable rule on its own, but to try to operate both rules at once leads to difficulties because the two rules in combination disobey both precepts—a long, low snail of under 5 g finds no home, while a tall, heavy snail could go in either class.

One may object at this point that no one would attempt classification with such unsuitable rules but the fact is that the division of rocks into *igneous* and *sedimentary* is defective in exactly the same way, as was mentioned in Chap. 2. To avoid such difficulties we must realize that a single criterion establishes *two* classes (the class of objects that satisfy the criterion and the class of objects that do not). Thus when we suggested (p. 13) that

> "Igneous rocks are those formed by solidification from a melt"

we established two classes of rock, igneous and nonigneous. When we went further and suggested

> "Sedimentary rocks are those formed by the settling of particles out of water or air under gravity"*

we divided both classes into two again:

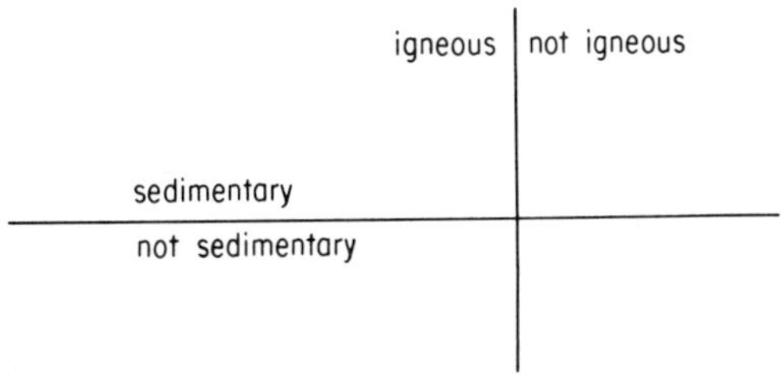

We should not therefore be surprised to find examples of four kinds of rock (excluding metamorphic effects), as indeed we did.

It is not the purpose of this section to advocate departing from the time-honored threefold division, but rather to show that attention to the classification process improves the operator's facility and understanding. A geologist who calls a coral reef "sedimentary" is extending the normal idea of a sediment for the sake of making his classification comprehensive and it is in his own interest to be aware of what he is doing.

The reason for four classes arising when only two were intended is that the criteria used were so different: one class was established by *mechanism of assembly* and the other by *source* of the constituents. The use of criteria of different kinds often leads to more classes than are intended, and similar difficulties. For another example, consider the terms *limestone* and *sandstone.* The former is defined by its composition—mainly $CaCO_3$ or $CaMg(CO_3)_2$—and the latter by its texture (most grains between $\frac{1}{16}$ mm and 2 mm diameter). Both are very useful terms, but they are not two equivalent classes in a classification of sediments. If we wish to classify sediments we must either:

Use texture only (conglomerate, sandstone, siltstone etc.)

or

Use composition only (limestone, aluminous sediment, siliceous sediment etc.)

or

Use both throughout (quartz sand, quartz silt, lime sand, lime silt etc.)

or

Use both, but sequentially. For example:
1. Use composition: *limestone* OR *not-limestone;*
2. If not limestone, use texture: *conglomerate* OR *sandstone* OR *siltstone.*

*See footnote on p. 13.

The last system can be set out more concisely thus:

limestone	conglomerate
	sandstone
	siltstone

It is probably the closest of the four to common usage, but implies an amended definition of a sandstone which is now "a rock *not mainly calcite or dolomite* whose grains are between $\frac{1}{16}$ mm and 2 mm diameter."

Again, it is not the purpose to recommend one system or another, but only to make clear how classifications can be kept trouble-free. The mixing of genetic criteria with observational criteria is a particularly common source of confusion; examples will recur throughout the following pages.

DEFINITION AND DESCRIPTION

Writing a definition is the same as establishing a class. For example, the same sentence has been used to establish the class of igneous rocks and to define the meaning of "igneous rock." The definition of an x— should thus provide the criteria for reaching a decision about any object in view, "Is this an x— or not?"—it must provide the *minimum* requirements for the object to be termed an x— .

The purpose of a description is quite different: it is to put a clear picture of the object or class of objects described into the reader's mind. In description, the more that can be said the better. A particularly important difference is the following: a definition should specify only properties that every member of the class defined has; but a description can state what properties most have, a few have, and so on. Words like "most," "a few" and "usually" have no place in a definition.

Since there is no point in giving a description without saying to what it refers, every description must be preceded by a definition, though sometimes the definition does not look like one. In the following examples the definition is in italic type.

> *A limestone is a rock whose main constituent is aragonite, calcite or dolomite.* Most limestones have densities between 2.4 and 2.8. They may be white, grey, black, yellow or other colors. Some limestones are

formed by inorganic chemical reaction, but most form from the skeletons of organisms . . . .

The rock at Regan's Quarry is a limestone of exceptionally low density and reddish color . . . .

Specimen 268 (location: Regan's Quarry). This specimen is a limestone with most grains under 1 mm . . . .

PART TWO

MAGMATIC ROCKS

The basic program of this part is to set up a problem and attempt to solve it. The problem is displayed in Chaps. 5 through 7, which summarize what is known about magmatic rocks from observation; the problem is to adapt theories from the physics and chemistry of artificial materials to furnish plausible explanations of what is seen. Chapters 8 through 11 are concerned with such theories. Chapter 12 is somewhat aside from the main stream, and seeks to answer the question "How is the knowledge discussed in the foregoing chapters actually gained?"

CHAPTER FIVE

FORM AND TEXTURE OF COMMON MAGMATIC ROCKS

If one were faced with the task of convincing a sceptic that some solid rocks form from molten rocks, one could well begin by taking him to an active volcano. At such a site, lava can be seen to solidify; the solid product can be seen to resemble older, coarser rocks; and by a series of comparisons, rocks far from known volcanoes can be linked to rocks whose origin as magma has been observed. Because of their position as the ultimate reference point in study of magmas, volcanoes are our first topic, but the discussion is not limited to description: the purpose is to discover what we can about the properties and powers of magmas. These discoveries are the basis for descriptions, toward the end of the chapter, of various non-volcanic rocks as magmatic.

VOLCANOES AND MAGMAS

Figure 5-1 shows the dimensions and profiles of some typical volcanoes. It is noticeable that steep-sided volcanoes are all small, though not all small volcanoes are steep-sided. The inclination of the flanks is

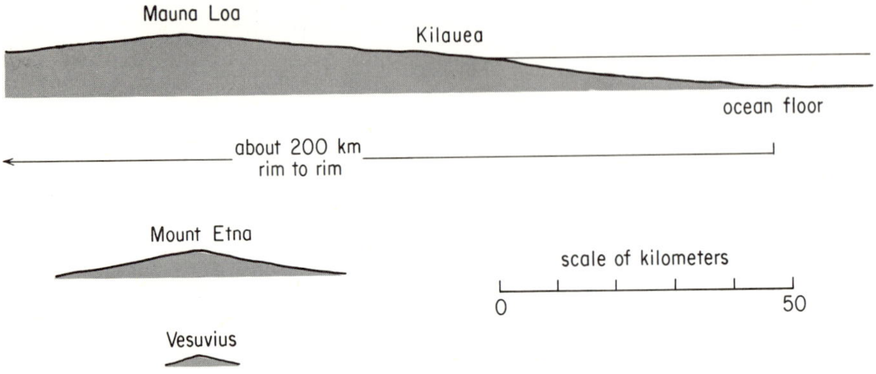

Fig. 5-1 Size of volcanoes. (Modified from Menard, 1964.)

determined by the nature of the material that pours out. The slope at Hawaii is about 5° on the average and is formed by basalt; even at this gentle inclination, the lava's fluidity permits it to flow several kilometers per hour. By contrast, the lava that escaped from Mont Pelée in 1903 was so stiff that it did not flow at all under its own weight (see Fig. 5-3); yet it had the structure of a viscous fluid and was squeezed vertically out of the vent as grease can be squeezed from a tube. Measurements made in the laboratory suggest that the difference in fluidity may have been as much as a factor of 10^8.

It does not necessarily follow that volcanoes with intermediate slopes—the volcanoes one normally thinks of such as Vesuvius, Fujiyama, Mount Etna, Lassen Peak—are made of lava of intermediate fluidity, because the angle of rest of dry volcanic fragments is also in the intermediate range. In fact, the cone shape of the volcanoes mentioned is primarily due to their being built of fragments: their slopes are similar to those of slag heaps or the piles of debris that accumulate beneath crumbling cliffs. Nevertheless, the difference in fluidity between the lavas of Mont Pelée and Hawaii has great petrological significance; the two fluidities represent opposite ends of a series, all intermediate members of which are theoretically possible.

The difference in fluidity is due partly to a difference in temperature. Although it is difficult to establish the temperature *within* a moving lava, it seems from surface measurements and laboratory tests that Hawaiian lavas are at about 1000°C when erupted, while the "spine" of Mont Pelée was 600°C or less. But there is in addition a chemical difference: other things being equal, the richer a lava is in silicon, the stiffer it is. As it happens, the custom in quoting chemical analyses of rocks is to state not the weight of silicon itself in the rock, but the weight of silicon dioxide there would be if all the silicon were in that form. Thus, records show that Hawaiian lava is about 50 per cent SiO_2 while Peléean lava is about 70 per cent SiO_2. Why does this difference control the fluidity? The other components of rocks are Al, Ca, Fe, Mg etc., and oxygen, and when the rock is thoroughly molten these are present mainly as the ions Al^{+3}, Ca^{+2} etc., and O^{-2}. As the rock approaches solidification bonds begin to form between these ions giving complex units such as $(AlO_2)^{-1}$. On account of silicon's electron configuration, however, it forms bonds more effectively than any other element: even in the liquid (magmatic) state, the lava contains quite complicated network structures (see Fig. 5-2A). The process is akin to polymerization in organic chemistry, and both result in the conversion of mobile fluids to stiffer ones. If we see a very viscous lava, we are therefore entitled to say, "There is a silica-rich lava."

TERMINOLOGY

SILICA: Silicon dioxide.

ACID, BASIC: Because enstatite, $MgSiO_3$, used to be regarded as a salt formed from an *acid*, H_2SiO_3, and a *base*, $Mg(OH)_2$, in the same way that NaCl is formed from HCl and NaOH, rocks rich in silica are called *acid* and rocks rich in Mg, Ca and Fe are called *basic.* These terms are more fully discussed in Chap. 6.

GLASS: Magma that has cooled to surface temperature without crystallizing is called glass.

A magma's ability to dissolve gases is also linked to its acidity. The idea of water dissolving in molten rock may be unfamiliar: if one dropped water on flowing lava, none would dissolve—it would spit straight back as steam. However, the dissolving of CO_2 in water by application of pressure is a familiar process for aerating drinks; and similarly, if molten rock and steam are confined together in a pressure chamber, some of the steam

Fig. 5-2 (A) Silicon-oxygen linkages in magma; (B) spicules suggesting an exploded foam.

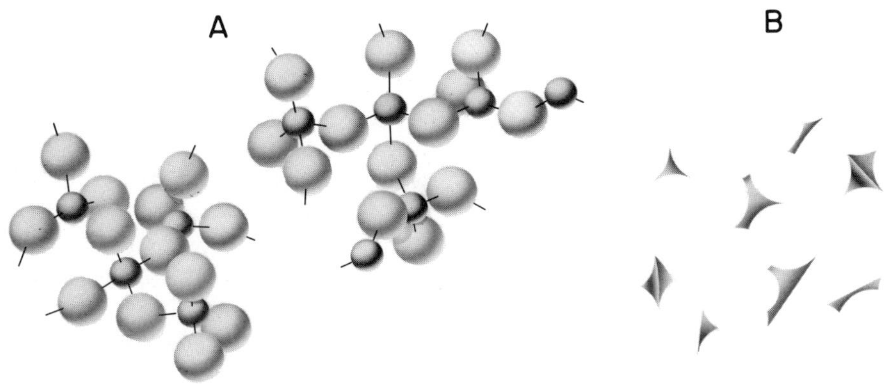

can be persuaded to go into solution. The earth's interior is such a pressure chamber where steam and molten rock coexist, so that lavas reaching the surface do contain steam, CO_2 and other gases in solution. Now we recall that water's content of HCl affects its ability to dissolve zinc, and so we may accept that a magma's content of SiO_2 affects its ability to dissolve steam; at a given pressure and temperature, the more acid the magma the more steam it will dissolve.

The combination of great viscosity and comparatively high content of dissolved gases, which may reach 10 per cent by weight, gives acid magmas an explosive power usually absent from basic ones. The energy of a volcanic explosion is of two kinds: (1) energy tied up in the strained surface rocks until the moment of rupture, when the individual fragments relax to a state of no stress, and (2) energy tied up in the bubbles of compressed gas that form in the viscous magma as it moves to the low-pressure surface environment. The second of the two is incomparably greater than the first. A fluid basic magma emits its sparse gases via a fairly peaceful froth, but in a brittle viscous acid magma, the froth stage is often not stable—the gases burst the froth, and the rock output of the volcano is found subsequently in the form of little cuspate fragments (Fig. 5-2B) scattered for miles around.

The approximate compositions of acid and basic magmas are shown in Table 5-1.

TABLE 5-1 COMMON COMPOSITIONS OF ACID AND BASIC MAGMAS

Acid Magma			*Basic Magma*		
		per cent by weight			per cent by weight
Nonvolatile fraction: (92–95% of whole)	SiO_2	65–75	Nonvolatile fraction: (96% of whole)	SiO_2	48–58
	Al_2O_3	12–16		Al_2O_3	13–17
	Fe_2O_3, FeO	4–8		Fe_2O_3, FeO	9–14
	MgO, CaO	4–6		MgO	5–8
				CaO	8–12
	Na_2O, K_2O	6–9		Na_2O, K_2O	3–5
Volatile fraction: (5–8% of whole)	H_2O	90	Volatile fraction: (2–6% of whole)	H_2O	80
	CO_2	3		CO_2	8
	halides	5		halides	5
	H_2S, SO_2	1		H_2S, SO_2	5

Note: Great departures from the proportions quoted are possible, particularly in the volatile fractions.

ROCK BODIES FORMED BY VOLCANIC ACTIVITY

In the preceding section, volcanic activity was discussed by reference to events in historical time, and attention was focused on magma in course of eruption. We turn next to the products of eruption, to rock bodies formed from ejected material that has settled on the ground (see Fig. 5-3).

The largest of these rock formations, the *basalt floods*, cannot be conceived as products of single vents; study of Iceland, for example, where sheets of basalt are particularly well exposed, shows that they are produced instead by magma flowing from fissures hundreds of kilometers long. No example of formation of a basalt flood has been recorded in historic time: the largest recorded flow, in Iceland in 1783, covered an area that would appear the size of a comma in Fig. 5-3A. The floods or plateaus are in fact structures of comparable size with the Appalachian mountain range; the emplacement of each one was a significant event even on a global scale. The composition of the basalts shows very little variation either within a flood or from one flood to another, in spite of their time span from Recent back to Triassic times or earlier. Clearly, a major basalt source available for tapping has been a permanent or at least recurrent feature of the earth.

(Evidence is accumulating that indicates an immeasurably more bulky volcanic formation even than these, on the Pacific ocean floor around the Marshall, Gilbert and Ellice Islands. The evidence, conveniently summarized by Menard (1964, Chap. 4) consists almost solely of echo-sounding profiles of the ocean bottom, and seismic profiles, i.e., estimates of seismic wave velocities at different depths below the ocean bottom. Clearly, circumscribing a unit, showing that its contents were emplaced in a single "geological event" and identifying the nature of the event is not as easy by such means as it is on a land surface; but the possibility of a unit measuring 10^7 km^3 being emplaced by a series of genetically related outpourings is to be kept in mind.)

The group of Hawaiian volcanoes, which are marginal to the hypothetical flood just mentioned, constitute the prime examples of *shield volcanoes*. They differ from the floods in being recognizably peaked (see Fig. 5-1) and from all other known volcanoes in being larger. Their distribution is linear, suggesting fracture control, but their contours form closed loops about individual centers or short fissures. The probability of any comparable shields existing in other oceans seems to diminish as ocean surveys progress; nor has any geologic structure been found that could have been a basalt shield of an earlier time. Basic igneous bodies of comparable size but greater age exist, but these are believed to have been emplaced subterraneously, so that the Hawaiian group is at present without parallels.

The next largest formations of volcanic origin are layers of volcanic ash, or *tuff*, as it is called when lithified. Some of these are as great in area

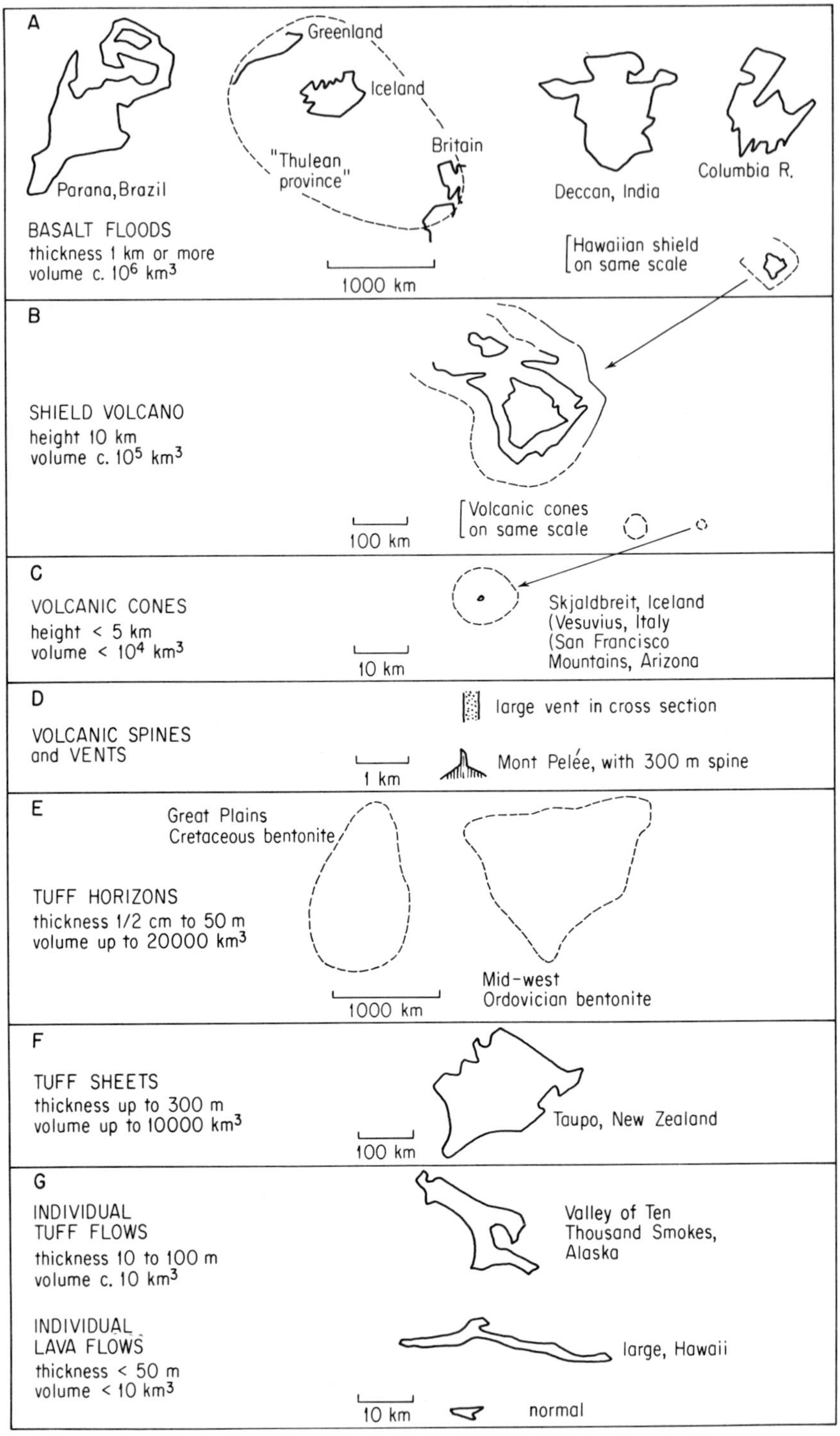

Fig. 5-3 Shape and size of volcanic features. (Modified in part from Rittmann, 1962; Ross, 1955; Cotton, 1952.)

as the flood basalts, though not one-tenth as thick. A layer in the Yukon Territory, in fact, can be traced over an area as great as that of the Columbia River basalts, without being anywhere more than a few inches thick. Estimating its volume as 50 km^3, we note that such a volume of ash could be generated by blowing off the top 4000 ft from Mount Etna (total height 11,000 ft), or by five explosions such as that in Kamchatka, 1955 (which is referred to rather for having been carefully analyzed in a quantitative way than for its size). The wide, uniform distribution of debris fits in with such an explosive origin, because it seems to require the ejection of very fine particles high into the air. The nearest modern analog is the explosion of Mount Katmai, Alaska, after which 12 in. of ash settled on Kodiak, over a hundred miles away. The bulkier formations, such as those in the Cretaceous of Wyoming-Manitoba, are composite, being built up from hundreds of thin layers in the same way that the shields of Hawaii are built up from hundreds of separate flows. Presumably the former area was one of more active vulcanism than anywhere in the world today, with large explosions going on as often as perhaps one a year. It should not be thought that all tuff deposits are thin; thicknesses of 1000 ft or more are known, but these are of quite small areal extent and so are more properly considered in a later paragraph.

The chemical composition and texture of tuff deposits deserve attention because of the possibility that we may be able to check by observation what has been presented as a hypothesis, namely, that acid magmas have a greater tendency than basic magmas to be explosive. If the hypothesis is correct, the abundance-ratio of acid tuffs to acid flows should be greater than the abundance-ratio of basic tuffs to basic flows. Many petrologists believe that observation confirms this suggestion, but it is instructive to note two obstacles that prevent a definite decision: firstly, there exist large bodies of rock whose texture does not now show whether they were tuffs or flows; secondly, there exist large bodies of rock whose composition does not now show whether they were acid or basic. Both kinds of uncertainty arise from the fact that tuffs are the most susceptible of all rocks to change of composition after formation. For example, the Wyoming deposits are no longer fragments of glass, rock chips and crystals, but are now wholly converted to powdery clay. Neither their original texture nor their original composition can be ascertained.

Overlapping in range of size with the tuff layers come the *volcanic cones*, among which the Andean volcanoes are the highest (e.g., Chimborazo, 20,640 ft). Most cones are lower (Lassen Peak, 10,453 ft; Mount Etna, 10,741 ft; Vesuvius, 3891 ft), and of course the height of the cone is only partly due to the volcano, since most land-based volcanoes rest on a non-volcanic basement some distance above sea level. Volcanic cones are the first strongly heterogeneous structures we have described. From eroded examples, we see that cones are like reinforced concrete—a mass of weakly coherent rubble laced by strong ribs and sheets of solidified lava. There is no restriction on composition, for in spite of the general rule that basic magmas are mobile and peaceful, explosive basic volcanoes do exist and form cones in the same way as acid ones.

Volcanic cones are composed of five kinds of individual unit:

1. Magma that has overflowed from the vent as lava streams and solidified;

2. Layers of fragments that have been shot from the vent high into the air and have fallen again;

3. Layers of fragments that have been lifted from the vent by rising gases but have then flowed over the rim and down the side in the same way that a dust cloud rolls over the land surface; such a mixture of hot gas and rock particles tends to separate under gravity but its turbulence is such that it remains a coherent body, in density heavier than air but lighter than water, long enough to roll several miles; the resulting rock bodies tend to fill valleys and are therefore elongated in plan like lava flows, in contrast to rock bodies of class (2) which are not topography-controlled;

4. Magma that, instead of flowing over the rim of the vent, has penetrated sideways into the cone of fragments before solidifying;

5. The vents themselves (a single cone may have several vents): the rock that eventually fills the vent depends on the history of the volcano in its closing stages—some vents are filled by a plug of solidified magma, some become choked with magmatic fragments and many remain open until collapse of the walls fills them with fragments of the surrounding nonvolcanic rocks. The vent itself is the smallest of all volcanic units—the volume of material extruded through it usually far exceeds the volume of the vent itself.

Figure 5-3 shows typical dimensions for these components.

TEXTURES OF VOLCANIC ROCKS

In the foregoing survey of volcanoes, the facts have been presented as if "known." The history of a volcanic deposit is usually reconstructed, however, by study of the texture of the rocks long after activity has ceased. In this section, therefore, some of the characteristic textures are described. It will be seen that the degree of confidence with which an origin can be ascribed to a prehistoric unit is highly variable. Figure 5-4 illustrates the terms introduced.

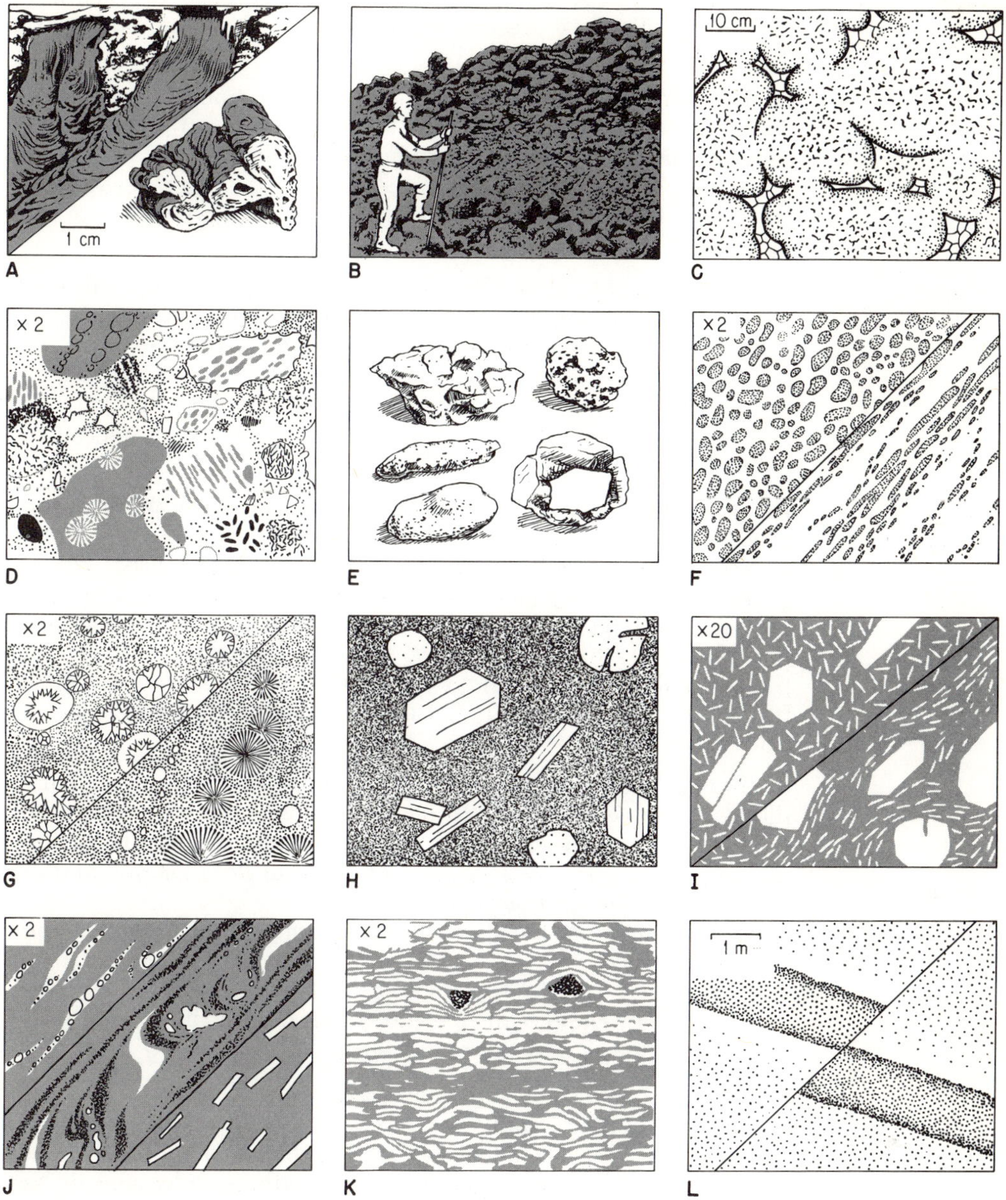

Fig. 5-4 Volcanic structures and textures. (A) Ropy lava—distant view and fragment; (B) blocky lava; (C) pillow lava; (D) agglomerate; (E) lapilli; (F) vesicles, normal and elongated; (G) amygdules and spherulites; (H) phenocrysts, euhedral and anhedral; (I) felty and trachytic textures; (J) flow banding or layering; (K) welded texture (tinted areas indicate glass); (L) baking and chilling effects.

A. ROPY LAVA: A term based on the appearance of the outer surface of a flow; not often recognizable once flow is buried by later deposits.

B. BLOCKY LAVA: A contrast with (A). Most modern flows belong distinctly to one category or the other. The blocks are fragments of cool solidified crust, broken and churned by flow of the hot interior. They may be sufficiently numerous to mask the interior completely, so that the flow looks merely like an advancing pile of cool rubble.

C. PILLOW LAVA: When lava flows under water, it sometimes advances by formation of a succession of pods; most pods eventually split, and lava flows through the split to form the next pod. The process has been observed in clear water, but can presumably occur also beneath water-laden sediments. In contrast to ropy and blocky lavas, pillow lava can be readily recognized in deposits as old as Precambrian.

D. AGGLOMERATE: Any rock emplaced by volcanic action and composed mainly of volcanic fragments, of which some are over 4 mm in size. Usually, "*x* is an agglomerate" is a hypothesis, whereas "*x* looks like an agglomerate" is a direct observation. The space between the larger fragments is usually filled by smaller fragments and ultimately by dust, but the fragments may be held together partly by flow material. Agglomerates often contain a minority of fragments of nonvolcanic rock torn from the vent walls during eruption.

E. LAPILLI: Strictly, any volcanic fragments with intermediate diameter between 4 and 32 mm; thus most agglomerates contain lapilli. However, the term is particularly used for somewhat rounded lumps that form by accretion, as drops of fluid magma and ash fly through the air together.
Ash: A collection of loose volcanic fragments whose mean diameter is 4 mm or less.
Volcanic ejecta over 32 *mm:* These have no special name. When composite they can be called *bombs;* when simple, they can be called fragments; *lumps* is also a useful term.

F. VESICLES (adj. *vesicular*): The holes formed when a lava containing gas bubbles solidifies: they may be rounded, irregular or elongated. Highly vesicular rock is called *pumice* or *scoria;* some authors refer to siliceous pumice and basaltic pumice and do not use *scoria*, while others use *pumice* for acid vesicular rocks and *scoria* for basic vesicular rocks.

G. AMYGDULES (adj. *amygdaloidal*): Vesicles that have been subsequently filled or almost filled. Since the material for the filling has to migrate via the aqueous pore-fluid in the rock, the common fillings are silica, calcite and zeolites, or hydration products like epidote.

Spherulites: Aggregates of crystals growing outward from a common center. They are found most commonly in glassy lava and, if growing in isolation, usually achieve more or less spherical exteriors.

Spherules: Since to call an object an amygdule or a spherulite implies a particular origin, the neutral term *spherule* is sometimes convenient, e.g., "The spherules in rock x seem to be amygdules, and so x is probably of volcanic origin."

H. PHENOCRYSTS: If a magma crystallizes to give some large crystals in a groundmass of much finer crystals or glass, the large crystals are called phenocrysts; it is usually supposed that the phenocrysts grew large while the magma was subterranean, and that rise of the magma to the surface later provoked quick, fine crystallization of the material between the phenocrysts.

Xenocrysts: Large crystals can be enclosed in a fine groundmass by other mechanisms, e.g., ash particles that happen to be single crystals can fall into a lava as it flows. When the large crystals and the groundmass come from separate sources, the former are called xenocrysts.

Megacrysts: Again, a neutral term is convenient for discussion of doubtful examples, e.g., "Most of the megacrysts seem to be phenocrysts."

Porphyritic: Any rock with megacrysts in a groundmass may be called porphyritic if it is thought to be of magmatic origin. However, similar texture can arise by sedimentary or metamorphic processes, and if a rock's history is not certain it should not be called porphyritic.

Euhedral, Anhedral: Crystals of any size can either impose their own characteristic shape on their surroundings or have some other shape imposed on them. The former are called euhedral, the latter anhedral crystals. Of the phenocrysts in the figure, the stippled ones are anhedral and the others euhedral.

I. FELTY, TRACHYTIC: A rock mainly composed of minute feldspar laths may be called felty if the laths are randomly oriented, or trachytic if they are arranged as if by flow. (The term *trachyte* is also in use and refers to a rock's composition, not its texture. It is therefore ambiguous to write "x is trachytic," but "x has trachytic texture" allows no confusion.)

J. FLOW TEXTURE (flow banding, flow layering): Any texture believed to have been formed during flow. Flow texture can be shown by particolored rocks, trachytic-textured rocks, or rocks with megacrysts, bubbles or dust arranged by flow.

K. WELDED TEXTURE: Rocks that look as if they are a mass of separate pieces that have been piled together while still plastic may be called welded, especially if the constituents are largely glassy. Such rocks are supposed to form by deposition of ejecta while they are still hot enough to be deformable and to stick to one another. Gas-charged ash flows are thought to be the main generators of welded rocks.

L. CHILLED MARGIN: The quickly-cooled outside of a lava flow is usually of finer grain than the interior, or even glassy. Chilled margins can develop where lava touches cold rock as well as where it is exposed to water or air. The rock next to a chilled margin should be examined carefully: if it shows alteration that can be ascribed to heating, it predates the lava; if it shows no alteration, it may simply be a heat-resistant rock (granite, for instance) but if it is a rock that *would* recrystallize on heating (shale, for instance) and yet has not recrystallized, then the chilling was against air or water and the neighboring rock emplaced later. The illustration shows a surface flow and a body emplaced subterraneously. Which surface was the original top of the surface flow?

RECOGNITION OF VOLCANIC ROCKS

From the textures described, it will be seen that flow rocks, tuffs and agglomerates are usually very different. Three separate problems are thus involved: (1) recognition of flows, (2) recognition of tuffs and (3) recognition of agglomerates. Success with such problems cannot be learned from a book, but there is a point which it is the purpose of the ensuing paragraphs to make: the recognition of genesis from observable properties and the prediction of observable properties from a given mode of genesis are quite separate skills. One is not the reverse of the other; different stores of knowledge are required for the two tasks. The following comments on the recognition process are designed to bring the difference to light.

Flows

Since solidified lava is of fine grain, it cannot be confused with magmatic rocks formed at greater depth in the earth, nor with coarse-grained sediments, nor with most highly metamorphosed rocks. The difficulty is to distinguish flows from highly compacted mudstones and from some fine-grained metamorphic rocks. Things to look for are:

1. The margins of the body, for chilling or baking;

2. Vesicles;

3. Megacrysts: These do not *prove* magmatic origin, but if they are euhedral, hard enough to scratch glass and distributed singly through the groundmass, they make magmatic origin likely.

More rarely, spherules, contained fragments of glass, or some kind of texture (flow or welded) will be seen; if spherules, they must be checked for structure—spherules formed by metamorphism often lack radial structure and merge more with the groundmass; if curvilinear bands are visible, they must be checked to see if they could be bedding in a sediment.

Tuffs

The main problem is to distinguish tuffs from nonigneous sediments. If the rock is pale in color, check its density, grittiness and reaction with acid: low-density noncarbonate rocks should be suspected as tuffs (but reaction with acid does not disprove tuff origin, because many tuffs pick up carbonate after deposition). Any large particles in the rock should be examined for glass, or euhedral outline, or signs of abrasion.

If the rock is of normal density and color, and contains small rock fragments and single crystals in a finer matrix, recognition must be given to the fact that it could be a tuff or a nonigneous sediment. A moment's reflection about rapid erosion of material off an active volcanic coast should remind us that tuffs are transitional into nonigneous sediments. It is therefore time wasting to ponder which "label" to assign, igneous or sedimentary. One's time is better spent estimating relative amounts of igneous and nonigneous components, degree of reworking by normal erosion and so on.

Agglomerates

If a rock is seen to be mainly composed of fragments, the problem is to decide between vulcanism, faulting, sedimentation etc., as the agent. If possible, the margins of the formation should be studied to find out whether they are sharp or gradational and whether or not they cut across the adjoining formations. The fragments themselves must be examined by the criteria given for flows and tuffs; if the fragments are largely volcanic, this is suggestive, although one should ponder circumstances in which a rock with 100 per cent volcanic fragments is produced by faulting or in which a rock with 30 per cent sedimentary fragments is produced by vulcanism. Thirdly, the relation between fragments and matrix is sometimes indicative: in agglomerates, the fragments are often ill-defined and the rock's fracture-surfaces run continuously across fragments and matrix indiscriminately, whereas in other fragmental rocks, the fragments may have well-defined surfaces and resist fracture.

It was mentioned that, even after the volcanic nature of a rock has been recognized, uncertainty may remain as to whether it is a flow or a welded tuff (Fig. 5-4K). Where the matrix is homogeneous and contained fragments are sparse, angular and equidimensional, origin as a flow is indicated. At the other extreme the matrix may be of glass that is patchy in characteristics (color, clarity, devitrification and so on) and the "fragments" may be abundant, rounded and with a common aspect of flatness; especially where the boundary of one lump appears to wrap around another, origin as welded tuff is indicated (In this case, the lumps may not be fragments in the normal sense, i.e., formed by fracture). If the rock under study is at neither extreme, the best way to a decision is to trace the unit in the field until it grades into a variant that *is* clearly a tuff or a flow.

These brief remarks are no substitute for field experience—"learning geology through the soles of one's boots." They are designed rather to bring to light the difference between the genetic approach and the observational approach. On the one hand we classify rocks by origin, as in Fig. 5-5A; on the other hand we classify them by observable features, as in Fig. 5-5B. Usually we wish to place a rock in a genetic class, but our

Fig. 5-5 Classifications by genesis and by observation.

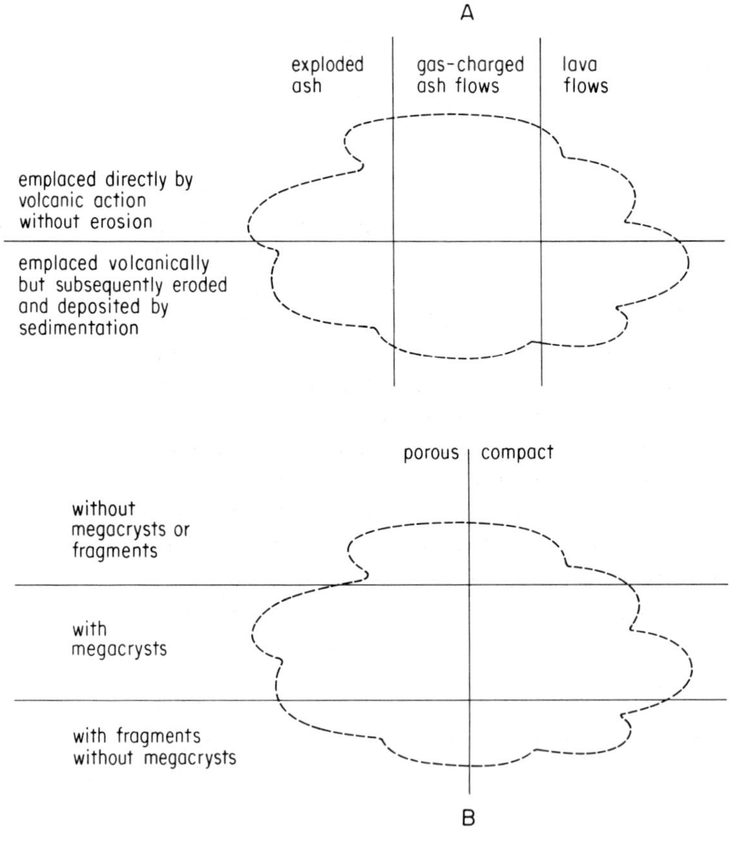

observations only put it in an observational class. There is no guarantee that when we have divided rocks into groups that differ in appearance etc., we have divided them into groups with different origins. In other words, no class in Fig. 5-5A necessarily corresponds with any class in Fig. 5-5B.

Unfortunately, the technical terms of geology are mainly genetic in meaning; there is a shortage of concise descriptive terms. Consequently, field records often read "Outcrop 44: basaltic agglomerate" when they should read "Outcrop 44: fine, dark, heavy rock with green lumps." Of course, the second version is not enough: it should be embellished with further details, and may be embellished with suitable conjectures. But the first version is not enough in a more serious way: it cannot be embellished at all. It does not record the rock; it records only a transient opinion in one person's mind. The time comes when details must be suppressed and conjectures summarized, but the field record is not the moment for this. To sum up, scientific study of rocks entails scientific discrimination between what is observed (the rock's condition today) and what is conjectured (the rock's condition at previous times).

ROCK BODIES FORMED BY SUBTERRANEAN MAGMATIC ACTIVITY

Reference has already been made to the formation of rock units by solidification of pools of magma within the crust. This section surveys the observable properties of these units. Again the dimensions are presented in diagrams (Figs. 5-6, 5-7) and comparison with Figs. 5-1 and 5-3 is interesting.

TERMINOLOGY

EXTRUSIVE ROCKS: Rocks emplaced in a flowing state on the land surface or sea floor.

INTRUSIVE ROCKS: Rocks emplaced in a flowing state below the land surface or sea floor.

(All magmatic rocks are thus intrusive or extrusive, tuffs being extruded in a broad sense, but not all intrusive and extrusive rocks are magmatic—wet sediments and bodies of rock salt can be intruded or extruded by suitable pressures.)

CONCORDANT BODY: A body of rock in a layered environment, whose boundaries do not cut across the layering.

DISCORDANT: Not concordant.

(Most extrusive rocks are thus concordant, although one should conceive circumstances in which they would not be. Where the adjacent rocks have no layering, tabular intrusions are usually called dikes, i.e., treated as discordant, or called sheets, a nonspecific term. Where the adjacent rocks have a strong cleavage and no layering, it is permissible to describe an intrusion as "concordant with the cleavage" if some good purpose is served by doing so, but it should *not* be described as a sill.)

Many terms in Figs. 5-6 and 5-7 have more exact meanings than a geologist's observations may justify. For example, it is often hard to decide whether a body is concordant or discordant; in such a situation a neutral term like *pod, lens, sheet* or simply *body* should be used.

Fig. 5-6 Shape and size of concordant intrusive bodies.

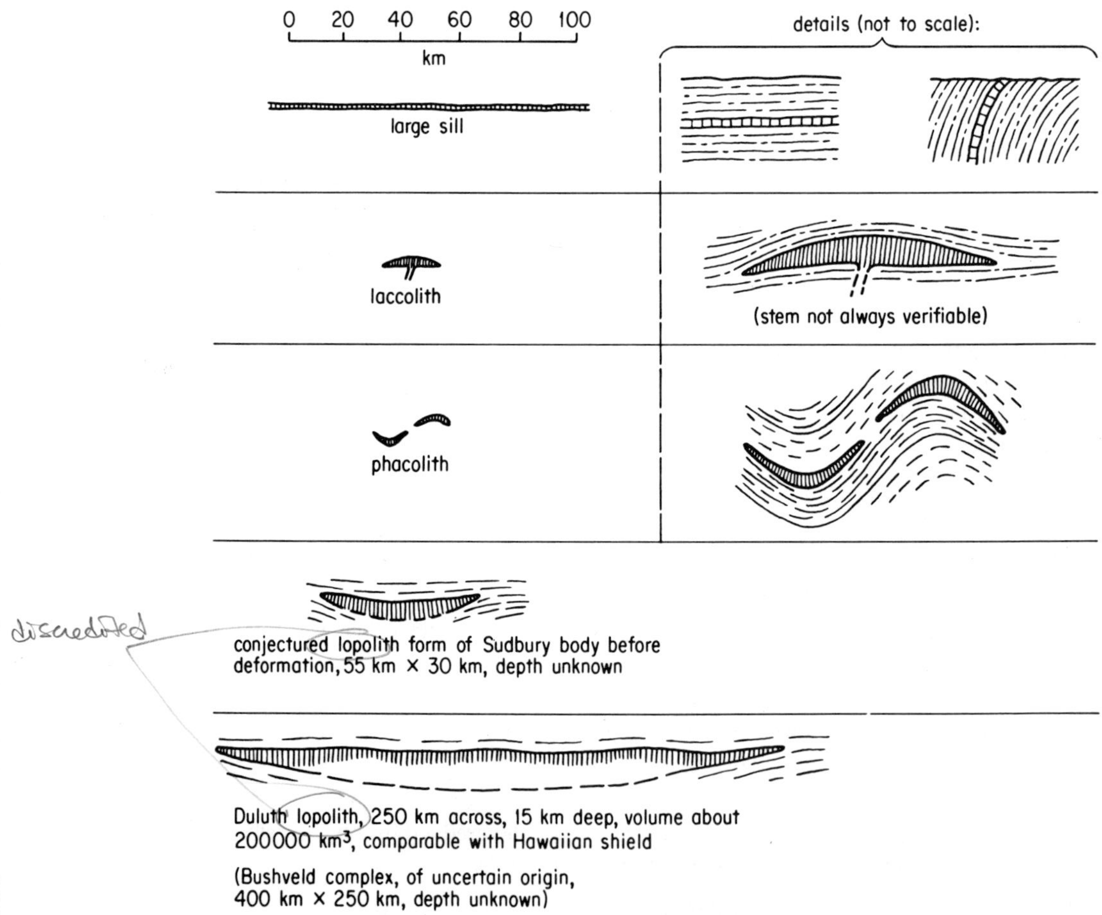

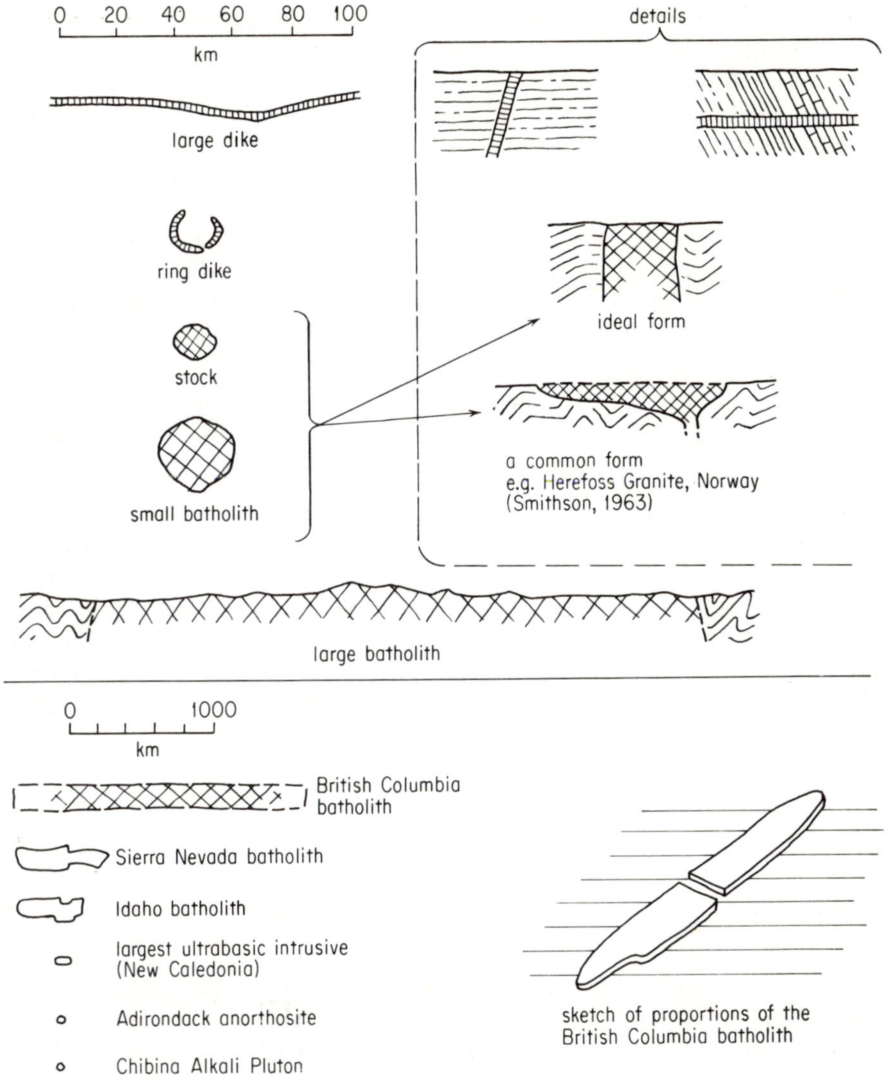

Fig. 5-7 Shape and size of discordant intrusive bodies.

PROVISO: Although small bodies occur as illustrated whose magmatic origin cannot be questioned, the larger the body the more doubt there exists about its formation. It is certainly unlikely that the Duluth lopolith (Fig. 5-6) or the Idaho batholith (Fig. 5-7) was ever all molten at once. It must also be remembered that doubt exists about the origin of many small bodies of the shapes shown.

Comparisons: The British Columbia batholith is comparable in volume with the basalt floods, being somewhat smaller in area but probably greater in thickness. (Some writers have emphasized the *depth* of batholiths, but the discussion in Chap. 1 shows that the depth of such a large batholith must be small in comparison with its length and breadth.) The floods and the batholith are composite structures with many distinguishable parts, but the batholith is by far the most variable from part to part in composition. No known concordant intrusive approaches these in size.

The Hawaiian shield, the Duluth lopolith and the Idaho batholith are similar in shape as well as in size, i.e., ovoid in plan, with thickness $\frac{1}{10}$ or $\frac{1}{20}$ of the breadth. The size range is continuous downwards—a lopolith can be alternatively described as a swollen sill or a large phacolith (at least insofar as its presence deforms the strata *beneath* it into a syncline), and batholiths come in all sizes down to 40 square miles in outcrop—at this point, they change, but only in name, to *stocks*. By tradition, an ideal stock is deeper than it is broad, but many discordant intrusives in this size range turn out to be as disc-like as laccoliths.

Compositions: As with extrusive rocks, there is a certain amount of correlation between the chemistry of intrusives and their geometrical form—large basic bodies tend to be concordant and large acid bodies discordant. The mechanism of emplacement of these bodies is much less certain than that of flows and tuffs, however, and further facts must be presented before possible hypotheses can be adequately discussed.

TEXTURES AND RECOGNITION OF SUBTERRANEAN MAGMATIC ROCKS

The terms *phenocryst*, *xenocryst*, *megacryst*, *porphyritic*, *euhedral* and *anhedral* (p. 41) are all applicable to intrusives as to extrusives. Intrusives can also show flow texture, either by orientation of megacrysts or by concentration of one mineral species in wisps, streaks or layers, as in Fig. 5-4J. In general, however, flow texture is less striking in intrusives than in extrusives.

Descriptions of some additional textures follow (see Fig. 5-8).

A. Poikilitic: If large grains of one mineral enclose many small grains of another mineral, the assemblage may be called poikilitic; the term distinctly implies a magmatic rock—if there is uncertainty on this point the crystals should be called "full of inclusions" or "sieved."

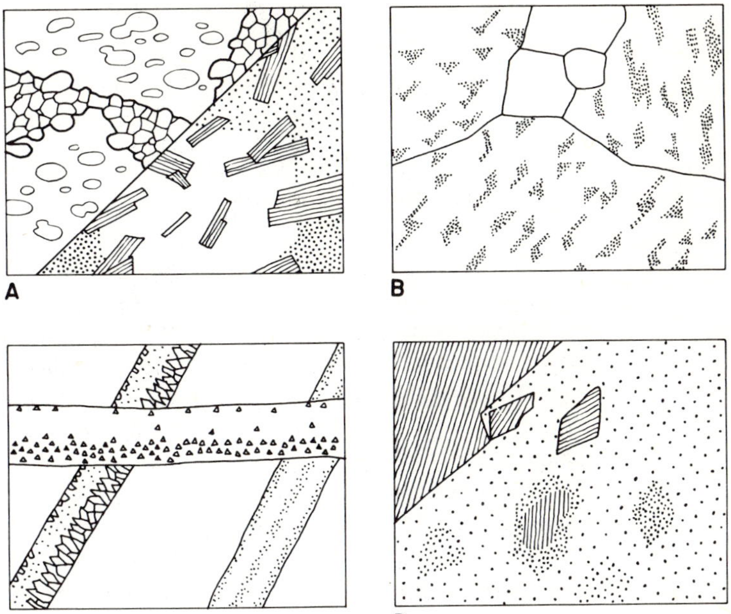

Fig. 5-8 Intrusive structures and textures. (A) Poikilitic and ophitic textures; (B) graphic intergrowth (called *granophyre* when the components are quartz and feldspar); (C) zoning in intrusions; (D) xenoliths, some partly digested.

Ophitic: The particular form of poikilitic texture where the host is pyroxene and the inclusions are laths of plagioclase.

B. Graphic Intergrowth: The particular form of poikilitic texture where the inclusions are of somewhat regular geometrical form and orientation. When the components are quartz and alkali feldspar, the intergrowth may be called granophyre, and the rock containing it also granophyre or granophyric. (Such intergrowth has been called pegmatite or micropegmatite, but pegmatite now has a quite different meaning and micropegmatite is best not used at all.) *Pegmatite:* A pegmatite is a body of particularly coarse grain, not necessarily magmatic. *Pegmatite*, as a rock name, means simply a coarse rock—most grains 1 cm or more across. *Pegmatitic* can be used for any portion of a nonsedimentary rock which is noticeably coarser than the rest (*Pegmatitic* is also used to specify a particular temperature range; for this usage, see Chap. 23).

C. Zoning (of intrusions): Concentration of some components in layers parallel to the walls. Zoning can be symmetrical or asymmetrical in both sills and dikes, although asymmetrical zoning due to gravity is confined mainly to sills. (The term must be used in such a way that it is clear whether this sense or zoning in crystals is meant.)

D. XENOLITH: A fragment of wall rock picked up by a magma in the course of its journey to its crystallization site. By reaction with the magma, xenoliths also may become zoned, as in **C**; by more extensive reaction they may be "digested" and merge imperceptibly into the magmatic rock.

On account of their emplacement beneath a rock cover, intrusive magmas crystallize more slowly than most extrusive ones, and so in general give more coarse-grained rocks. These cannot be confused with fine sediments and are distinguished from coarse sediments by the compactness with which their grains are intergrown. On the other hand, some metamorphic rocks and some igneous rocks look alike, and there are many rocks to which neither of these labels can be attached with confidence. Why is this?

First, we note that most metamorphic rocks are either metamorphosed igneous rocks or metamorphosed sediments; naturally, a slightly metamorphosed coarse igneous rock will differ but little from an unmetamorphosed one, particularly because igneous minerals are stable at elevated temperatures; therefore, we shall have difficulty distinguishing these two groups. However, it is usually not important that we should do so since many problems in geology can be solved without ever deciding whether a rock has or has not undergone slight metamorphism. The questions that much more often need answers are:

1. Could rock x be a highly metamorphosed volcanic rock?
2. Could rock z be a highly metamorphosed nonvolcanic sediment?

Most techniques for answering such questions depend on a knowledge of metamorphism as well as of magmatism. All that we can do in this chapter is to list some properties which, *if shown*, indicate subterranean magmatic origin; if they are not shown, we do not assume metamorphism—we merely suspend judgment:

1. Sharp contact against unmetamorphosed rocks;
2. Chilling on inside of contact;
3. Heat effects localized against outside of contact;
4. Xenoliths inside contact, especially if they do not match the rocks outside the contact;
5. Discordant outline against unmetamorphosed rocks;
6. High degree of homogeneity over a large area, e.g., several square miles;
7. Ophitic texture.

All but two of these require examination of the margin, and only one can be observed in a hand specimen. In short, magmatic rocks are better recognized by their relations than by their intrinsic properties.

CHAPTER SIX

CLASSIFICATION AND NOMENCLATURE OF MAGMATIC ROCKS

Two purposes are served by establishing classes of rocks; it becomes easier to summarize knowledge from the past and it becomes easier to accumulate knowledge in the future. But a system well suited to one purpose is not necessarily well suited to the other; therefore it is advisable before selecting a system to be clear what one's purpose is. The purpose of the following two sections is to summarize knowledge from the past, and the classifications used are selected with that end in view.

CLASSIFICATION OF MAGMATIC ROCKS

The main classification used is set out in Fig. 6-1. In addition to the classes established in Fig. 6-1, there are three other useful ways in which rocks may be divided:

1. Oversaturated / saturated / undersaturated;
2. Acid / intermediate / basic / ultrabasic;
3. Alkalic / nonalkalic.

The first system depends on the existence of undersaturated minerals, that is, minerals that would react with quartz if placed in contact with it, such as the feldspathoids and most olivines. Then we may say:

> If a rock contains quartz, the rock is oversaturated;
> If it contains any undersaturated mineral, the rock is undersaturated;
> If it contains no quartz nor any undersaturated mineral, the rock is saturated.

All saturated rocks fall within the shaded area in Fig. 6-1. The commonest such rocks are made of feldspar with pyroxene or amphibole.

The second system is chemical rather than mineralogical, being based on the weight percentage of SiO_2 in the rock; thus, if SiO_2 percentage is:

over 66, rock is acid;
52–66, rock is intermediate;
45–52, rock is basic;
under 45, rock is ultrabasic.

The third system is also chemical, but is based on relative numbers of molecules rather than on weights. Let the number of molecules of Na_2O in a given quantity of rock be n_{Na_2O}, and similarly for other oxides. Then, if

$$n_{Na_2O} + n_{K_2O} > n_{Al_2O_3}$$

or

$$n_{Na_2O} + n_{K_2O} > \tfrac{1}{6} n_{SiO_2}$$

the rock is alkalic. The effect of this definition is that an alkalic rock cannot contain all its Na_2O and K_2O in the form of feldspar or mica; some has to be accommodated in other minerals. The class of nonalkalic rocks has no special name; if a rock is not named as alkalic, one is usually correct in assuming it to be nonalkalic.

It is desirable to notice at this point the difference between absolute and relative terms. A rock is not *more saturated* or *less saturated*—it is either *saturated* or not so; but *acid* and *basic* are sometimes used as relative terms. For example, one may say of an intrusion, "The margins are more basic than the center" even though the SiO_2 percentage is everywhere over 52.

NAMING OF MAGMATIC ROCKS

The four independent classifications introduced all satisfy the basic requirements—in each one, every rock has a place and no rock has two places. Further, they help us to summarize past discoveries: we may now state "Most undersaturated alkali rocks carry nepheline" or "Granodiorites hardly ever contain muscovite" and expect our meaning to be unambiguously understood. But when the ambition is to convey information about a particular specimen that has been examined, are these classifications

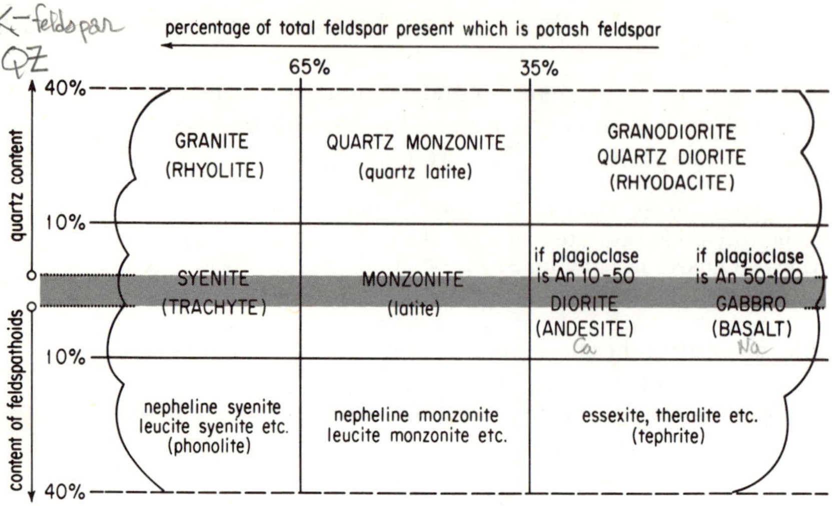

Fig. 6-1 A classification of igneous rocks. See explanatory notes following (modified from Nockolds, 1954, and Travis, Copyright 1955, Colorado School of Mines).

1. If plagioclase present is albite, Ab_{90-100}, rock is put in column at left-hand end, however much plagioclase is present.
2. Total amount of feldspar present does not affect class of rock unless there is over 90 per cent or under 10 per cent; these extreme cases are discussed elsewhere (Chap. 7 and Appendix 2).
3. Names in parentheses are fine-grain equivalents of names without parentheses; *micro-* may be prefixed to any of the latter for stricter specification of grain size, but *diabase* is often used in place of microdiorite or microgabbro.
4. Names in lower case are less commonly used; by contrast, names in capitals are widely used, not always with exactly the meaning given in this chart.
5. Rocks occur with more than 40 per cent quartz or feldspathoids, but the names given here should not be applied to them—the rocks with quartz are unlikely to be magmatic and the ones with feldspathoids (very rare) have names of their own.
6. The points at which the vertical boundaries are drawn vary with different authorites in the ranges 30–40 per cent and 60–70 per cent.
7. Some alternative terms are:
   adamellite = quartz monzonite
   dellenite = quartz latite
   tonalite = extreme granodiorite (potash feldspar < 10 per cent of total feldspar)
   dacite = extreme rhyodacite (potash feldspar < 10 per cent of total feldspar)
   trachyandesite = latite
8. A careful and authoritative attempt to serve all purposes simultaneously, and to give precision to modern trends in usage, has been made by Streckeisen, whose recommendations form Appendix 2.

any use? The answer is that the usefulness of any term is always dependent on the user's purpose. The describing of rocks as an end in itself is a somewhat artificial exercise, practised mainly in schools. More commonly, the describer has a specific purpose in mind, such as to describe the suitability of rock x as a building stone, or to describe the relation of rock x to rock y. For each such purpose, different terms will be preferred. It is the describer's duty, not to pick his terms from a special set of nine or sixteen "names," but to find, somewhere in the language, terms that embody exactly the meaning he wishes to convey.

At once, we remark that there are two common purposes: (1) to convey observations and (2) to convey hypotheses. It is usually also desirable to let the reader know which is which; hence the usefulness of terms which are purely observational or purely genetic. Many things in petrology are subject to uncertainty beside genesis, however, and in particular the mineral composition of a fine-grained rock; even in a coarse-grained rock, it is often impossible to say with confidence which of the three vertical divisions a rock belongs to. Thus "x is a quartz monzonite" is often not an observation, but a hypothesis, more properly phrased, "I think x is a quartz monzonite." Now this statement, while permissible, fails to give the reader any *facts* whatever: are two feldspars distinguishable? is either visibly plagioclase? The record fails as description unless supplemented by some such phrases as, "A few grains of plagioclase and quartz recognizable; plagioclase estimated as 20–40 per cent of total feldspar." Having been obliged to say this much to present any *facts* at all, what does the author convey by adding a name?—nothing but his guess as to which side of the granite/quartz monzonite divide the rock lies on. In accordance with the precept above, if it is the author's purpose to express this guess, then to say "I think x is a quartz monzonite" is in order; but if his purpose is to express his uncertainty as to where the proportion lies between 20 per cent and 40 per cent, his record is perfect as it stands and is only marred by the addition of a "name."

The upshot of the foregoing discussion is that the terms in Fig. 6-1 are usually unsuitable for use in describing particular specimens. The observations usually need several phrases or sentences for their exact expression, and any term, however unscientific, that is exact is preferable to a technical term that means not quite what is intended. In conclusion, two more examples are given:

> Compare "x is a fine, dark, heavy rock with green lumps" with "x is a basaltic agglomerate." Either may serve one's purpose, but to give the reader an accurate mental image, the former is more reliable.

> Compare "outcrop 23: coarse quartz-feldspar rock similar to outcrop 22" with "outcrop 22: granite . . . outcrop 23: granite." Suppose that laboratory study shows the rock from outcrop 22 to be a gold-bearing granodiorite; did the writer of the second record make the same misidentification at both outcrops, or is outcrop 23 truly granite and therefore probably not a valuable extension from outcrop 22? In this instance, the more direct record of what was observed also gives more exactly the information of economic consequence.

CHAPTER SEVEN

MINERAL VARIETIES, TRENDS AND ASSOCIATIONS IN MAGMATIC ROCKS

In the preceding chapter, some classes of rock were defined; in this chapter, the same classes will be described. The theme of the descriptions is that a rock's mineralogy and its chemistry are related, because of the restricted compositions a mineral can have.* The order of procedure will be first to describe the trend of variation among common rocks, and then to describe less common rock types by stating their deviations from the main trend.

MAJOR TRENDS IN THE COMPOSITION OF MAGMATIC ROCKS

Figure 7-1 shows the trend of variation among common rocks, and also indicates the predominance of rocks with silica content just over 50 per cent or just over 70 per cent (basalt and granite/quartz monzonite respectively). Any estimate of relative abundance in a population such as

*Mineral data are given in Tables A1-3 and A1-4.

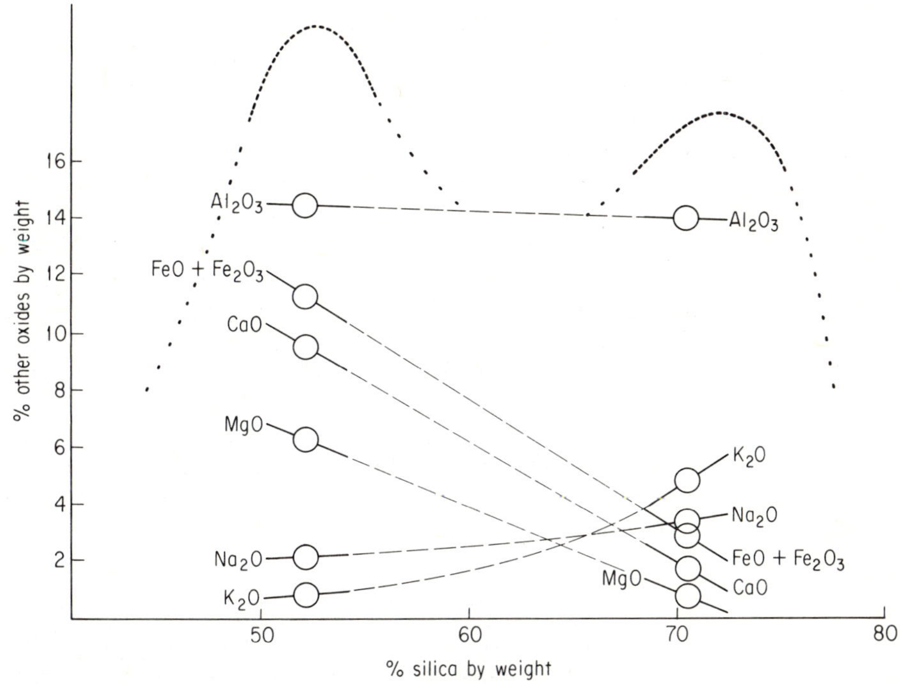

Fig. 7-1 Approximate compositions of the most abundant igneous rocks. The dotted line is an indication, of uncertain precision, of the relative abundances of rocks with different silica contents (modified from Richardson and Sneesby, 1922).

"all magmatic rocks" is greatly affected by the method used for sampling the population. Almost certainly, the curve drawn does not show the predominance to be as striking as it actually is in terms of volumes because the curve is based on the relative abundance of analyses in a population of analyses, and such a population almost necessarily includes a higher proportion of rare varieties than their volume justifies. There is then, in terms of volumes, a marked scarcity of rocks with silica percentage between 58 and 66.

If there were in fact two wholly independent populations, we might well expect the trends in one to lack continuity with the trends in the other; but no discontinuity in trend appears in the data. Relatively scarce though they are, rocks in the middle range are numerous enough to show that the "other oxide" percentages change moderately smoothly. In passing from 50 per cent SiO_2 to 75 per cent SiO_2 we pass from 50 per cent "other oxides" to 25 per cent "other oxides." Purely arithmetically, therefore, we might expect each individual oxide at 75 per cent SiO_2 to be half as abundant as at 50 per cent SiO_2. In fact, CaO, FeO, Fe_2O_3 and MgO all fall by more than half, while Al_2O_3, Na_2O and K_2O do not fall by as much as half; in other words, relatively, they increase. Na_2O and K_2O increase absolutely as well, but the relative increase in Al_2O_3 is as important; from being less than $\frac{1}{3}$ of "total other oxides" at 50 per cent SiO_2, it is over $\frac{1}{2}$ of "total other oxides" at 75 per cent SiO_2.

The mineral trends which parallel these chemical trends are shown in Fig. 7-2. It cannot be stressed too strongly that minerals do not actually occur in the proportions shown: the mineral compositions of real rocks are far too variable to be represented. The procedure is comparable to drawing a single line to show how men's weight is related to their height. Though the following properties are evident in the diagram, their importance justifies further description. Trends are noted as if from left to right:

1. Free quartz appears when the rock's content of SiO_2 reaches 55 to 60 per cent by weight; the amount of free quartz is greater in more acid rocks, and it often forms 30 per cent of the rock by volume when the weight percentage of SiO_2 is over 70.

Fig. 7-2 General trend of mineral compositions and silica content of rocks. Horizontal axis: percentage of silica by weight as in Fig. 7-1. Vertical axis: mineral content in percentages by volume. Example: the diagram indicates that rocks with 58 per cent silica by weight often contain some potash feldspar, much plagioclase (somewhat sodic), some pyroxene, hornblende and biotite, possibly a little quartz and not much, if any, olivine or muscovite. The diagram makes no attempt to show how widely rocks with 58 per cent silica can vary in mineralogy, and to read "pyroxene content = 12 per cent" is misleading; "pyroxene content in range 0–25 per cent" is a reading more consistent with the intention behind the diagram.

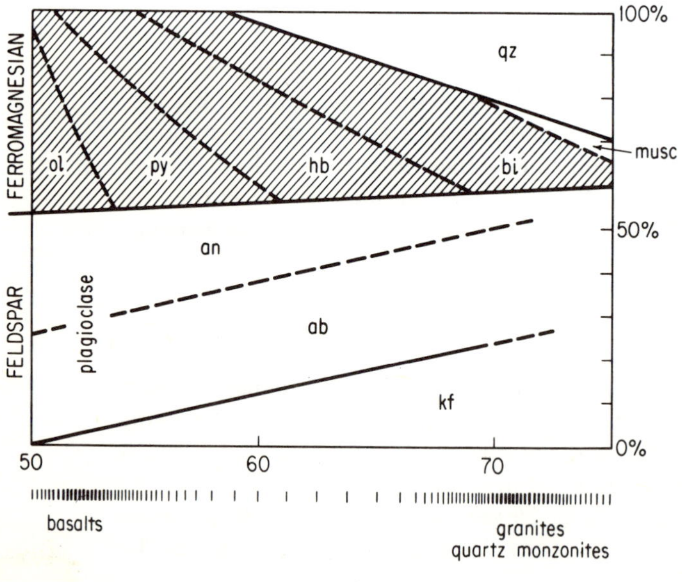

2. The total feldspar content of a rock is often about 60 per cent by volume, although especially in basic rocks it can be much less or much more according to the aluminum content. Acid rocks carry more potash feldspar and their plagioclase is poorer in anorthite, in comparison with less acid rocks. (Anorthite molecules contain more aluminum than the potash feldspar that comes in to take their place, but the diminution in anorthite is accompanied by the appearance of aluminous ferromagnesians, and total aluminum changes little, as in Fig. 7-1.)

3. The main aluminous ferromagnesian minerals (micas and amphiboles) are also hydroxyl-bearing; not only do acid magmas dissolve more water than basic magmas, but on crystallizing, they generate hydrous minerals in larger quantities.

4. Potassium ions, as well as aluminum and hydroxyl ions, enter the ferromagnesian minerals in rocks with higher contents of SiO_2 by weight; thus even where potash feldspar merely equals soda feldspar in amount (at the acid end of the range illustrated), the K_2O content exceeds that of Na_2O, as in Fig. 7-1.

5. The disappearance of olivine is not followed immediately by the appearance of free quartz; over a small interval, increase of silica is absorbed by the formation of more amphibole in place of pyroxenes and more potash feldspar in place of anorthite. This interval corresponds to the group of "saturated" rocks.

6. As the total content of iron and magnesium oxides decreases (see Fig. 7-1), the ratio of iron to magnesium increases; thus in a given rock suite, the pyroxenes of the more acid members are less magnesian than those of the less acid members.

(If minerals had simple invariable formulas, there would of course be a very close arithmetic relation between the oxide amounts shown in Fig. 7-1 and the mineral amounts shown in Fig. 7-2. Even as it is, a chemical analysis in oxide amounts can be used to calculate a *possible* set of mineral amounts, and this process often reveals the mineralogic significance of quite small changes in chemistry. Such a chemically-possible set is called a normative analysis or *norm*, in contrast to a set of observed volume percentages which constitutes a modal analysis or *mode*. Examples of a simplified form of norm calculation are given in Appendix 3.)

The behavior of some minor constituents is sufficiently systematic to be included in this summary: Ni, Cr, Co and S are more abundant in basic rocks than in acid; Li, Rb, Zr, Mo and F are more abundant in acid rocks than in basic. Abundance of Ti and P is not confined to any particular degree of acidity, but is noticeable in rocks off the main trend in the alkali class.

The foregoing remarks and Figs. 7-1 and 7-2 summarize *trends* that are present in probably 90 per cent of continental magmatic rocks (from beneath the oceans, sampling has so far been less informative). The discussion of departures from the main trend will necessarily occupy more space, but should not mask the transcendent importance of the trends just introduced.

DEPARTURES FROM THE MAIN TREND

Reference to Fig. 6-1 shows that the main trend of variation in magmatic rocks runs diagonally across the diagram, cutting off a small corner of granodiorites and a larger corner of syenites, nepheline syenites etc. Independently of the divisions in Fig. 6-1, a class of alkali rocks was distinguished from the commoner nonalkali kinds. Rocks exceptionally rich or poor in feldspar were also noted, and finally we need to consider rocks exceptionally rich or poor in hydroxyl ions. The different deviations are necessarily described one by one, as if there were a finite number of ways in which a rock can depart from "normal"; but of course there are not. If we could conceive a diagram in enough dimensions to fully portray the composition of all magmatic rocks, it would be a continuum like Fig. 7-3A, not like Fig. 7-3B.

Granodiorites

Granodiorites can be regarded as transitional in two ways into rocks of the main trend. A diorite or gabbro can grade into a granodiorite if the quartz content of the rocks is below 10 per cent at one point and over 10 per cent at another; such a granodiorite might have as much as 40 per cent of dark minerals, with appreciable pyroxene. On the other hand, a quartz monzonite can grade into a granodiorite if the ratio of plagioclase to potash

Fig. 7-3 The nature of the variation among rocks. As far as can be discovered, rocks form a continuous population as in A; the concept of central types and discrete deviation paths, as in B, is a descriptive convenience only.

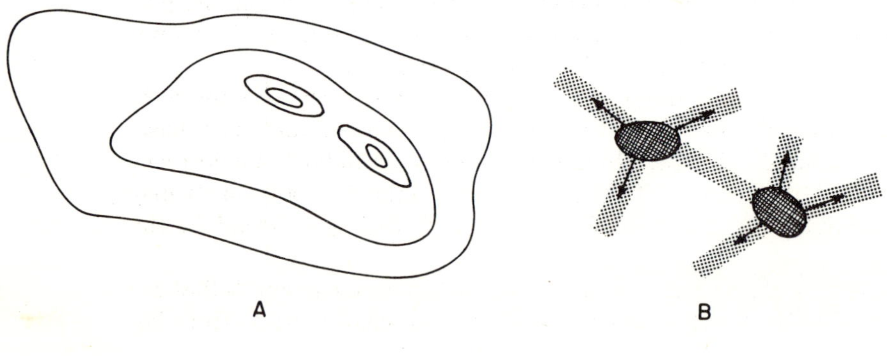

feldspar is higher at one point than at another; a granodiorite with such affinity is likely to be poorer in dark minerals and to have biotite rather than pyroxene. Granodiorites of both these kinds occur naturally, as well as ones of intermediate character.

Syenites

The field of syenites may similarly be regarded from two points of view: a syenite may resemble a granite except for being quartz-poor, or it may resemble a monzonite except for being anorthite-poor. Again the difference would be in the ferromagnesians—the former would carry biotite and perhaps muscovite, with little hornblende; the latter would carry more ferromagnesians, especially more hornblende, and little muscovite.

The difference between *rock suites* including granodiorites and *rock suites* including syenites can be expressed by the Alkali Lime Index. A *suite* is a group of rocks whose field relations and composition characteristics make it appear that they have a common source. For any suite, a diagram such as Fig. 7-1 can be drawn and, from the diagram, a silica percentage can be located at which the total $Na_2O + K_2O$ percentage equals the CaO percentage. This SiO_2 percentage is called the Alkali Lime Index for the suite (Fig. 7-4A; Peacock, 1931). It will be seen that for a rock suite to include syenites, it must have a low Alkali Lime Index and to include granodiorites, it must have a high Alkali Lime Index (Fig. 7-4B). Suites with low indexes are called alkali suites, but it is not correct to say, on this basis alone, that a syenite is an alkali rock (see Chap. 6 and next paragraph).

Alkali Rocks

The criteria in Chap. 6 generate three subdivisions of alkali rocks:

1. $(Na_2O + K_2O) > Al_2O_3$. In this condition, there is insufficient alumina to make feldspar, feldspathoid or mica out of all the alkali present; the excess therefore enters either an amphibole such as riebeckite or a pyroxene such as aegirine.

2. $(Na_2O + K_2O) > \frac{1}{6}\ SiO_2$. In this condition, shortage of silica prevents all the alkali from going into feldspar, but if alumina is sufficient nepheline or leucite forms instead.

3. If both criteria are satisfied at once, the rock is likely to carry nepheline and riebeckite or aegirine.

The tendency of alkali rocks to be rich in Ti and P also appears in their mineralogy: the pyroxene is often titanium-rich as well as soda-rich, and the phosphorus leads to relatively abundant apatite. Whereas

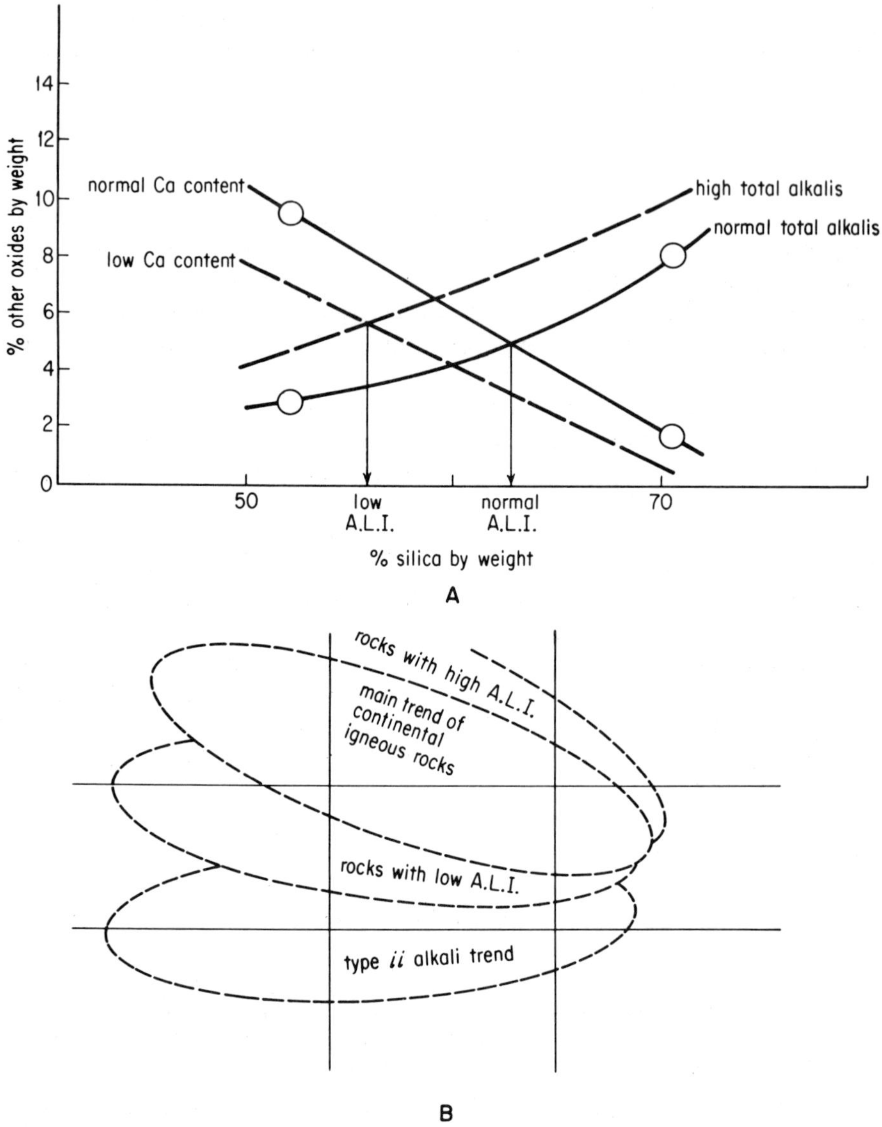

Fig. 7-4 The Alkali Lime Index and alkali rock suites. (A) Construction of the Alkali Lime Index (A.L.I.); (B) suites of different A.L.I. compared with the grid from Fig. 6-1; a suite with low A.L.I. may be called an *alkali suite*.

alkali rocks of types (2) and (3), with silica shortage, are confined to the lower parts of Figs. 6-1 and 7-4B, rocks that are alkalic through alumina shortage can occur in any of the classes. It is left to the reader to satisfy himself that an alkali rock in the class of granodiorites can be conceived, but that it is highly unlikely that any such rock will occur in nature.

Hydroxyl Content Variations

One of the important equations of petrology is:

$$\underset{\text{HYPERSTHENE}}{3\ \binom{Mg}{Fe}SiO_3} + \underset{\text{POTASH FELDSPAR}}{KAlSi_3O_8} + H_2O = \underset{\text{BIOTITE}}{K\binom{Mg}{Fe}_3Si_3AlO_{10}(OH)_2} + \underset{\text{QUARTZ}}{3\ SiO_2}$$

Figure 7-2 shows that potash feldspar and hypersthene usually coexist in rocks up to about 60 per cent silica and that, beyond that point, biotite and quartz usually appear instead. However, deviations from this pattern must be expected, and in fact the crust does contain some silica-poor rocks carrying biotite and some silica-rich rocks carrying hypersthene. The former usually fall into the class of biotite diorites or granodiorites; the latter would fall into the class of hypersthene monzonites or granites, if they were indeed igneous. As a matter of fact, the igneous origin of many such rocks is in doubt and the term *charnockite* may be preferred; *charnockite* implies the same minerals as hypersthene granite but does not imply anything about genesis.

A second equation of great importance is:

basalt + water = hornblende gabbro (hornblende + plagioclase)

Exact equations in terms of minerals are very cumbersome but are of the type:

$$\underset{\text{PYROXENES}}{CaMgSi_2O_6 + 3\ MgSiO_3} + \underset{\text{ANORTHITE}}{CaAl_2Si_2O_8} + H_2O = \underset{\text{AMPHIBOLE}}{(Ca_2Mg_4Al)Si_7AlO_{22}(OH)_2}$$

Neither the amount nor the pressure of water needed to drive the reaction completely to the right is great—2 per cent by weight at 2 kb pressure is sufficient (Yoder and Tilley, 1962). Consequently, beside normal gabbros with olivine and pyroxene, we expect to encounter hornblende gabbros formed by crystallization in deep, moist situations. As with charnockites, there are other ways of producing a coarse-grained amphibole-plagioclase rock beside direct crystallization; these will be discussed in Part Four.

Anorthosites

Rocks containing more than 90 per cent of plagioclase do not belong in any class in Fig. 6-1, but form the separate class of anorthosites. The feldspar they consist of is usually andesine or labradorite, less often a more calcic plagioclase. Their small content of other minerals is usually pyroxene, so that they are closer to a gabbro than to any other class of rock. Some anorthosite bodies are in fact sheets that alternate with sheets very rich in pyroxene, so that the assembly differs from a diorite or gabbro only in the segregation of the components. Other bodies have no visible pyroxene-rich associates, however, and the origin of these bodies is a more difficult problem (see Chap. 9).

Ultrabasic Rocks

Of rocks containing less than 10 per cent of feldspar, most are clearly not magmatic, but there are a few whose origin may at least be sufficiently connected with magmatic processes to justify their mention at this point. These are the dunites, peridotites and pyroxenites (dunite = over 90 per cent olivine; peridotite = over 90 per cent olivine and pyroxene; pyroxenite = over 90 per cent pyroxene). They may be regarded, like anorthosite, as deviations from the class of gabbros in which the ratio between feldspars and ferromagnesians has gone to an extreme. Of the three kinds, pyroxenite is by far the least abundant, and peridotites are usually so close to dunite in composition that many geologists do not distinguish between the two. The accessory minerals are often of importance, because the enrichment in Ni, Cr, Co etc. that was mentioned as a characteristic of gabbroic rocks is sometimes even more marked in the ultrabasics—the nickel in the world's great nickel reservoir at Sudbury, Ontario, is concentrated in the ultrabasic parts and so is the chromium in the Stillwater complex, Montana.

Carbonatites

Rocks are known whose field relations indicate solidification from a melt and yet whose grains are mostly carbonates rather than silicates. Such rocks could form by fusion of limestone, but they could also conceivably form from a large volume of silicate magma, if the carbonate content of the magma could all be concentrated in a small space. Because they *may* be extreme deviations from a silicate family, carbonatites deserve mention here.

The main constituent is either dolomite or calcite, and one or the other usually forms over 80 per cent of the rock; less commonly, both calcite and dolomite are important. The remaining fraction can be made up from a very wide variety of minerals—silicates, oxides, sulfides and more rare varieties. The commercial importance of carbonatites rests, in fact, on their tendency to carry rare metals, such as niobium, in economic concentration in the minor constituents. Rocks rich in sodium carbonate are also known from recent eruptions in Africa, but on account of this mineral's solubility in water, it is far less likely to be preserved in the geological record.

FORMS AND ABUNDANCE OF THE LESS COMMON MAGMATIC ROCKS

The description in Chap. 5 of forms in which magmatic rocks occur was centered on the common varieties, granite, quartz monzonite, basalt etc. A statement on the mode of occurrence of the less common varieties is now in order.

Granodiorites, syenites and their volcanic equivalents have the same modes of occurrence as the commoner rocks, except that they do not, by themselves, form bodies of large size (over 1000 km³). They form small stocks, dikes and flows by themselves, and in addition occur extensively as variations within large bodies of more normal composition. For example, the Hawaiian rocks include andesite, trachyte and rhyodacite flows; quartz monzonite batholiths may contain pockets of granodiorite, and dominantly gabbroic lopoliths are sometimes layered in such a way that one layer treated in isolation might conform to syenite or granodioritc in composition.

Alkali rocks can be divided by their occurrence into two groups, those that are associated with abundant carbonate rocks and those that are not. The latter occur, like granodiorites, (1) as variants, of comparatively small bulk, associated with larger, more normal bodies—for example, in the Pacific basalt islands and in the lavas of East Africa, and (2) as small bodies in their own right—for example, at Chibina in the northwest extremity of the U.S.S.R. This body measures about 40 km across, but the great majority of alkali intrusives are far smaller. The former, carbonate-associated alkali rocks, are less extensive still: both intrusions, as at Alnö, Sweden, and extrusions, as at Vesuvius, Italy, are limited to a few cubic kilometers.

Charnockites, anorthosites and ultrabasic rocks have in common the fact that no corresponding lavas are known (hence the uncertainty as to their genuinely magmatic origin). Charnockites are known only in regions that have been severely metamorphosed. Hence, although rocks now having charnockite composition can be mapped over irregular areas of many square kilometers (possibly 400 km² in the Adirondack Mountains), there is no agreement on the size or shape of any hypothetical charnockite magma-body.

Anorthosite occurs in two ways, (1) in association with comparable amounts of gabbro and more basic rocks, in layered lopoliths and other intrusions, and (2) as the major component of large, roughly homogeneous bodies, where the relative volume of other rock types is very small. The size ranges of the two groups are similar—the Adirondack anorthosite body (type 2) is similar to the Sudbury body in area, and larger bodies of type 2 occur in Labrador and Quebec. The Adirondack body also seems to be lopolithic in form, its underside, as reconstructed by geophysical methods, being concave upwards for the most part; but it may differ from an ideal lopolith in being discordant and in this respect resemble more the Herefoss body (Fig. 5-7). Charnockite and anorthosite often occur together, with gabbroic intermediaries also present. The assemblage is known only from shield areas of metamorphosed Precambrian rocks, as in India, Australia, Scandinavia and North America.

Ultrabasic bodies are also of two kinds: (1) subdivisions of layered intrusions, and (2) independent intrusions (cf. anorthosite). The ultrabasic layers are concentrated mainly in the lower parts of the layered bodies; pyroxene concentrations are as common and as massive as olivine concentrations (700 m or more in the Stillwater complex, Montana). The independent intrusions, on the other hand, are dominantly olivine or serpentine, with pyroxenite and gabbro as minor variants mainly on the margins. The ultrabasic layers and anorthosite layers of the lopoliths are in a way complementary, and combine to give an average composition for the whole assemblage somewhere in the gabbro range. The independent ultrabasic intrusions, however, have no connection whatever with the independent anorthosites; they are mostly much smaller, and, instead of forming broad flat bodies in a fairly stable crust, they form small steep-sided intrusives strongly concentrated along lines of mountain-building activity. The best-known such line is that of the Appalachian Mountains, with its hundreds of separate bodies each a few kilometers long, or less; but in other such lines, the individual bodies are larger, and reach batholith dimensions in New Zealand and New Caledonia (110 by 40 km).

The occurrence of carbonatites brings us back to the field of unquestionably magmatic rocks, because lavas of carbonate composition have been seen solidifying (Dawson, 1962). However, the volume of carbonatite known is tiny. Magmatic origin has been ascribed to 30 or 40 bodies but none is more than a few kilometers across. They do not differ in form from silicate bodies of comparable size (mainly dikes and irregular plugs) except for a somewhat more marked prevalence of ring structures. Whereas *some* alkali rocks are associated with carbonatites, *all* carbonatites have alkali rocks in association.

The observable facts about the magmatic rocks of the earth's crust have now been summarized. Clearly, the summary is an extreme reduction of the body of present knowledge, which is in turn a circumscribed part of what advancing technology will one day reveal. On the other hand, the facts that have been presented are more than sufficient to start speculation ranging. It is the purpose of succeeding chapters to see how many of these facts can be given mutually consistent explanations.

CHAPTER EIGHT

PETROGENESIS 1: BINARY SYSTEMS

The formation of a magmatic rock is an extended process: long before crystallization begins, the ions in the magma form many bonds with each other, so that the magma has an internal structure which differs from a crystal more in the degree of disorderliness than in any absolute property; and long after crystallization is more or less complete, an interstitial fluid permits change of crystal outline and adjustment of crystal composition to falling temperature. Nevertheless, the key to the understanding of magmatic rocks is the freezing process, and it will be our first topic of study.

The freezing of water is a sufficiently well-known process. The freezing of brine is slightly more complicated, because the composition of the solid that forms is not the same as the composition of the liquid it forms from. The fact that brine contains two components, H_2O and NaCl, gives it this flexibility—in the partly-frozen state, there is no reason for the ratio $NaCl/H_2O$ to be the same in the floating ice as in the liquid. When a magma with nine major components and many minor ones is studied, it is not surprising that the complications are greater. In fact, magmas are so complex that it is surprising to find that much of their behavior can be understood by comparison with systems with only two components; but this is so, and consequently we shall make a thorough study of two-component systems our first objective.

BINARY SYSTEMS

A *system* is a quantity of material defined by weights or numbers of molecules, atoms etc., not by volume. (It is usually a convenience to imagine some kind of container around the system but the container is not part of the system.)

A *phase* is any physically homogeneous part of the system. Ten g of water sealed in a tube with no bubbles is a system with one phase; the same 10 g sealed in a larger tube, so that a bubble is present as well as liquid water, is a system with two phases. The same 10 g sealed in an expandable container will consist of one phase (vapor) if the external pressure is less than the vapor pressure of the water; if the external pressure is raised, however, the container will shrink as the water condenses, until the system consists of another single phase (water) at much smaller volume. Many geological situations are of the latter kind, where the environment exerts a pressure on the system greater than the vapor pressure of the constituents. In this circumstance no vapor phase remains and the volume occupied by the system is determined by the density of the constituents.* Some further examples of systems in this vaporless condition are:

> 10 g of water and 1 g of powdered glass: two phases—one liquid and one solid;
>
> 10 g of ice and 1 g of powdered glass: two phases, both solid;
>
> 10 g of water and 1 g of oil: two phases, both liquid;
>
> 10 g of water and 1 g of alcohol: one phase, a solution;
>
> 10 g of water and 1 g of salt: one phase, a solution;
>
> 10 g of water and 10 g of salt: two phases—a solution (saturated) and some excess solid;
>
> 10 g of anorthite and 10 g of albite: one phase, a solid solution (homogeneous plagioclase);
>
> 10 g of potash feldspar and 10 g of albite: two phases—two solid solutions (albite with a small potash feldspar content and potash feldspar with a small albite content).

The number of components in a system is the smallest number of constituents by means of which the compositions of all phases can be expressed. The examples above are all two-component or *binary* systems. There is sometimes an element of choice over what constituents to use to state the compositions of the phases; for example, if a mixture of enstatite and forsterite is melted to a homogeneous liquid, the liquid's composition can be expressed as A g of enstatite and B g of forsterite, or as X g of

*The term "condensed system" is in use, but it does not refer to systems in the vaporless condition here described.

MgO and *Y* g of SiO_2. There is no choice over the number of components—the number is two, however the specifications are made. It will be noted that the number of components is often less than the number of atomic species: in the above example it is not necessary to state the amounts of oxygen, magnesium and silicon separately, because of the implication that each magnesium atom is accompanied by exactly one oxygen atom and each silicon atom by two. A similar implication was made when brine was described as a two-component system: if the brine in question is made by adding salt to water, it is not necessary to state the quantities of Na and Cl separately; with sufficient precision for most purposes, the two can be taken as equally abundant. However, if part of such a brine is allowed to evaporate, this assumption is no longer safe: H ions and Cl ions may enter the vapor, forming HCl molecules and leaving Na and OH ions behind. If this happens, it will no longer be possible to express the composition of the brine as x of H_2O plus y of NaCl—three components will be necessary instead. It is therefore unwise to attempt categoric statements such as: "The system MgO–SiO_2 *is* a two-component system"; it is less misleading to say: "Under such-and-such conditions, combinations of MgO and SiO_2 behave as two-component systems with sufficient precision for the purpose in hand." Under many geologic conditions, and for geologic purposes, this is true; but in semiconductor research, for example, one has to recognize the fact that even crystalline MgO need not carry *exactly* equal numbers of Mg and O atoms, so that for such purposes, the assertion is not true.

Clearly, the number of components has some effect on the number of phases that are possible; an exact relation between these two will emerge in the following pages (especially Chap. 20).

BINARY DIAGRAMS AND THE BALANCE PRINCIPLE

Because the behavior of a system depends only on the proportions of the components and not on their absolute amounts, binary systems are usually discussed in terms of percentages, rather than in grams. The contents of a system can then be specified by a point on a line as in Fig. 8-1A or, more usually, as in Fig. 8-1B.

Next consider the mixing together of two feldspar crystals whose compositions are represented by the points B and C in Fig. 8-2. We wish to know what overall proportions would result upon mixing b grams of feldspar B with c grams of feldspar C. It can be shown algebraically that the overall proportions are given by A if we make $AB = [c/(b + c)] \times BC$. In this condition it is also true that $AC = [b/(b + c)] \times BC$ and that $b \times AB = c \times AC$.

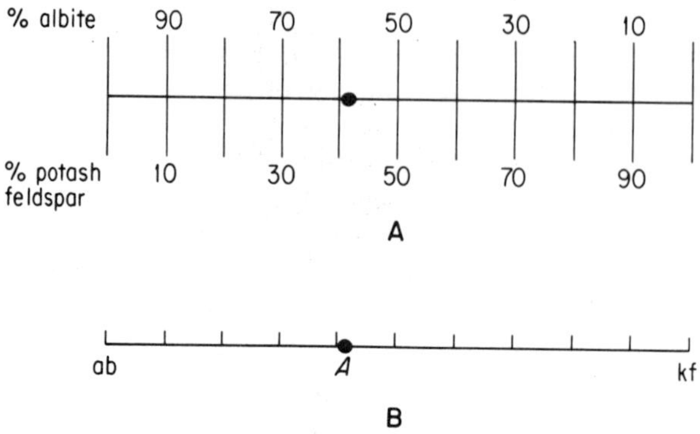

Fig. 8-1 Binary scales.

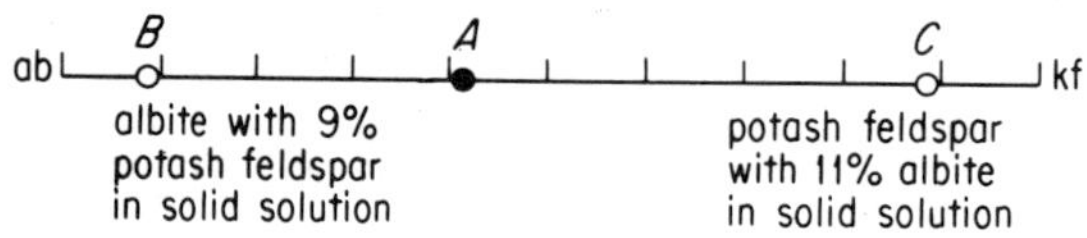

Fig. 8-2 A binary partition.

Commonly in geology we wish to consider unmixing, and to answer the question: If a grams of mixture A separates into two phases with compositions B and C, how much of each phase will result? The same relations are used for this problem.

EXAMPLE: Let A be 41 per cent potash feldspar; then

$$b \times (41 - 9) = c \times (89 - 41)$$

whence

$$2b = 3c$$

Suppose the total weight a of the system is 100 g, i.e., $b + c = 100$; then $b = 60$ g, $c = 40$ g.

Check: Content of albite molecules in 60 g of $B = 60 \times 91\% = 54.6$ g
Content of albite molecules in 40 g of $C = 40 \times 11\% = 4.4$ g

Total $= 59.0$ g

Content of albite molecules in 100 g of $A = 59$ g, a satisfactory check.

The statement that the mixing and unmixing of phases obeys the rule $b \times AB = c \times AC$ is known as the balance principle, by analogy with a balance arm supported at A. Other statements about the same behavior are: (1) the more a phase differs from the bulk in composition, the less of it there can be, and (2) the amount of one phase is proportional to the composition difference of the other phase from the bulk composition. The reader will notice that both of these need qualification to make them exact; they are merely short verbal forms to serve as reminders of the exact algebraic relations; Fig. 8-3 is an ideographic reminder.

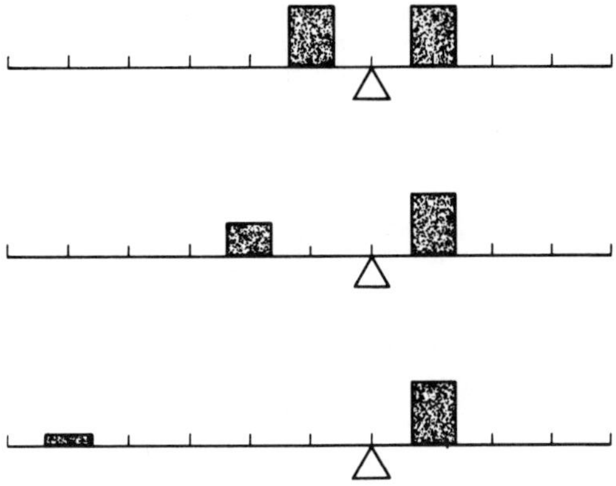

Fig. 8-3 The balance principle. The further from the pivot the left-hand weight is set, the smaller the weight required for balance.

THE QUARTZ-ALBITE SYSTEM*

We are now in a position to examine the cooling process in some liquids containing two components. The first to be discussed are albite-quartz liquids.

Imagine a melt containing the appropriate ions—Na^{+1}, Al^{+3}, Si^{+4}, O^{-2}—with equal numbers of Na and Al, far more Si than either, and O in the appropriate quantity for neutrality. Even before crystallization begins, many Si—O bonds are present: they are simply longer, weaker and more liable to disruption than in crystalline quartz. Above the crystallizing point, the vigorous particle motion and interference from the Na and Al ions prevent the Si and O from locking into the rather rigid system that constitutes a crystal, but with falling temperature, the particles lose some vigor; a moment comes when the Na and Al ions can no longer prevent the Si—O bonds from locking firm, and some crystals of quartz appear. The growth of these quartz crystals changes the ion proportions in the remaining liquid—it is impoverished in Si and relatively enriched in Na and Al. Because the latter ions hinder the congealing of the disorderly Si—O linkages, no more quartz will form unless the temperature continues to fall. Thus, in Fig. 8-4, if X_1 is one point at which crystallization is just possible, X_2 may be another.

In ideal circumstances, the process does not go on in jumps, so that there is a line of points between X_1 and X_2 at all of which crystallization is

*In anhydrous experiments with mixtures of SiO_2 and $NaAlSi_3O_8$, quartz never crystallizes—its place is taken by tridymite or cristobalite. Quartz does crystallize from hydrous melts of silica and albite, but the system is then not strictly binary. To speak of quartz crystallizing in the binary system is therefore inaccurate. However, for the purpose of making intelligible the behavior of magmas, this usage is the most convenient.

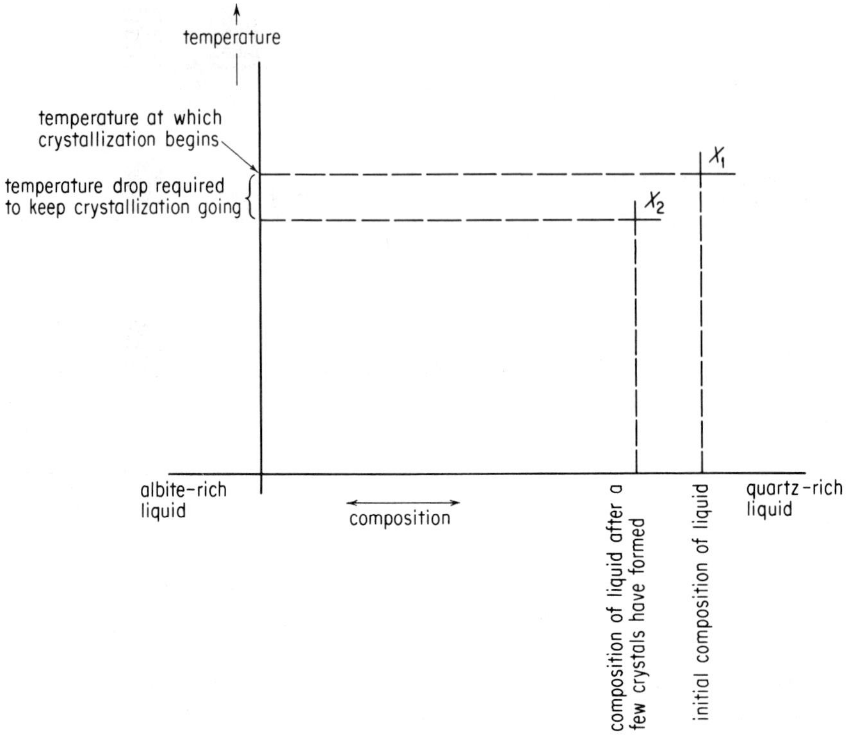

Fig. 8-4 Behavior in the quartz–albite system (1).

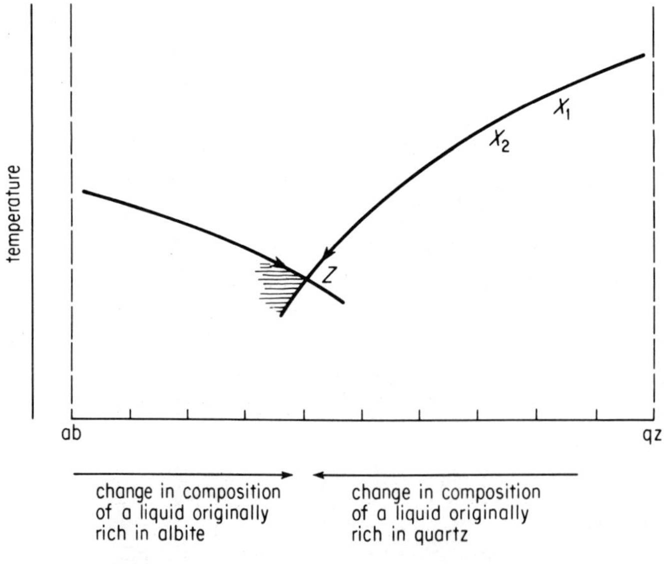

Fig. 8-5 Behavior in the quartz–albite system (2). Lines show change of temperature required to just keep crystallization going in a liquid of changing composition.

just possible and through which the liquid will pass as quartz separates; experiment shows that, in most cases, the line gets steeper as crystallization proceeds. Similar effects control the crystallization of albite from liquids at the other end of the composition range, so that the complete diagram for cooling of liquids in this system is as shown in Fig. 8-5.

What is the significance of the point of intersection, Z? All along the line X_1–X_2 and on downward, Na and Al have been becoming more abundant in the liquid. Eventually the condition must be reached where collisions between these ions and Si—O units become so frequent that albite crystals begin macroscopic growth. The point Z is this point. Whichever side Z is approached from, it represents the liquid's condition when crystallization of the second mineral begins to occur, in conjunction with continued crystallization of the first mineral that appeared. The two minerals crystallize at Z in the same proportions as they exist in the liquid at Z. Suppose the liquid at Z has a composition 40 per cent quartz, 60 per cent albite and that a cluster of crystals were to form containing quartz crystals and albite crystals in the ratio 45:55 by weight. The liquid would be relatively enriched in albite and would move into the shaded area in Fig. 8-5. Here albite would continue to crystallize but quartz would not, and the composition of the liquid would be driven back to Z. Thus once the composition corresponding to Z is reached, the liquid changes no more; crystallization continues until all liquid has been used up.

The quantities involved are considered next. Let the initial liquid at temperature T_0 be equivalent to 15 per cent albite, 85 per cent quartz (A_0 in Fig. 8-6). We read from the figure that upon cooling, crystallization begins at T_1, and may ask: What has happened by the time the temperature falls to T_2? We read that at T_2, the phases present are a liquid of composition B and solid quartz crystals. To make the notation match Fig. 8-2, pure quartz is referred to as composition C and the starting mixture as composition A. Then application of the formulas given shows that $b = c$; i.e., of every 100 g of initial liquid A, there have formed 50 g of liquid B and 50 g of quartz crystals (*check:* albite content of liquid phase = 50 g × 30% = 15 g of albite = amount of albite in initial mixture). An exactly similar calculation for temperature T_3, infinitesimally above that of Z, shows that at this point 75 g of quartz have crystallized and 25 g of liquid remain (still carrying 15 g of albite component but by now only 10 g of quartz component). After the liquid crystallizes at Z in 60:40 proportions, thus delivering quartz and albite simultaneously, the final quantities of solid that result—85 g of quartz and 15 g of albite—match the quantities in the initial liquid. This state is reached by T_4, infinitesimally below the temperature of Z.

The progress of the *system* can be followed, as well as the progress of the separate phases. The initial state is: bulk composition A, temperature

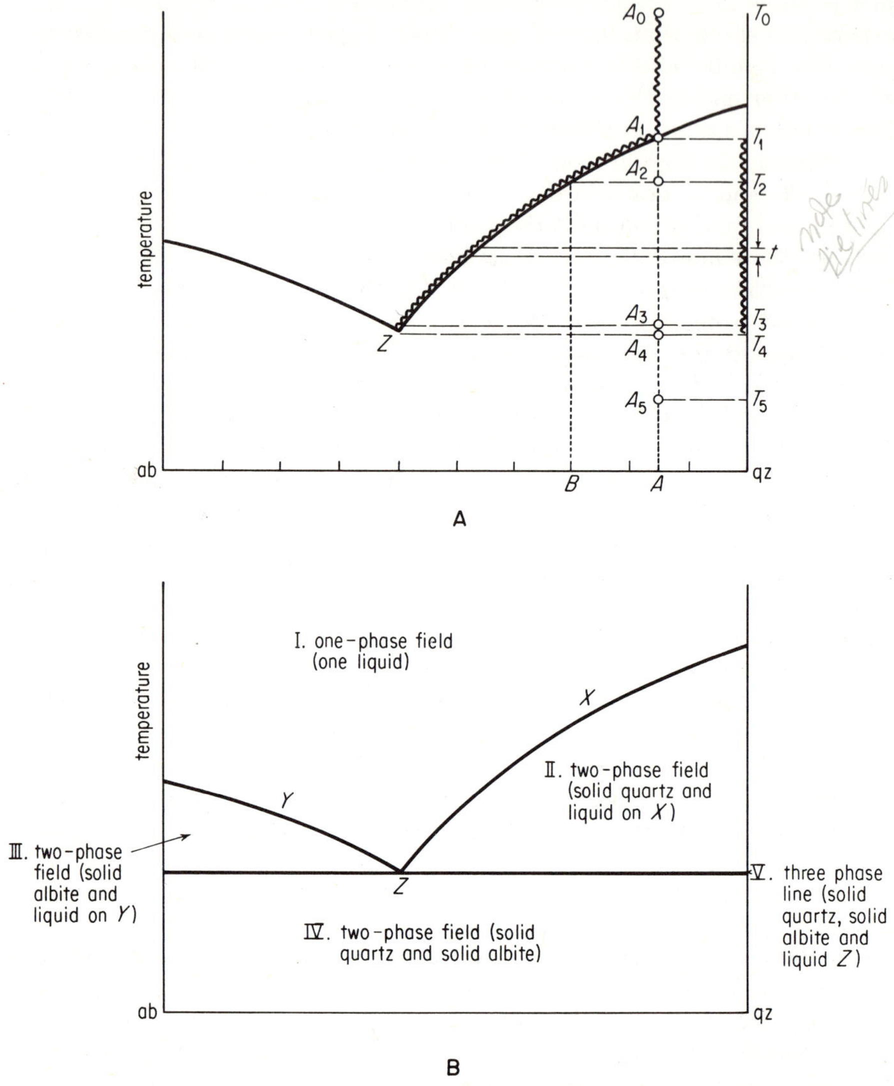

Fig. 8-6 Behavior in the quartz–albite system (3).

T_0, represented by the point A_0. Subsequent states are:

bulk composition A, temperature T_1—represented by A_1
bulk composition A, temperature T_2—represented by A_2
bulk composition A, temperature T_3—represented by A_3
bulk composition A, temperature T_4—represented by A_4
bulk composition A, temperature T_5—represented by A_5

Because nothing is added or taken away, the system as a whole is constrained to travel down only one line of constant composition. In so doing, however, it changes from a one-phase state (liquid only), through a two-phase state (liquid and one solid) and a three-phase state (liquid and two solids), to a final two-phase state (two solids). The states of all possible such systems made of quartz and albite can be summarized as in Fig. 8-6B; for example, whenever the point representing the system is in Field II, the phases present at equilibrium will be some liquid on line X and solid quartz; hence Field II is a two-phase field.

TERMINOLOGY

PHASE DIAGRAM: Fig. 8-6B is called a phase diagram at constant pressure.

LIQUIDUS: The line X in Fig. 8-6B is called a liquidus, the line Y is another.

EUTECTIC POINT: The intersection of lines X and Y is called a eutectic point, because, at the pressure for which the diagram is drawn, a solid mass of this composition can be completely melted at a lower temperature than a solid mass of any other composition.

The heat changes in the system are next studied. We shall need the following symbols:

CL = specific heat of liquid
CQ = specific heat of quartz
CA = specific heat of albite
HQ = latent heat of crystallization of quartz
HA = latent heat of crystallization of albite
ML = mass of liquid
MQ = mass of solid quartz
MA = mass of solid albite

(The first five quantities change as cooling proceeds, but so slightly that they may be treated as constant. The last three of course change radically as cooling proceeds.) Over the path A_0–A_1, the heat loss is $ML \times CL \times (T_0 - T_1)$. Over the path A_1–A_3 the heat loss is of three kinds. Consider cooling through a small temperature interval t within T_1–T_3, during which a small quantity of quartz mQ crystallizes. The heat given out is:

1. $ML \times CL \times t$ due to cooling of liquid
2. $MQ \times CQ \times t$ due to cooling of solid quartz
3. $mQ \times HQ$ due to crystallizing of quartz

At the temperature of Z, a quantity of heat

$$(NQ \times HQ) + (MA \times HA)$$

is given out, where NQ = mass of quartz that crystallizes at Z. No heat

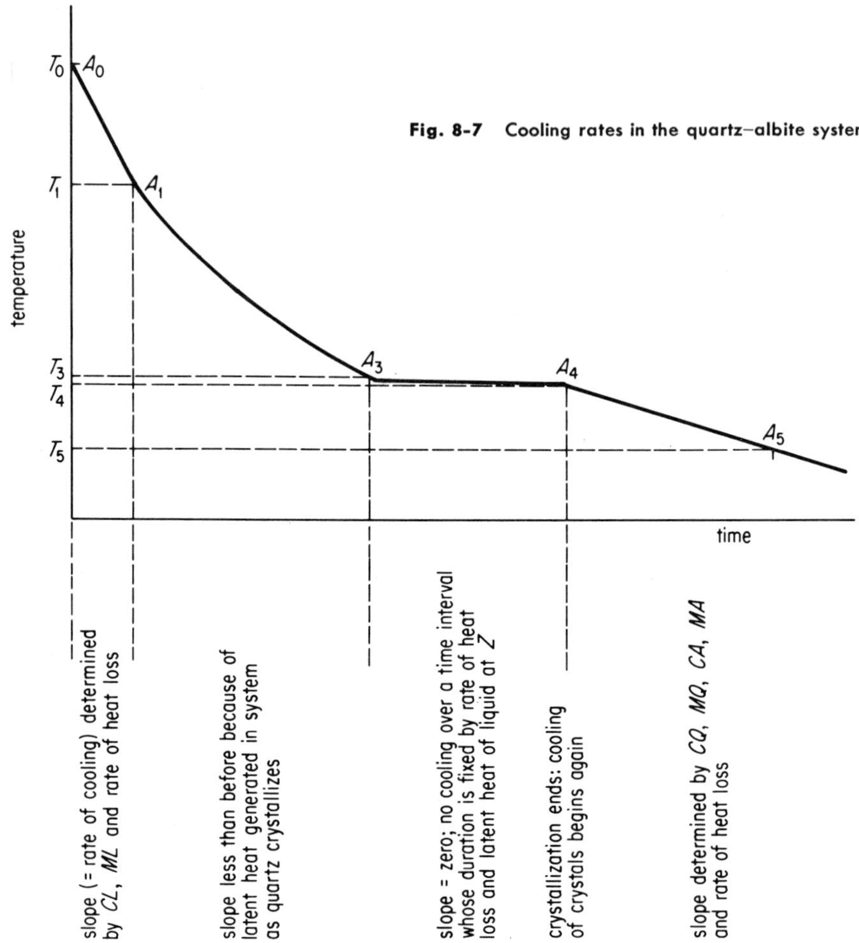

Fig. 8-7 Cooling rates in the quartz–albite system.

loss due to cooling occurs during this crystallization, because the temperature does not fall below that of Z until crystallization is complete. Finally, over the remainder of the journey to A_4, heat is lost by the cooling solids:

$$MQ \times CQ \times (T_3 - T_4) + MA \times CA \times (T_3 - T_4).$$

The heat loss can be summed up in a diagram if it is assumed that the system loses heat at a constant rate with time (see Fig. 8-7). It is important to note that the system remains on the three-phase line (V in Fig. 8-6B) for a period of time which is certainly finite in the idealized example and could be thousands of years in an analogous geologic situation.

The hypothesis was advanced in Chap. 2 that some magmas are produced by the melting of crystalline rocks. It is therefore also important to understand what happens if heat is supplied at a constant rate with time to a beach sand or sandstone of, say, 15 per cent albite crystals and 85 per cent quartz crystals. Fortunately, no new processes are involved; the result is an exact reversal of the steps that have been described. Important points are:

1. The first liquid to form has composition far from the composition of the solid it forms from (first liquid carries 60 per cent albite, at Z);

2. After this liquid begins to form, a considerable amount of heat can be supplied without its composition changing—the additional heat merely generates more liquid of composition Z;

3. The liquid formed at Z extracts *all* the albite from the solid mixture: by the time state A_3 is reached, only quartz is left in the solid state. Thus *partial melting* can generate a marked composition contrast between the melt and the solid residue;

4. *Total melting*, to a fluid equivalent in composition to the initial solid, requires far more heat and far more time. Whereas crystallization usually proceeds to completion, melting is probably only partial in most geological instances.

Lastly, we consider the effect of pressure on the cooling process. If Fig. 8-6 exactly represents behavior for a system that is maintained at a constant pressure P_1, then a system maintained at constant pressure P_2 will behave according to a diagram of similar but slightly different shape. This means that crystallization or melting can be provoked by change of pressure as well as by change of temperature. Consider a system in the state D, Fig. 8-8. At low pressure P_1, it will be a homogeneous liquid, but if the pressure rises to P_2, two phases will coexist—solid quartz and a more albitic liquid. Whether increase of pressure causes melting or crystallization depends on the components. In ideal binary systems of silicates, increase in pressure usually causes crystallization, but in analogous geological situations, the presence of water usually interferes with ideal binary behavior, and increase of pressure is just as likely to cause melting. This very important effect is more fully discussed in the next section.

The behavior of quartz-albite liquids has been discussed in detail for two reasons:

1. If one system is understood in detail, brief outlines of other systems are adequate. The reader should practise on the systems in the following pages: he should be able, from the outline given, to fill in the details of what is happening at any point in any diagram.

2. Many geologically important systems either are binary systems of the same type, or resemble them sufficiently to be understood in the same terms; for example, the systems quartz-anorthite, nepheline-albite, quartz-diopside and diopside-anorthite behave almost as if they were binary over a wide range of conditions.

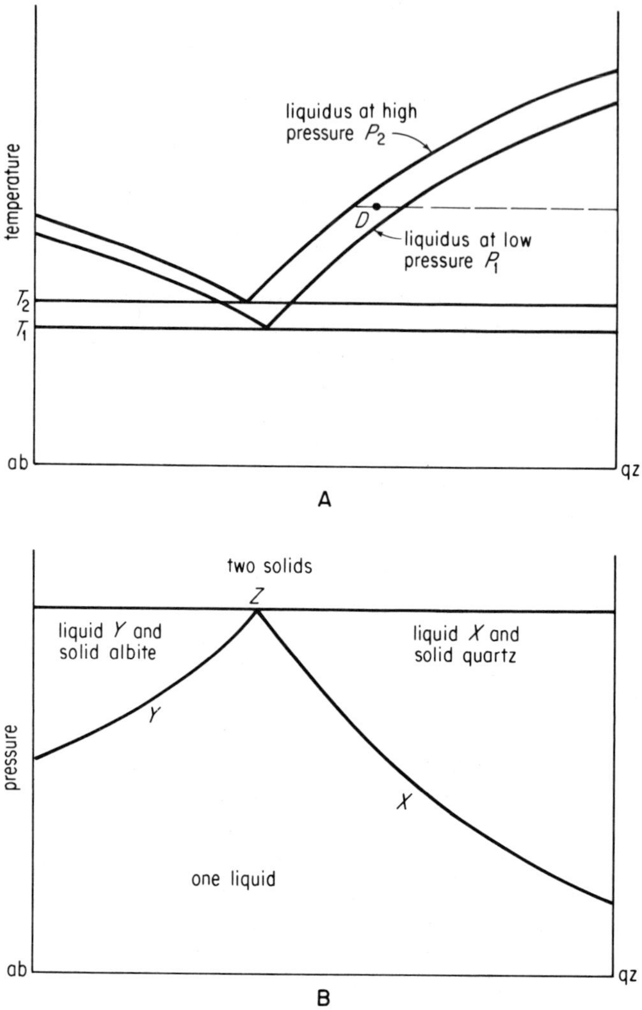

Fig. 8-8 Changes with pressure in the quartz–albite system. (A) Phase diagram at constant pressure P_2 superimposed on diagram for constant pressure P_1; (B) phase diagram at constant temperature (between T_1 and T_2 in diagram A).

(Further details follow in Chap. 9. It may be noted that even the system Pb–Zn departs from binary behavior to a slight extent on account of isotopes, and no crystal is truly stoichiometric—the albite that crystallizes from a quartz-albite melt is not identical with the albite that crystallizes from an alumina-albite melt. Whether we say a system is or is not of a given type therefore depends on the magnitude of the deviations from ideal conditions that we are prepared to neglect.)

THE QUARTZ-WATER SYSTEM

To understand this system, it is necessary to remember the "critical behavior" of water, summarized as follows: when water boils under a pressure of 1 atm, its relative density changes from 0.96 to 0.0006; when it boils at 100 atm, its density changes from 0.69 to 0.054; at 200 atm, from 0.48 to 0.18; and under 222 atm, there is no sudden change of density at all. With rising temperature, the water simply becomes more and more diffuse in a completely continuous way. We shall consider the quartz-water system only at pressures greater than 222 atm and shall refer to H_2O in this state as *aqueous fluid* in preference to more specific terms such as *water* or *steam*. A partial phase diagram for the system at some pressure P above the critical pressure of 221 atm is shown in Fig. 8-9, which is far from correct dimensionally.

The diagram contains two portions similar to Fig. 8-5. At the left-hand end, we read the cooling of a fluid of composition A thus: it cools to T_1 without change of state; at T_1, when the fluid can be described as a saturated solution of quartz in the aqueous fluid, quartz begins to crystallize; the liquid cools and loses quartz until either it reaches B, where quartz and ice crystallize simultaneously, or, more probably, it reaches the temperature of its environment and stops cooling. This process is wholly analogous to the crystallization of sugar or any other solute from aqueous solution; it also resembles the behavior of quartz-rich liquids in the quartz-albite system. A fluid of composition C, on cooling, forms ice as the first solid and becomes enriched in the solute; this behavior again is similar to

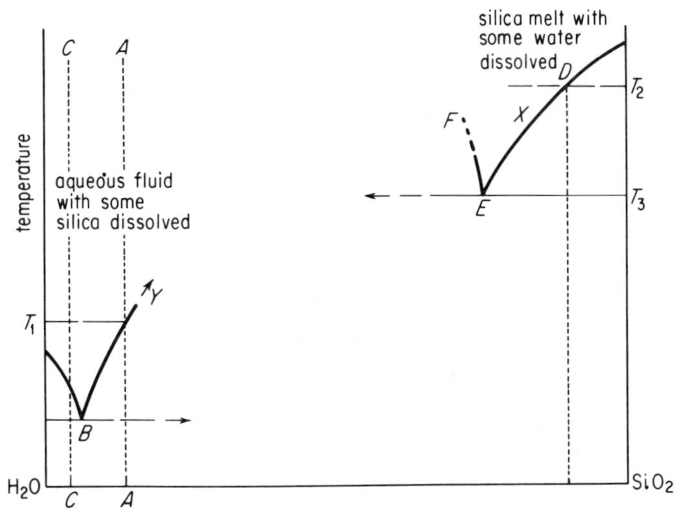

Fig. 8-9 Behavior in the quartz–water system (1) (not to scale).

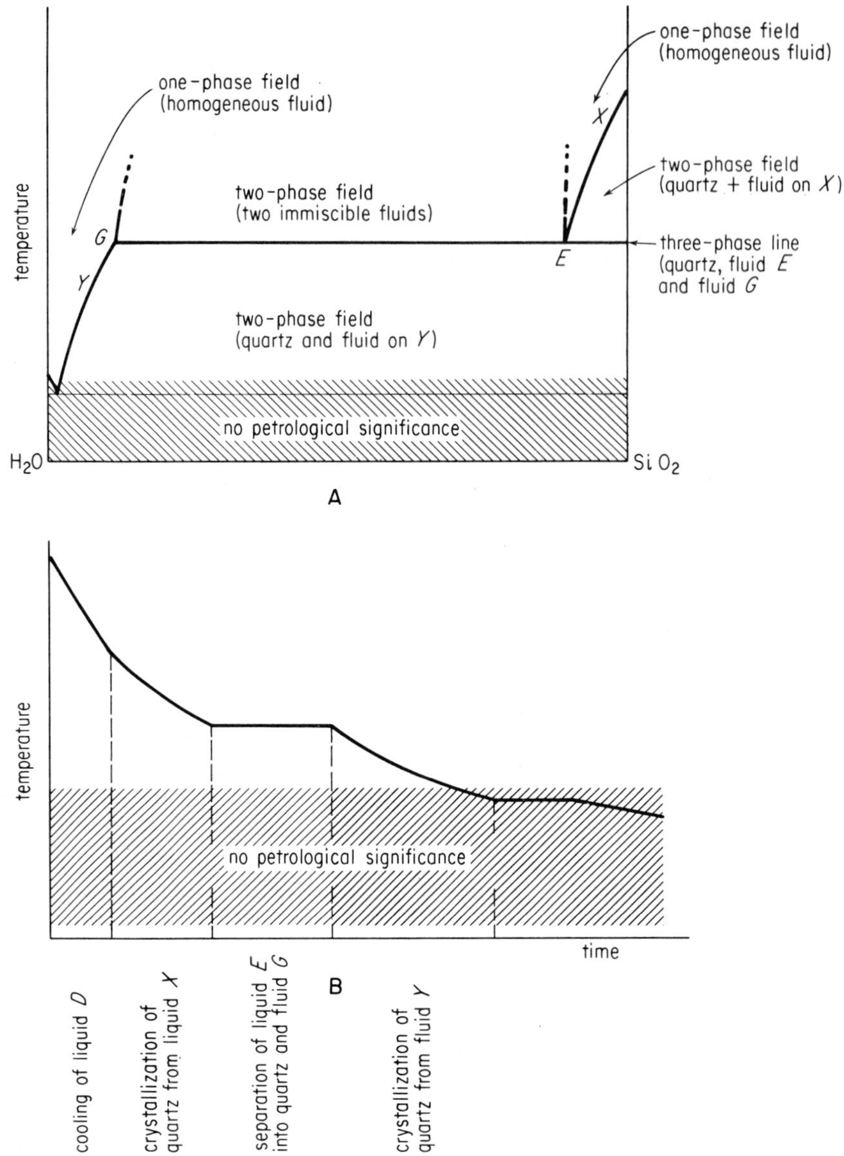

Fig. 8-10 Behavior in the quartz–water system (2).

everyday fluids (seawater etc.) and to albite-rich liquids in the quartz-albite system, but it has no geological importance.

At the right-hand end, a liquid of composition D can be described as a solution of water in molten quartz. At T_2, quartz begins to crystallize, the liquid cools and is enriched in the other component. The liquid reaches E where, by analogy with the quartz-albite system, we expect a second phase to begin to separate, simultaneously with quartz. But the second phase which appears, corresponding to albite, is not pure water; it is an aqueous solution represented by the point where the two arrows intersect. Thus, above T_3, we have quartz in contact with a water-saturated silica melt on curve X; below T_3, we have quartz in contact with a silica-saturated aqueous fluid on curve Y. A more complete phase diagram at constant pressure and a heat-loss diagram are shown in Fig. 8-10. As far as is known, E is over 90 per cent quartz in all conditions up to 8 kb and 1000°C, and G is under one per cent quartz at least up to 4 kb and 550°C (Weill and Fyfe, 1964; Kennedy *et al.*, 1962). At higher pressures and temperatures, G can move to much higher values (details are given in Chap. 10) and the presence of contaminants such as hydrofluoric acid (HF) also increases the solubility of quartz.

The behavior of many silicate minerals is expected to resemble that of quartz. In particular the albite-water and potash feldspar-water systems are known to be of this kind, and the gap EG of similar size (Goranson, 1938). At temperature T_3, the two fluids E and G coexist without mixing, like oil and water, and the gap is consequently known as an *immiscibility gap*. The gap gets wider at lower pressure (Fig. 8-11) and the temperature at which it occurs gets higher. Here is one reason for the effervescence of lavas. Let a magma that has reached the point R be elevated from a deep high-pressure site to a shallower low-pressure site: in the high-pressure situation, R is in the field where the system exists as a single homogeneous fluid, but in the low-pressure situation, R is in the field where the system consists of two fluids, one siliceous and one aqueous. In passing from one situation to the other, the magma excretes aqueous droplets, which either escape toward the surface or form vesicles. The process is closely analogous to uncorking champagne.

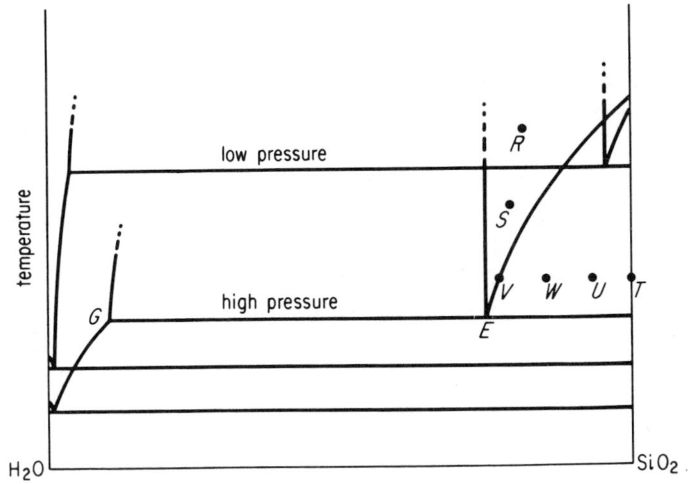

Fig. 8-11 Changes with pressure in the quartz–water system. The figure shows partial phase diagrams at two different pressures, both above the critical pressure for water.

If the magma is initially in condition *S*, the state approached at low pressure is again a two-phase state, but now the two phases are an aqueous fluid, as before, and solid crystals. That is to say, as the pressure falls, crystallization proceeds (with vesiculation). This behavior is different from that of anhydrous systems, where decrease in pressure usually discourages crystallization (Fig. 8-8).

As before, it is important to think in terms of melting solids as well as crystallizing fluids. A rock consisting of crystals with a few percentage parts of water can be made to begin melting by raising its temperature. Once it has reached a certain temperature, it can also be made to begin melting by raising the pressure. When a rock is buried beneath a growing pile of sediments, pressure and temperature usually rise together and cooperate in encouraging melting. Changing the water content can also either initiate or extend the melting. For example, dry crystals may reach *T* in Fig. 8-11; if then water creeps into the system and changes its overall composition to *U*, at high pressure the second phase that forms will be a melt *V*—thus the water initiates melting. If water continues to creep in and moves the system to *W*, no new phase appears but the proportion of melt *V* increases at the expense of solid *T*—thus water extends melting. The geological importance of these effects is unknown, because knowledge of the movement of water in the crust below 10 km is nil. However, there is no doubt that water-bearing sediments are sometimes carried to great depth before being metamorphosed to an anhydrous state, so that water migrating back to the surface is a geological agent whose potency should not be overlooked. To reinforce this point we need only note that when *V* is at 90 per cent, the balance principle shows that every gram of water added to *T* causes ten times its weight of melt to form.

THE ALBITE-ANORTHITE SYSTEM

This differs from the quartz-albite system in containing two solids that can form homogeneous solid solutions of all proportions. A mixture of solid quartz and solid albite can be ground until it looks homogeneous, but it still remains an aggregate of some particles of quartz and some particles of albite; short of melting the mixture to a glass, nothing will make the two combine. If, however, a mixture of albite and anorthite is ground fine and then kept hot, the two will diffuse into each other without ever leaving the solid state, and form a *solid solution*. The process is exceedingly slow; it is much quicker to melt the two together and allow the melt to crystallize, but the result is the same—a homogeneous crystal with a uniform distribution of Ca ions and Na ions throughout.

The phase diagram for the system is shown in Fig. 8-12. The line X is a liquidus as in Fig. 8-6 and shows the temperature at which any liquid cooling from the field above will begin to crystallize. The crystals that form, however, are not pure anorthite. Their composition is shown by the *solidus* W, thus: a liquid of composition A, on cooling to crystallization point T_1, develops a first few crystals of composition B. These crystals,

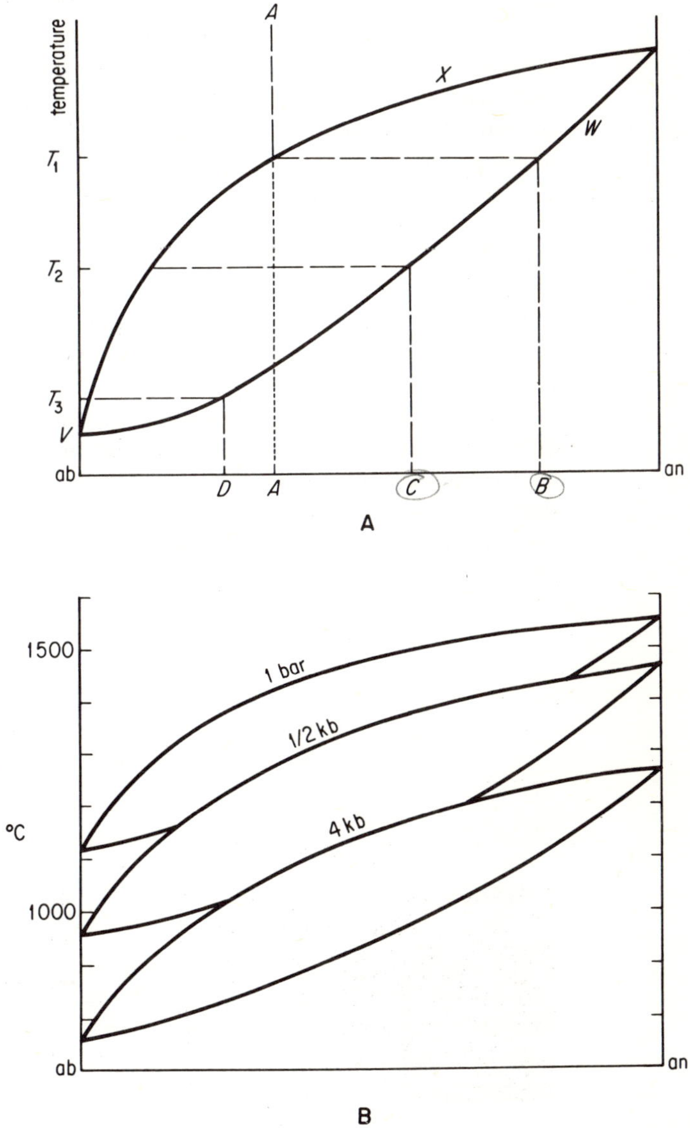

Fig. 8-12 The albite–anorthite system. (A) At one bar; (B) showing depression of the liquidus at ½ kb and 4 kb water pressure (Bowen, 1913; Yoder, 1958; Tuttle and Bowen, 1958).

while not pure anorthite, are still much richer in anorthite than the liquid from which they formed; consequently they leave behind in the liquid an increased proportion of albite. Exactly as in the quartz-albite system, crystallization continues if, and only if, the liquid cools down the line X as its composition changes. By the time it cools to T_2, the crystals forming have composition C, and, if all the crystals, once formed, play no further part in the changes, the liquid may cool to temperatures such as T_3, when crystals form with composition D. In theory, the sequence ends with the last drop of liquid just reaching the composition of pure albite, but in the real world, some kind of interference always prevents this ideal from being attained. Instead the process stops when continued accumulation of crystals more albitic than A makes the overall composition of the entire crop of crystals match the initial composition A. When this happens, there can be no liquid left.

In reality the solids B, C, D etc. do not all form separate crystals; more commonly, plagioclase of later stages grows like a skin on earlier crystals. The result is called *zoning*. If the environment changed smoothly, the composition change would be smooth, from B at the center to D or so at the rim; but naturally-occurring plagioclase crystals are more often zoned in steps, the trend from calcic to sodic being subject to fluctuations (Fig. 8-13).

As was mentioned at the start of this section, plagioclase crystals of different compositions are not, for the most part, stable in contact with each other, and diffusion tends to homogenize most such assemblages. The same is true of the contrasted zones: if kept hot long enough, the zones will merge, and ideally the end product of cooling a liquid A would be a mass of homogeneous solid plagioclase of composition A. Such homogenization

Fig. 8-13 **Zoning in plagioclase.**

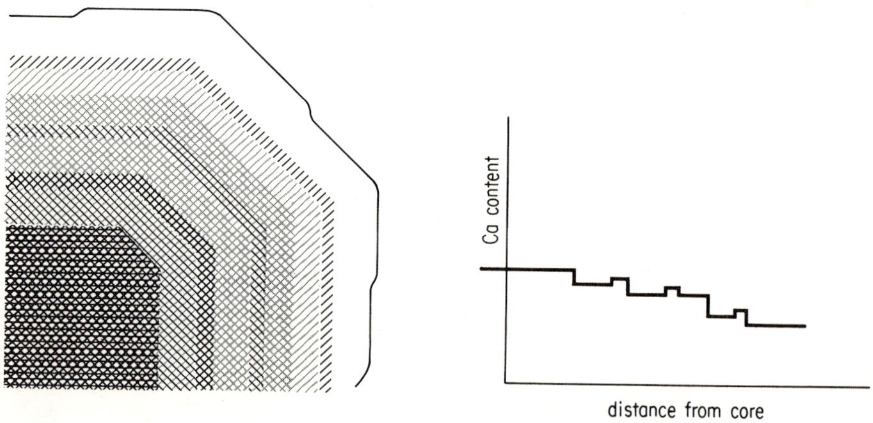

is rarely complete in igneous rocks. The ones that cool most quickly (flows and dikes) show the sharpest zones; larger intrusions cool more slowly and the zones are blurred; in igneous rocks subsequently subject to metamorphism, the plagioclase is often quite thoroughly homogenized. There is also the possibility of diffusion between early-formed crystals and the sodic liquid of a later stage; thus a crystal of B might be changed to C even before D begins to form. Nothing, however, can prevent the final crystal aggregate from having bulk composition equal to A.

(Discussion of such processes is often carried forward in terms of two extreme situations. In one, it is supposed, as above, that any crystal once formed remains thereafter inert; i.e., adjustment of crystal composition by diffusion is nil. In the other, it is supposed that adjustment by diffusion is instantaneous and complete, so that, by temperature T_2, for example, all crystals that formed earlier in the composition range B–C have adjusted by diffusion to composition C. On the latter supposition, no crystal more albitic than A ever forms. The two sequences are known as the *perfect fractionation path* and the *equilibrium path* respectively, and any actual path is bound to lie somewhere between them; that is, some crystals or parts of crystals more sodic than A are bound to form.)

The marked contrast in composition between the plagioclase liquids and solids that are stable in contact with each other is also important in remelting—heating labradorite delivers oligoclase to the melt in the early stages, and the solid residue becomes more calcic. The effect of pressure on the system is as for quartz-albite mixtures: in the ideal binary situation (anhydrous), increase of pressure raises all the temperatures by about 10°C per kb; in presence of water vapor, increase of pressure has a more marked effect in the opposite sense—a lowering of 30°C per kb for most crustal pressures, but up to 200°C during the change from surface conditions to 1 kb (see Fig. 8-12B).

CHAPTER NINE

PETROGENESIS 2: APPLICATION TO ROCKS

The preceding chapter was about simple mixtures of pure chemicals. To explain the variety and abundance of natural rocks we need to know how the behavior of quartz and albite in a multi-component magma differs from their behavior in an ideal binary situation. In brief, the same things tend to happen, but in different quantities and at different temperatures. In this chapter, we shall attempt to construct the behavior of a multi-component magma by adding extra components te a binary system one at a time.

POLYCOMPONENT SYSTEMS AND THE FRACTIONAL CRYSTALLIZATION OF BASALT

As an example of the effect of a third component on a binary system, let us consider a melt formed from some diopside, some anorthite and some albite.

When approaching any system more complex than a binary one, it is well to begin by considering the behavior of the end members in pairs; that is, in this case, by considering the systems diopside-anorthite, diopside-albite and albite-anorthite (the last was studied in Chap. 8). At once we stumble on the fact that a system of two minerals is not necessarily a system of two components, or binary system. The reason lies in the facility with which minerals depart from stoichiometric proportions or, in other words, form solid solutions. For example, in the system diopside-albite, some sodium and aluminum can dissolve in the diopside leaving excess quartz:

$$\underset{\text{DIOPSIDE}}{CaMgSi_2O_6} + \underset{\text{ALBITE}}{NaAlSi_3O_8} = \underset{\substack{\text{PYROXENE}\\ \text{SOLID SOLUTION}}}{\begin{Bmatrix} CaMgSi_2O_6 \\ NaAlSi_2O_6 \end{Bmatrix}} + SiO_2$$

and at the same time, some calcium can dissolve in the albite to form a plagioclase. The result is that after a liquid equivalent to a mixture of pure diopside and pure albite develops a few crystals, the remainder is no longer exactly a diopside-albite mixture (see Fig. 9-1); the situation resembles the partial evaporation of brine, discussed in Chap. 8. In spite of these complications, much can be learned by discussing such systems

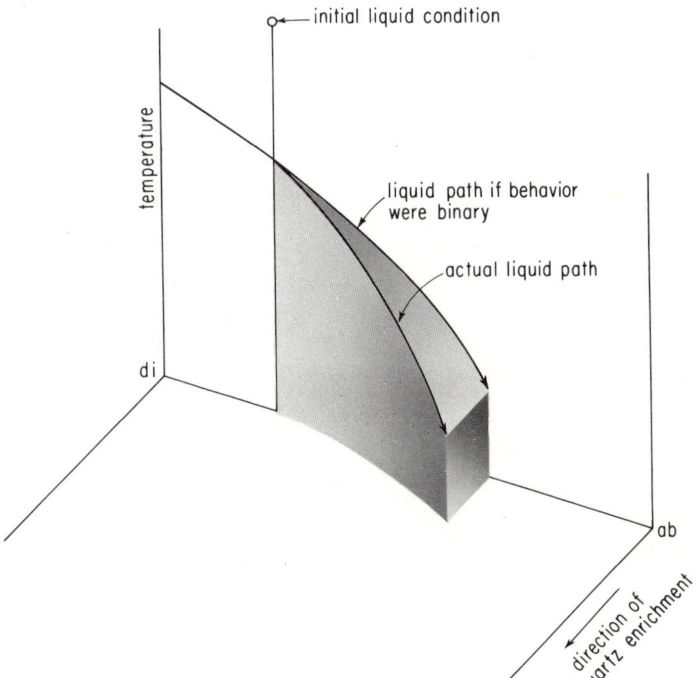

Fig. 9-1 Departure from binary behavior in the diopside–albite system. Beside the one shown, there is another important departure from binary behavior—instead of albite, a noticeably calcic plagioclase forms (Schairer and Yoder, 1960).

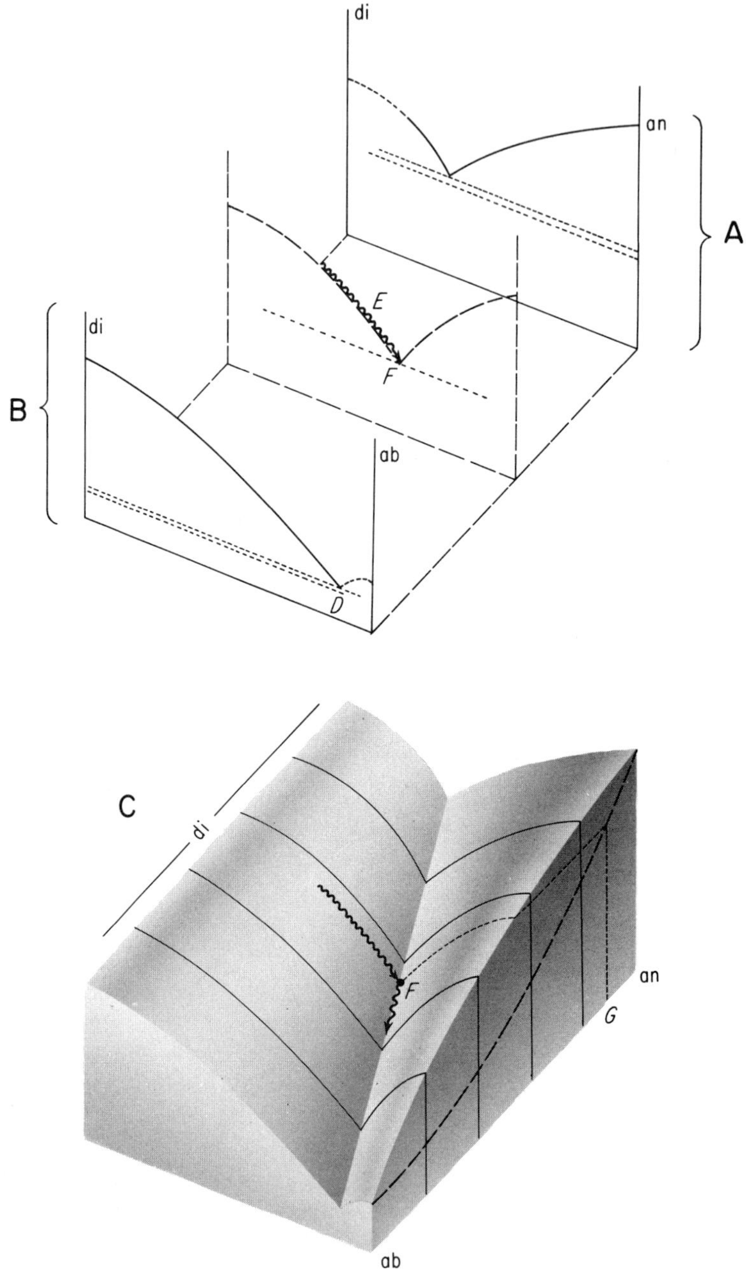

Fig. 9-2 The diopside–plagioclase system. In (A) and (B) the phase boundaries are dotted where departure from binary behavior is most marked; in other parts of the diagrams, simple behavior is assumed. The plagioclase that crystallizes from liquid *F* will not be exactly *G* but will be close to *G* (Schairer and Yoder, 1960; Bowen, 1915).

as if they were binary. In Fig. 9-2 and elsewhere, full lines have been replaced by dotted lines where departure from binary behavior becomes most conspicuous; where the line is full, it may be assumed that the hypothetical binary behavior is a useful guide to actual behavior.

Except for these deviations (the calcium effect is actually the more conspicuous) the system diopside-albite resembles the system quartz-albite in general shape (compare Fig. 9-2B with Fig. 8-6). If a diopside-rich liquid cools, diopside crystallizes alone until the composition of the liquid residue reaches *D*; from this moment on, diopside and plagioclase crystallize simultaneously until no liquid remains (a little quartz seems likely to appear finally by virtue of the deviations mentioned, although no experiment has yet demonstrated this effect). The system diopside-anorthite also departs from binary yet approximates behavior of the quartz-albite type over a wide range of compositions (Fig. 9-2A); again, from a diopside-rich liquid, a pyroxene close to diopside is the first solid to form. We may therefore suppose that such a pyroxene will be the first solid to form from diopside-rich liquids containing both albite and anorthite components. Suppose a liquid initially contains albite and anorthite components in equal amounts. When crystallization begins, the liquid will be relatively enriched in both albite and anorthite but these will remain equal to each other in amount; the change in such a liquid can be represented by an arrow *E*. Eventually the point must be reached where a plagioclase begins to crystallize in conjunction with diopside, and we may ask what kind of plagioclase it will be. We can read from Fig. 8-12A that a liquid containing equal amounts of albite and anorthite, and no diopside, will deliver crystals close to An_{80}. In the presence of diopside, the first plagioclase crystals to appear will not have exactly this composition, but will be somewhere in the bytownite range.

If the liquid originally had equal amounts of anorthite and albite, and bytownite separates out, more albite than anorthite must be left behind. The composition of the liquid therefore changes from (diopside + plagioclase Ab_{50}) to, say, (diopside + plagioclase Ab_{55}). At the same time, the liquid is relatively impoverished in diopside; in the composition diagram it moves away from the anorthite end and away from the diopside face—in colloquial terms, the liquid "runs down the trough" (Fig. 9-2C). The behavior below *F* is similar to eutectic behavior in that two minerals crystallize simultaneously; but it differs from eutectic behavior in that change of liquid composition and fall of temperature continue. The journey from *F* down the trough is referred to as *cotectic* crystallization of diopside and an ever-changing plagioclase; the line of the trough is called a *cotectic curve*. The end product is an aggregate of diopside and plagioclase crystals of various compositions. The process is therefore a kind of combination of pseudobinary systems in which the crystallization of plagioclase accompanies the crystallization of diopside as in diopside-anorthite mixtures and the composition of the plagioclase changes as in the albite-anorthite system.

Of course, solid solution with $NaAlSi_2O_6$ is not the only kind that diopside can be involved in—solid solution with $CaFeSi_2O_6$ is also conspicuous geologically. Over much of the range from 100 per cent Mg to 100 per cent Fe, this solid solution resembles that between albite and anorthite (at the iron-rich end, at high temperatures, the crystals take a different structure, that of wollastonite instead of pyroxene, so that the pyroxene solidus and liquidus do not continue into this region). Thus, if both Fe and Mg are available in the melt, early-formed Mg-rich pyroxene will be followed by later, more Fe-rich pyroxene; but this effect does not alter the behavior already discussed—it is merely superimposed on it. Thus, crystallization delivers a succession of different pyroxenes simultaneously with a succession of different plagioclases. Let us add another component, a slight surplus of silica in the initial liquid: as in the systems diopside-quartz, anorthite-quartz and albite-quartz, silica concentration in the liquid eventually leads to the appearance of crystalline quartz. Let us add another component, water: as in the system quartz-water, separation of an aqueous fluid eventually begins, at the same time as crystallization of the last of the silicates continues. In the same way that extra components turned the three-phase temperature in Fig. 8-6B into a three-phase interval, from F downwards, they also turn the three-phase temperature in Fig. 8-10A into a three-phase interval. We therefore recognize the following stages as possible on theoretical grounds:

1. Crystallization of pyroxene alone;
2. Crystallization of pyroxene and calcic plagioclase;
3. Crystallization of lower-melting pyroxene and medium plagioclase;
4. Crystallization of low-melting pyroxene and sodic plagioclase; quartz appears;
5. Aqueous fluid separates and coexists with magmatic fluid; crystallization continues;
6. Magma exhausted; crystals cool with aqueous fluid interstitial.

In natural rocks this sequence would of course be modified. We have to remark first that olivine may occur with or in place of magnesian pyroxene; that at low temperatures in a hydrous environment, hornblende or biotite will occur in place of low-melting pyroxene; and that where albite occurs, potash feldspar is likely also. Thus stage (1) generates crystals which, settling under gravity, could give peridotite layers in lopoliths. Stage (2) generates pyroxene-labradorite rocks, such as basalts and gabbros, grading into stage (3), andesites and diorites. Stage (4) gives quartz monzonites, where hornblende or biotite often takes the place of pyroxene; by now much Fe, Mg and Ca have been used up, so that dark minerals are scarcer and the feldspar becomes alkaline. Stage (5) is when the effect of an igneous body on its environment is most striking: it is still hot, and the aqueous fluid is highly active both chemically, as a catalyst, and physically—it penetrates the wall rocks far more freely than the viscous silicate melt, carries solutes, aids diffusion and is generally crucial to the formation of economic mineral deposits. By stage (6), the solvent power of the aqueous fluid is much diminished; it can still turn olivine to serpentine or feldspar to sericite, and can fill wall-rock fissures with vein-quartz, but these are the final acts in the life of a magma body.

If a single, well-confined magma body passes through these stages successively, of course the different products will all accumulate in the magma chamber and, except for those components that are able to penetrate the walls, the final solid body will retain the composition of the initial liquid. It is when the magma moves while cooling that the possibility arises of generating rock bodies with one composition at one site and a different composition at another site; but a necessary condition is that the early-formed crystals should be separated, at least partially, from the surrounding fluid. Fortunately, mechanisms for achieving such separation can be conceived, using either the density contrast or the viscosity contrast between the crystal-rich and crystal-poor parts of the magma. It is thus possible to conceive the generation of different rock types as follows (we presuppose a large pool of basaltic magma within the earth):

1. A part is ejected to the surface and crystallizes en masse to flood basalt etc.;

2. A part begins to crystallize within the earth;

3. The early crystals segregate, giving peridotite;

4. Much of the surrounding liquid crystallizes, giving gabbro or diorite;

5. The remaining liquid moves to a new site where it crystallizes further to monzonite and quartz monzonite or rhyolite-rhyodacite;

6. The last magmatic liquid and the aqueous fluid between them form veins, especially pegmatites, around the quartz monzonite.

The process as a whole is referred to as the *fractional crystallization of basalt.* It is a hypothetical process—no one has seen it occur. As a hypothesis, it is conspicuously successful in explaining many observed facts, although there are some it is quite unsuited to explain.

It must at once be emphasized that the order of events indicated is not the only one possible. Within the range described as basaltic, magmas can vary in the proportions of the different components, and in addition, the sequence followed at one pressure can differ from the sequence followed at another pressure. The following ten ways for the crystallization sequence to begin all seem to be possible, although the five within the boundary have been more thoroughly demonstrated in both field and laboratory:

1	2	3	4 high pressure H_2O sufficient	5 high pressure H_2O deficient
ol ol + py	ol ol + plag	py py + plag	am am + plag	py py + gt
py ol + py (low Mg, high Ca, Fe)	plag ol + plag	plag py + plag	plag am + plag	gt py + gt

It is well to recall at this point what is meant by basaltic magma: it is merely magma in which elements are present in the same proportions as in basalt. Basaltic magma can crystallize to form other rocks than basalt—obviously gabbro is a possibility, including hornblende gabbro as noted in Chap. 7; another possibility is eclogite, a rock dominated by sodic pyroxene and garnet.*

If crystallization begins with any of the sequences inside the boundary, some kind of basalt or pyroxene gabbro is likely to result, and one of the trends in Fig. 7-4B will be followed; but if crystallization begins with amphibole (column 4) the late liquid may be enriched in plagioclase and tend toward anorthosite, and if it begins with either of the eclogite sequences (column 5), the effect of separation of the late liquid will be different again. Thus when we speak of the *fractional crystallization of basalt* and refer to columns 1–3, we are discussing a more restricted range of possibilities than the *fractional crystallization of basaltic magma*, which embraces in addition columns 4 and 5.

Even within columns 1–3, the scope for variation is rather great.

*The equivalence of olivine basalt and eclogite can be briefly indicated thus:

	Basalt		*Eclogite*	
olivine	$\binom{Mg}{Fe}_2SiO_4$	$\rightleftharpoons$	$Ca\binom{Mg}{Fe}_2Al_2Si_3O_{12}$	garnet
plagioclase	$CaAl_2Si_2O_8$			
	$NaAlSi_3O_8$	$\rightleftharpoons$	SiO_2 + $NaAlSi_2O_6$	sodic
pyroxene	$Ca\binom{Mg}{Fe}Si_2O_6$		$Ca\binom{Mg}{Fe}Si_2O_6$	pyroxene

Firstly, the sequence (ol $\rightarrow$ ol + py) has two variants: in one, the amount of olivine continues to increase as pyroxene begins to crystallize; in the other, the crystallization of pyroxene, once started, consumes some or all of the preformed olivine (see Appendix 4 for details). A similar pair of possibilities exists at a later stage with pyroxene and amphibole—they may increase in amount simultaneously, or an amphibole may form partly from magma and partly at the expense of preformed pyroxene. The resorption of olivine has been thoroughly explored in the laboratory but is rarely indicated by observations in thin section. The resorption of pyroxene has been explored with more difficulty in the laboratory, because of the high water vapor pressures needed; in thin section, amphibole that appears to be growing at the expense of pyroxene is very often seen, but in thin section, of course, it can rarely be *shown* that magma was also involved in the reaction. The quantitative importance of such resorptions during crystallization is therefore somewhat uncertain. Because of their possibly great influence on crystallization trends, they are further discussed in Appendix 4.

Secondly, in addition to the variety of ways in which the crystallization sequence may begin, there are different possible subsequent trends (Fig. 7-4B). More detailed study of polycomponent systems shows that to produce a suite with high alkali lime index or one with low alkali lime index, or even a suite of type 2 alkali rocks, only a very slight change in the initial magma may be sufficient. The extent to which the iron is divalent or trivalent also influences the late products more noticeably than the early ones.

The fractional crystallization theory is thus very versatile in explaining a wide variety of rocks; it was in fact a classic example of success in explaining field observations by treating rocks just as any other polyphase chemical system. Beside its power to explain the variety of known rocks, there are three further points in its favor:

1. Great quantities of basalt magma are known to have existed—the basalt floods (Fig. 5-3A) show that recurrently, at different places and different times, enormous quantities of basaltic magma have been available for injection. By contrast, latite floods for example are unknown, so that any theory involving large subterranean pools of latite magma is less acceptable.

2. The balance principle states that if a system splits into separate phases, the phases most removed from the bulk composition of the system are likely to be the least in quantity. Treating the "main trend" (Fig. 7-4B) as the balance arm, this implies that if a basaltic magma solidifies to some basaltic rocks (*close* to the magma in composition), it can only generate fewer granitic or rhyolitic rocks, for these are farther from the initial magma in composition. In the oceanic islands, this is exactly the quantitative relation found—

much basalt, little rhyolite. The same relation appears again in the Duluth body and other lopoliths (much gabbro, little granite), and in the lopolith-size sills, such as those of the Karroo region of South Africa.

3. Some localities have distinctive minor features such as richness in apatite. Where such a feature recurs in all igneous rocks of a locality from basic to acid, its presence supports the general idea of differentiation from a common source, though it does not inform us directly about what the differentiation mechanism might be. Richness or poorness in Na or K is a feature that often distinguishes one suite from another, particularly when expressed in a comparison with Ca or Si or in an alkali lime index.

On the other hand, there are three limitations on the applicability of the theory outlined for columns (1), (2) and (3) on p. 92:

1. If the magma contains two per cent of water by weight at over 2 kb pressure, crystallization will be as in column 4 rather than as in 1, 2 or 3 (Yoder and Tilley, 1962);

2. Even if the magma is dry, at over 18 kb (and possibly as low as 10 kb) crystallization will be as in column 5 rather than as in 1, 2 or 3 (ibid);

3. If the magma is more granitic than basaltic, as it is likely to be if it forms by fusion of continental rocks, nothing that has been written in this chapter so far is adequate to describe what may happen.

Because of limitation (3), the fractional melting of sediments is the next topic to be discussed.

FRACTIONAL MELTING OF SEDIMENTS TO GRANITE

The introduction mentioned that rocks of the crust, under favorable conditions, may melt to magma of granitic composition. We shall now consider in more detail the products that could result. Since granite and quartz monzonite are composed principally of quartz, potash feldspar and albite or oligoclase, we again have three mineral pairs whose behavior is relevant, as in diopside-plagioclase mixtures. The quartz-albite system has been discussed; the quartz-potash feldspar system is closely similar, except for some complexities which appear at pressures lower than those we shall need to consider; the albite-potash feldspar system is shown in Fig. 9-3, as it appears under several kilobars pressure.

The liquidus resembles that in the quartz-albite system in having a

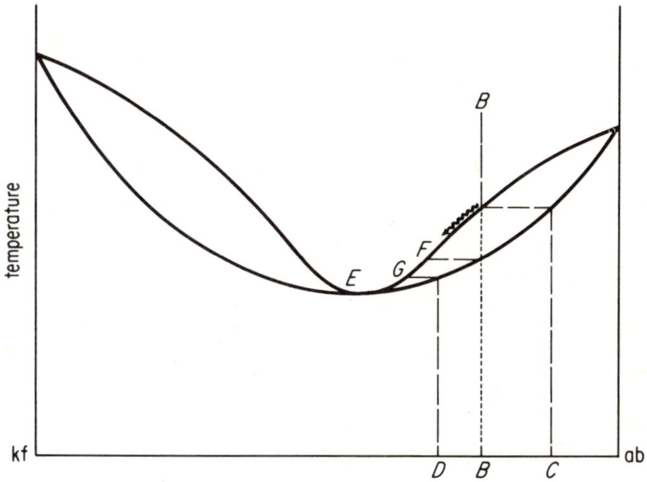

Fig. 9-3 The albite-potash feldspar system, in conditions where neither leucite formation nor solid-state immiscibility interferes with the melting behavior (after Bowen and Tuttle, 1950).

marked central depression, but here the lowest point, *E*, is not a point of discontinuity. The system can alternatively be likened to two plagioclase systems with a common low point. The low point, however, is not the melting point of any ideal pure chemical, as the albite point is in the plagioclase system; the low point is merely the melting point of the mixture of potash feldspar and albite that melts most easily at whatever pressure the diagram illustrates—a point that occurs at different temperatures and different compositions for different pressures. The cooling history of a liquid *B* is very similar to that of the plagioclase liquid *A*, Fig. 8-12A—the solids that form range from composition *C* at the first moment of crystallization to compositions in the region *D* as the last liquid disappears. The exact manner in which the albite and potash feldspar are mixed as solids does not concern us here (they can in fact coexist as separate grains, as perthite intergrowths or as a homogeneous solid solution). Of more consequence at the moment is the question: If an assemblage of grains of the alkali feldspars is heated, what kind of melt will form? The answer is:

1. If the heated grains contain some whose composition is to the left of *E* and some whose composition is to the right, the first liquid will form exactly at *E*;

2. If the heated grains are all to one side of *E*, for example within the range *C–D*, the first liquid to form will be close to *E* but not at *E*. (Suppose (a) that in being heated, the grains in range *C–D* all diffuse into each other to form solid grains of *B*—then the first liquid to form will have composition *F*; suppose (b) that in being heated the grains do not diffuse into each other—then the first liquid to form will form from a crystal *D* and have composition *G*. Which occurs depends on the rate of heating.)

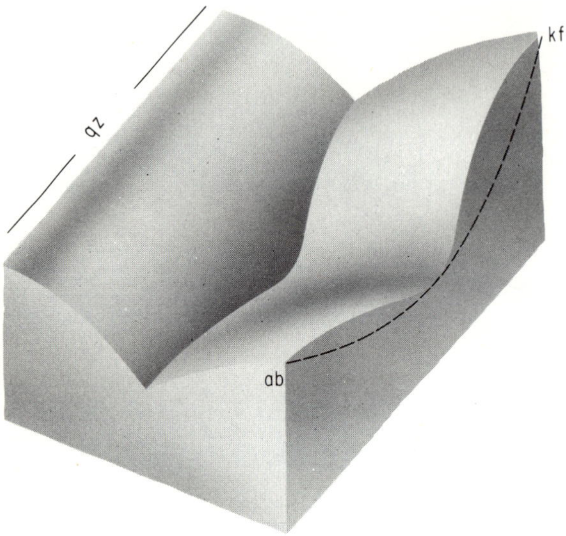

Fig. 9-4 The liquidus in the quartz-alkali feldspar system. A generalized picture of behavior under several kilobars' water pressure, based on work by Tuttle and Bowen (1958). The water present is a fourth component, but the first solids to form need only three components for their description.

Details aside, the first liquid to form will certainly be much closer to E than the bulk composition of the assemblage from which it forms.

We can now consider the three components in conjunction (Fig. 9-4). The figure combines the three two-mineral systems, as did Fig. 9-2, and shows that the three-component liquid with lowest melting point of all has about twice as much feldspar as quartz and equal amounts of the two feldspars—in other words, about equal amounts of all three components.

At this point the reader should remind himself of the melting process in the quartz-albite system (Chap. 8, especially Fig. 8-6): even when the solid is 85 per cent quartz, the first liquid to form is 60 per cent albite (subject to the approximations made in the diagram). Similarly, when a solid is 70 per cent quartz, 15 per cent albite and 15 per cent potash feldspar, the first liquid to form is, by the above, about one-third of each in composition. As such a mixture is heated, liquid of this composition will continue to form as long as the two feldspars remain, i.e., until, of the original mixture, 55 per cent remains solid (pure quartz) and 45 per cent is liquid. This is the reverse of fractional crystallization; it is *fractional melting,* and can produce a liquid fraction with granitic composition from a wide range of initial solid mixtures (Winkler, 1965, Chap. 16).

In the natural situation, fractional melting may reverse the trends of fractional crystallization in considerable detail. Thus, if the temperature of the first-formed liquid rises, minerals can be taken into the melt in the order of their crystallization (but the last first). The actual order will of course depend on the composition of the rock that is being heated, and the amount of variety in the sequences that are possible was emphasized in the preceding section. If the environment were basalt, the reversal could be pursued in theory right back, through one of the ten possibilities on p. 92, to homogeneous magma; but two facts argue against this happening in nature: (1) melting is less likely to go to completion than crystallization, at least at the crustal level discussed and (2) the environment being melted is usually not basaltic in bulk composition.

What *is* the environment being melted likely to be, at 20–30 km depth? There is great doubt about this question, but two generalizations are probably permissible:

1. It is likely to be richer in granitic components (Si, Na, K) than basalt itself;
2. It is likely to deviate from a simple mixture of granite on the one hand and basalt on the other, on account of sedimentary processes.

The first idea is based on observations of the Precambrian shields of Canada, Northern Europe, Australia etc.—the minerals occurring in these shields show that they have been buried to depths of over 10 km and are thus perhaps similar to what is 10 km or more below us now. The second idea is based on the general tendency of sedimentary processes to sort the magmatic components out. Even if a land surface were bare granite and bare basalt, the rivers running off it would deliver quartz to the beach sands, calcium to the shelf limestones, sodium to the sea and aluminum to the deep-water sediments—or, admitting that this is a caricature of the sedimentary process, at least we can suppose that any body of sediments is likely to be enriched in *something*, in comparison with its igneous forebears. The rocks being melted are likely to be sedimentary, at least in part: thus the melting of a limestone-rich succession will give a calcic melt, a quartzose succession will give an acid melt, and a clay succession will give an aluminous melt—all slight variations on the dominant granitic theme.

A very important matter is the behavior of sodium and potassium. In weathering, alkali feldspars go to hydrated aluminum silicates and illite, with a great loss of alkali metals from the crystal lattice. However, by the time a mudstone has recrystallized under metamorphism to grains big enough for microscope study, it contains feldspars and micas again. The mechanism and timing of the alkali metals' return from the sea to the silicates is an outstanding problem of petrology. Ocean sampling is beginning to give data on the alkali metals adsorbed on, and interstitial between, sediment particles, but how these particles change as they are buried under first 100 and then 1000 m of overlying marine sediment is very little known: are they usually squeezed dry while still sodium-poor, or do they absorb much sodium before losing their interstitial brine? If we knew the factors that encourage absorption, we might be able to explain better the association between geosynclines and granite magmas. Pending this knowledge, we can already conceive that poor retention of alkali metals will lead to an aluminous melt and good retention to a more normal melt. These facts are summed up by a modification of Barth's epigram (1962, p. 379): *the diversity of granitic rocks is caused by sedimentary processes.*

Like fractional crystallization, fractional melting may or may not be followed by segregation of the fractions. A small, homogeneous stock whose composition is close to the low point in Fig. 9-4 is usually supposed to have formed from magma that has been completely segregated from its solid residues. On the other hand, a heterogeneous batholith may contain some segregated magma, some segregated solid residues, some pockets where melting occurred but no segregation, and all intermediates between these. The extent to which batholiths contain material that did not melt during the batholith's formation is much argued, and will be discussed in later sections. Here, we are content to summarize the foregoing as follows:

1. Fractional melting of sub-continental rocks can lead to a variety of magmas basically granite or quartz-monzonite in composition;
2. Of such magmas, those efficiently segregated from their solid residues will be close to each other in composition and have roughly equal amounts of quartz, potash feldspar and sodic plagioclase;
3. Magmas less efficiently segregated will give rocks of more varied composition and texture;
4. The melted and unmelted parts of a body will not necessarily be easily distinguished, on account of the power of the interstitial aqueous fluid to promote rearrangement of crystal outlines etc., during the cooling episode.

GENESIS OF LESS ABUNDANT ROCKS

To complete our discussion of the genesis of the rocks described in Chap. 7, we turn to the less abundant varieties.

Carbonate Magmas

The possible mechanisms for generating carbonate magma can be divided into (1) organic and (2) inorganic. Organic mechanisms are those involving the concentration of CO_2 in the form of calcite or dolomite by animals; the alternative, inorganic mechanisms, must depend on the concentration of CO_2 by some fractionation process, from whatever CO_2 was dispersed among the components of the primitive earth. The latter mechanisms are hard to visualize because of our ignorance, but the former mechanisms are impossible, at least through that major fraction of earth history when carbonate-forming animals did not exist. Thus age comparisons of carbonatites with early fossils are important. Many carbonatite plugs are emplaced in Precambrian surroundings and may themselves be Precambrian, but their dating vis-à-vis the dating of the earliest algal limestone reefs has not been established with sufficient assurance to show that they *could not* be remelted limestones.

If it is shown likely that at least some carbonatites predate carbonate-forming animals, the inorganic hypotheses have to be thought of. They are attractive even at present, on account of isotope and trace-element differences between many carbonatite plugs and ordinary limestone—the richness of some carbonatites in niobium for instance. No present-day observations can completely inform us about the Precambrian environment, but there are two main reasons for believing the crust and mantle to have been richer in carbon formerly than they are now: (1) all the carbon now existing near the surface as limestone must, in pre-algal times, have been more widely dispersed, and (2) carbon is among the lighter elements concentrated by gravity in the atmosphere. The escape of carbon into the atmosphere is a continuing process, even now incomplete and presumably less complete 1500 million years ago. The Precambrian crust could then have been rich in carbon in comparison with the present crust. Given a quantity of carbon in the crust, nothing prevents our believing that carbonate magmas may coexist with silicate magmas, or that a silicate magma may generate a carbonate magma by late-stage concentration of CO_2. Even if experiment were to show these possibilities unlikely, this would be compatible with the facts since we already know that the conditions for formation of a carbonate magma are rather peculiar; otherwise carbonatites would cover more than 1/100,000 of the present land surface. To sum up: (1) carbonate magmas are known to exist, having been seen flowing in Tanzania (Dawson, 1962), and (2) both organic and inorganic hypotheses for their origin must be entertained.

Alkali Rocks and Assimilation

At the interface between a magma and a solid rock, the possibility of interpenetration exists (Fig. 5-8D). If a xenolith is partly digested, some of its constituents are *assimilated* by the magma; if it is completely digested, all are assimilated (wall-constituents can also be assimilated, of course, without fragmentation into xenoliths). Where the environment is solidified magma from an earlier stage of the same magmatic event, the process is called *hybridization;* when, as is more common, the environment is some independent solid, the process is *contamination* (assimilation *of* limestone = contamination *by* limestone). The result of assimilation is not always an alkali rock; it usually is not. In fact, all magmas are likely to be contaminated to some extent—once a little magma is present, further fractional melting can alternatively be described as *selective assimilation.* Hence assimilation of sediments is probably very extensive in the formation of ordinary batholiths. It is discussed here in conjunction with alkali rocks only for the sake of a clear example of the process.

The reaction of interest is:

limestone +	=	calcium silicates +
silica-rich magma		silica-poor magma
		(with escape of CO_2 as gas)

Two possibilities must be distinguished: (1) the calcium enters the magma phase; (2) the silica enters the solid phase. In (1) the calcium silicates produced will be lime-feldspar, lime-pyroxene and perhaps lime-garnet, at the expense of quartz, biotite, hornblende and hypersthene; in (2) the calcium silicates produced will be nonaluminous ones—wollastonite, pyroxene and possibly larnite and others. Of course, the processes may be concurrent, and give a lime-enriched magma within silica-enriched wall rocks. In any case, there is apt to be insufficient silica left to form feldspar from all the sodium and potassium, so that nepheline or leucite appears. The leucite lavas of Vesuvius and the nepheline lavas of the Hegau, West Germany, have both been attributed to this mechanism, although their true genesis remains in question.

Since there is no point in defining a term exactly unless one uses it exactly, it should be noticed that only process (2), the escape of silica to the walls, leads to the formation of alkali rocks as defined in Chap. 6. Even when entry of calcium into the magma causes the development of nepheline, it cannot change the ratio $(Na + K)/Si$ in the rock. Thus, not all feldspathoid rocks are alkaline, any more than all alkali rocks have feldspathoid.

Beside feldspathoid rocks associated with limestones, there are feldspathoid rocks associated with carbonatite; in fact, no carbonatites occur without associated feldspathoid rocks. Of course, if carbonatites are fused limestones, the explanation already given for the limestone-feldspathoid assemblage is applicable. If, however, a carbonate magma is produced by segregation in the crust or mantle, we may ask: Does the carbonate magma form and then part of it react with normal magma to give a feldspathoid magma, or is the differentiation process such that carbonate-rich magma and feldspathoidal magma both tend to segregate in the same localities? In other words, is one the cause of the other or do both arise from a common cause? At present, there is no known means of answering this question, but it remains an interesting one, of a kind recurrent through all science.

Peridotite Bodies

Peridotite in layered complexes has been explained as a concentrate of early-formed crystals from a basaltic magma of less extreme composition. Here we are concerned with the other kind of peridotite body, small, discordant and apparently deep in comparison with its length or breadth, whose typical location is in fold-mountain belts (included with these, though by no means small, are the peridotite batholiths of New Zealand and New Caledonia). Briefly, there is no accepted explanation of these bodies. Nevertheless, the problems they pose deserve consideration.

Two questions must be kept separate: (1) How does a body of this composition come to exist? and (2) How is such a body emplaced—as magma, solid, mixture etc.? At once, we distinguish hypotheses where the body becomes peridotite after emplacement from those where it is peridotite in composition when emplaced. We note also that there is no need for all peridotites to have one origin. (Much trouble has arisen in petrology from the wrong supposition "one rock-type—one origin." It was pointed out, with regard to vulcanism, that rocks with similar properties can have different origins and, with regard to fractionation, that rocks with similar origins can have different properties. The same mistake is made whenever a genetic term is used to specify an observational class—for example, flow folds, wrench faults etc.)

Hypotheses where peridotite is introduced as such are divided into (1) those which involve a worldwide peridotite substratum, and (2) those which do not. Hypotheses where peridotite is generated by modification of material already present at the site include (3) generation from pyroxenite, (4) generation from limestone, and others. Clearly the question of whether there is a periodotite substratum is crucial; but it is fruitless to discuss it on purely petrological grounds, and the evidence from seismic studies, gravity studies, high-pressure experiments and so on is out of place here. Briefly, the problem concerns the material below the Mohorovicic discontinuity, in which the velocity of the compressional wave is about 8 km/sec. At present, popular conjecture is that the material is either peridotite or eclogite. If it is peridotite, the uniformity and recurrence of basalt floods requires explanation; if it is eclogite, the uniformity and recurrence of peridotites is the difficulty. In size, the basalt floods are a larger problem than the peridotites, but to balance this we find the recurrent generation of basalt from peridotite a little more plausible than the recurrent generation of peridotite from eclogite or basalt. This is the sort of problem which can be resolved *locally* by deep drilling; but the hypothesis that the rock beneath the Mohorovicic discontinuity is homogeneous is nothing but a mental convenience, so that even a drilled sample or two is an insufficient basis for a worldwide extrapolation.

Whichever source for the peridotite material is favored, its emplacement is a separate problem. The conflicting evidence is:

1. Many bodies have discordant boundaries as if emplaced in a magmatic state;

2. The temperature of the corresponding anhydrous magma is over 1600°C;

3. The rocks around some of these bodies show heating effects that look far too mild to be compatible with such high temperatures.

The conflict can be resolved by (1) thinking of ways to emplace the material at a lower temperature, or (2) thinking of ways to protect the environment from the effects of heating. A high pressure of water vapor at the site goes far toward meeting both requirements: as in the quartz-water system, if the water vapor pressure is high enough to keep several parts per cent of water dissolved in the magma, the magma will remain liquid far below its dry freezing point. At the same time, the high water vapor pressure minimizes the dehydration effect of heating on the environment (as will be seen in the section on metamorphism, some rocks can be heated to 800°C in a hydrous state and still undergo only changes that would occur at 400°C in the dry state). However, if an explanation is sought along these lines, it then has to be explained how the olivine sometimes remains unserpentinized; in this matter, the silica content of the ambient aqueous fluid may be the critical factor.

Another helpful observation is that not all of the peridotite body needs to be liquid on emplacement; a mixture of 90 parts solid and 10 parts liquid may be mobile enough to form the discordant boundaries observed. However, the two phases should then be recognizable in the texture of the final solid, whereas actual textures often contain no relics of a second phase. Thus the idea is reinforced that the second phase consists of "fugitive" components, mainly water. Perhaps these do their job of lubricating the movement of olivine crystals, and then slip from between the grains leaving no trace of their presence.

The peridotite problem is not a good example of a problem successfully solved, but it is a good example of the need to *think back*. No problem of genesis, structural or mineralogical, can be solved merely by observing properties of rocks as they are now; it has to be asked—What were the properties of the rocks *then*, when formed or deformed? The fugitive components—present then, now lost—are crucial; two per cent of water, whose presence or absence we have to imagine, may be more effective than 20 per cent of a mineral whose presence or absence we are able to observe.

Anorthosites

The great independent anorthosites seem to pose problems similar to those posed by peridotites; they appear to be intrusive, and yet most magmas of anorthosite composition (over 90 per cent plagioclase) would melt at over 1400°C when dry. Again, facts to remember are: (1) the problem of emplacement is separate from the problem of generation, and (2) it is not necessary that all large anorthosites have the same origin. It is also noteworthy that large parts of, for example, the "Adirondack anorthosite body" are not in fact anorthosite, but carry over 10 per cent of dark minerals, and that the more closely the body is studied, the more heterogeneous it turns out to be.

As with large independent peridotite bodies, it seems likely that the presence of water is important. At low water pressures, upon melting a potash-poor crustal rock such as basalt or some calcareous shales, plagioclase and pyroxene components enter the liquid fraction together; but at water pressures over 5 kb, the tendency becomes strong for plagioclase components to enter the liquid fraction while iron and magnesium remain in the solids. Such a liquid fraction, if separated and crystallized in a new location, could generate a rock containing at least 80 per cent plagioclase, which is as much as many so-called anorthosite samples actually contain. Other features favoring an origin on these lines are: (1) the large size of some of the plagioclase crystals—in other situations, e.g., pegmatites, large crystal size is attributed to abundant water; (2) the ophitic texture shown by some pyroxene crystals in the anorthosite. The question of generation is by no means closed, but experimental work since 1950 has certainly made magmatic genesis more plausible than it appeared in earlier years (Yoder and Tilley, 1962, p. 461). Rigorous examination of hypotheses will continue to be impossible as long as data is so sparse on what the dark mineral content's mean value and range of variation actually is.

CHAPTER TEN

THE VOLATILE FRACTION

In the preceding chapter the presence of water and other volatile components in magmas was frequently mentioned, together with their separation into a discrete phase (*volatile fraction* = *aqueous phase* = *fugitive components*). Knowledge of this phase is negligible because in its natural state it is inaccessible. Observations can be made only after it has reached the surface or cooled or undergone reactions. Three possible kinds of sample are:

1. The gases that bubble out from fairly peaceful lavas such as exist intermittently at Hawaii—these can be drawn off through a long tube for analysis;
2. The gases and liquids that escape from fissures in areas that have been volcanic but are now quiet, such as Yellowstone Park;
3. The gases and liquids that exist in cavities, in solid crystals such as quartz—the existence of tiny cavities is fairly commonplace and cavities yielding analyzable samples are not scarce.

Clearly, none of these is representative of volatile fractions as a whole. Data from them must be pieced out with experimental measurements on fluids kept under high pressure in equilibrium with hot silicates. Measurements of this kind being scanty, data must also be pieced out by extrapolation, and also with pure exploratory speculations.

A rough picture of two possible compositions was given in Table 5-1. Since there is no reason to suppose the volatile fraction to be any less variable than the silicate fraction, all kinds of deviation from these models must be conceived, in the same way that all kinds of deviation exist from the dominant rocks, granite and basalt. However, almost always, water is the main component. A diagram of the density of water under geological conditions is given in Fig. 10-1, and some deductions from it follow.

1. The field where water and steam coexist as separate phases is very small; it corresponds to hot, shallow localities in the crust, such as occur only where magma approaches or reaches the surface (hatched line in figure).

2. The density of the aqueous fluid is very sensitive to pressure. Consequently, the density we hypothesize at a given depth depends on whether we suppose

$$P_{H_2O} \cong P_{overburden}$$

or

$$P_{H_2O} < P_{overburden}$$

However, air-filled holes in the crust are rare, so that it is usually safe to assume P_{H_2O} at depth 1 km is at least that due to a head of water 1 km high. On this basis a density-depth curve can be drawn for any assumed change of temperature with depth, and the density shown will be a minimum density. The dotted line is such a curve for an average temperature gradient, 30°C/km (cf. Fig. 1-5). It is seen that as far as density goes, the aqueous fluid is closer to water than it is to steam.

3. A density of 0.5 is not unreasonable to expect, for example in a magma chamber at 700°C and 7–10 km depth (Fig. 10-1B). Then if a melt of density 3.0, containing five per cent excess dissolved water by weight, separates to solid crystals and aqueous fluid, the relative *volumes* will be about one of fluid to three of crystals—a very dilute "mush." (*Excess water* here means water that is not taken up on crystallization in forming hydrous minerals such as mica. The volume ratio may be derived by considering 10 g of mixture, 9.5 g of silicate plus 0.5 g excess water; these will occupy about 3 cc and 1 cc respectively.) Of course, immediately upon separation, the aqueous fluid will tend to leave the crystals and escape through the walls of the magma chamber, but at a depth of 5 km or more, these cannot be expected to be freely permeable. We thus have to imagine considerable volumes of watery fluid striving to reach the surface but taking a long time to do it. During the escape interval, they not only carry a quantity of passive solute with them, but also act as a diffusion medium, allowing solution of a mineral at one point and deposition at another.

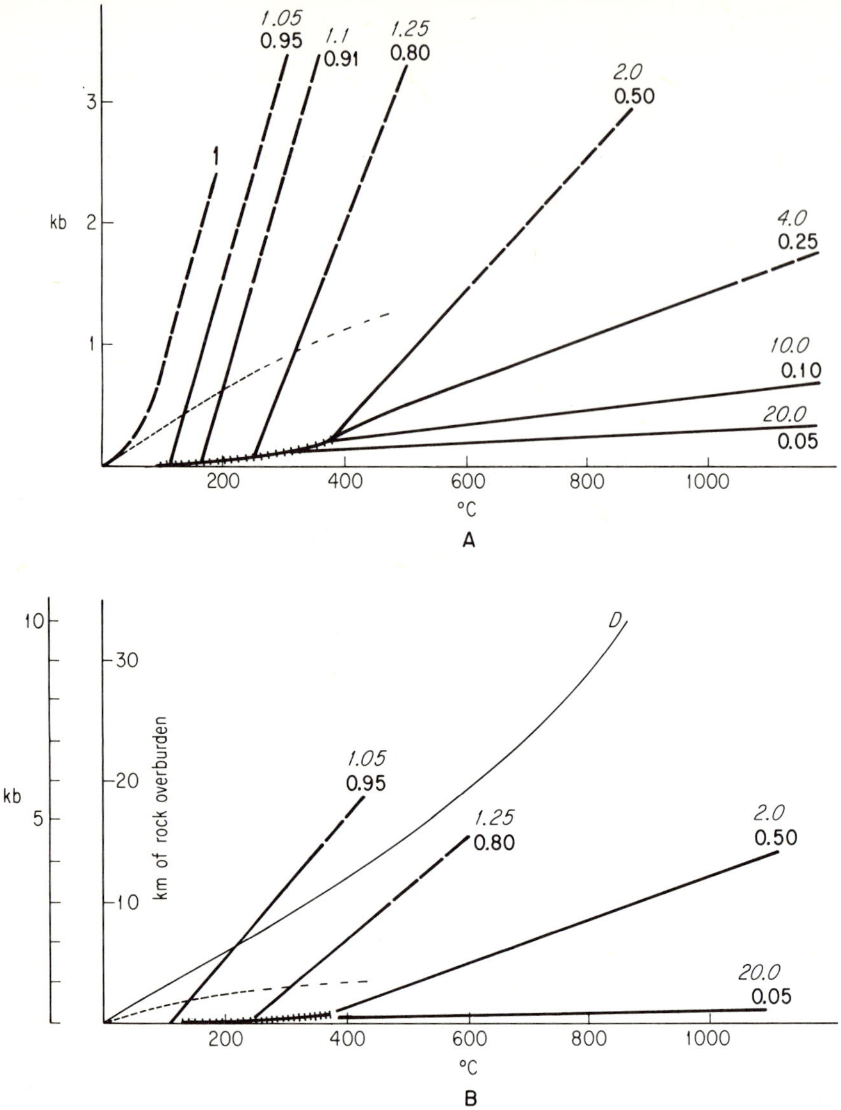

Fig. 10-1 The density of water within the crust. (A) Water has a density of 0.5 g/cc at 600°C and 1.5 kb, and in other conditions represented by points on the line marked 0.5. Other lines show the conditions necessary for water to have other densities, as marked in boldface numbers. The italic numbers show the volume in millilitres that 1 g of water occupies under the different conditions. The hatched line passes through conditions where water boils (where a liquid of one density and a vapor of another density coexist in equilibrium). The dotted line is calculated for a water-filled hole or fissure in the earth. If the temperature rises 30°C/km (cf. Chap. 1), at 4 km depth the temperature will be 120°–140°C and the pressure will be due to 4×10^5 cm of water of density close to 1, i.e. 0.4 kb; but over the next 4 km, the pressure will not rise to 0.8 kb because the water at that depth is less dense—0.95 to 0.91 g/cc. The dotted curve therefore flattens—where it crosses a density line, its gradient is proportional to that density. (Density data from Holser and Kennedy, 1958, 1959.) (B) The data of Fig. 10-1A are plotted on the scales used as standard throughout the book; hatched line, dotted line and numbers as in A above, curve *D* as in Fig. 1-5. If water occurs not in a fissure but in a closed space where $P_{water} = P_{overburden}$, its condition is likely to be above curve *D*, and its density above 0.8 g/cc.

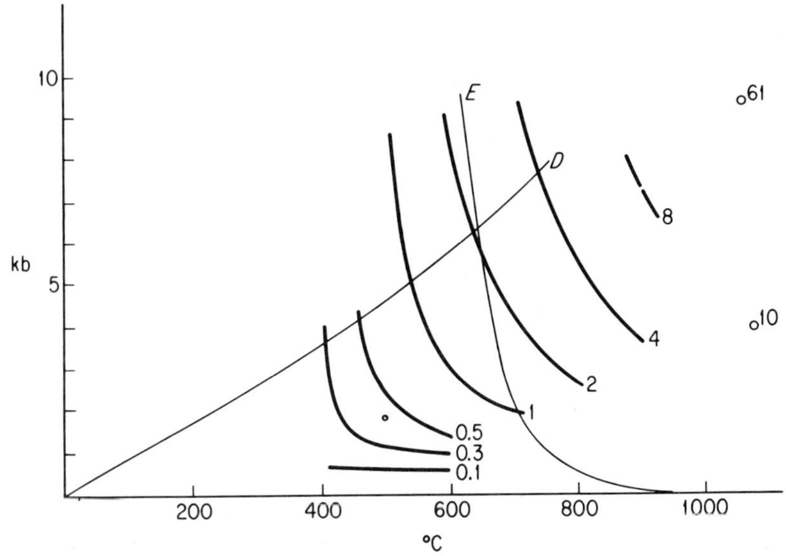

Fig. 10-2 The solubility of quartz in water (in weight per cent). At small circle, the solubilities of albite and potash feldspar are 0.27 and 0.25 per cent respectively (from data of Morey and Hesselgesser, 1951, Weill and Fyfe, 1964, and Anderson and Burnham, 1965). Curves *D* and *E* as in Fig. 1-5. (The two larger circles are data points from the work of Kennedy *et al.*, 1962; they show that very large solubilities are reached below 10 kb. Above about 1050°C and 10 kb, molten quartz and water are miscible in all proportions.)

Aside from material whose diffusion is thus made possible, the amount of passive solute that can be carried depends on the solubility of silicates in the aqueous phase, data on which are shown in Fig. 10-2. Most of the data is for pure SiO_2 in pure water; the two figures for albite and potash feldspar are in the same range. Allowing for the fact that a natural fluid's solute power is apt to be enhanced by presence of acids, we take one per cent as an adequate rough figure for a body at 10 km depth and 600–800°C. A sample calculation then runs as follows (we suppose a small stock 5 km in diameter and 2 km in depth):

Weight of rock = volume × density = 12×10^{16} g

Weight of aqueous fluid generated at solidification of the above = 5% × 12×10^{16} = 6×10^{15} g fluid

Solute content, at 1% by weight, = 6×10^{13} g = 2×10^{13} cc

This is enough to form twenty veins 1 m thick and 500 m long stretching 2 km from the stock—just about one to every square kilometer of its top surface (Fig. 10-3). If the erosion level later came within 2 km of the stock, all these would be exposed and would give an intensity of veining comparable with that found in real situations, e.g., quartz veins in mineralized

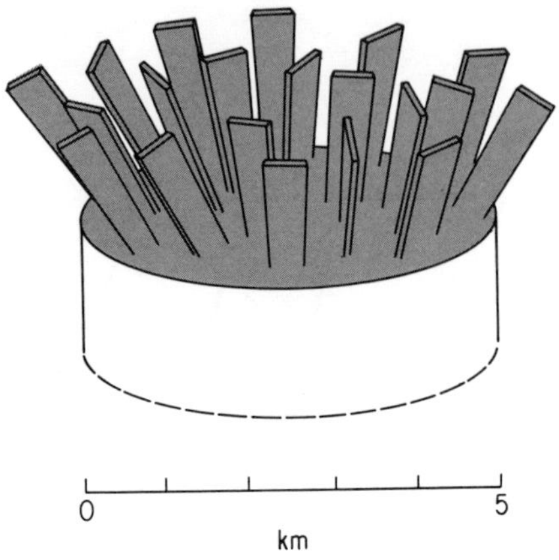

Fig. 10-3 Veins emanating from a stock by the agency of quartz-saturated water.

aureoles. The calculation does not, of course, show that veins *do* form by this mechanism, only that perhaps some could. In addition, it gives a concrete picture; without some such estimate, "the solute content of a hypothetical vapor phase" is a rather elusive concept to discuss.

The veining mechanism outlined is only one of several possibilities. If we summarize it as

> **1.** Transport in an aqueous fluid of magmatic origin, of vein-filling minerals from the same magmatic source,

then other possibilities are:

> **2.** Transport in an aqueous fluid of magmatic origin, of vein-filling minerals from a local source;
>
> **3.** Transport in an aqueous fluid of local origin, of vein-filling minerals from a local source;
>
> **4.** Transport of vein-filling minerals as a melt.

The last gives veins that are simply small dikes. In possibility (3), the hypothesis is that as a crack opens, pore fluid runs into it from the walls carrying solutes which crystallize in the crack. In possibility (2), the fluid which acts as solvent and diffusion medium is magmatic but the solids that form the vein are leached from the vein walls. Undoubtedly, some veins do form by mechanisms (1) and (4), i.e., are purely magmatic in source, so that no survey of magmatism is complete without their mention. However, the mechanisms distinguished here can mingle in nature, and thus many veins are at least partly metamorphic. Details of texture and composition are therefore reserved until later (Chaps. 22, 23).

CHAPTER ELEVEN

THE PHYSICS OF MAGMATIC PROCESSES

It is as impossible to describe the physics of a magmatic event in all its details as it is to describe its chemistry. The purpose of this chapter is to introduce just a few basic topics and to make the point that magmatic processes are subject to the same laws of physics that govern blast furnaces, the molding of plastics and so on.

THE COOLING OF INTRUSIONS

A solid body that is hotter in one part than another will conduct heat from the hot part to the cooler part. If we imagine a small area A within the body, oriented across the heat flow, the rate of flow of heat across this area is KAg, where K = heat conductivity of the material and g = temperature gradient normal to the area A. Let us now think of two areas A at distances x_1 and x_2 measured in the x-direction, normal to A, from some arbitrary zero-point (Fig. 11-1). If the temperature gradient is g_1 at x_1 and g_2 at x_2, the prism or cylinder between the two areas will suffer

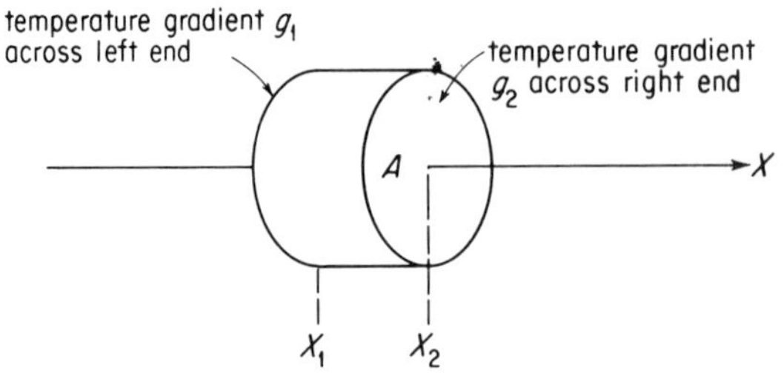

Fig. 11-1 A simple heat-conduction situation.

inflow of heat KAg_1 and outflow KAg_2. If we suppose that no gain or loss of heat occurs over the "sides" of the prism and that g_1 is greater than g_2, the prism will get warmer. The rate at which heat accumulates in the prism is $KA(g_1 - g_2)$; its volume is $A(x_1 - x_2)$; if its heat capacity is S calories per cc, its temperature will rise R°C/sec where

$$A(x_1 - x_2)SR = KA(g_1 - g_2)$$

or

$$SR = K\frac{g_1 - g_2}{x_1 - x_2}$$

Since $g_1 - g_2$ is the change in temperature gradient in passing from x_1 to x_2, we may phrase this as: $S \times$ change of temperature with time $= K \times$ change of gradient with distance. This fundamental law can be applied to heat flow in all kinds of situations, and very similar laws govern the diffusion of solutes in liquids and the movement of compressible fluids in porous media—both processes of geological interest. The application here is to the escape of heat from the margin of a magmatic body into the surrounding rocks.

The approach used is to consider the simplest possible situation first and then set about modifying it to bring it as close as possible to a geological situation. The simplest possible situation is an infinitely extended plane interface between an infinitely thick cold body on one side and an infinitely thick hot body on the other, S and K being the same in each body. As soon as this situation is created, heat will begin to flow across the interface. If the initial temperature distribution created is as in Fig. 11-2A, a short time later it must be of some form such as Fig. 11-2B. Reasons are:

1. At a great distance out into the cold body, the warming influence of the hot body will hardly have begun to be felt; at such a distance the temperature distribution must be close to that in Fig. 11-2A. The same argument holds for the hot body.

2. If K and S are the same in both bodies, the heating effect of the hot body will equal the cooling effect of the cold body. The interface thus remains at $T + U$ and the temperature-distance curve is

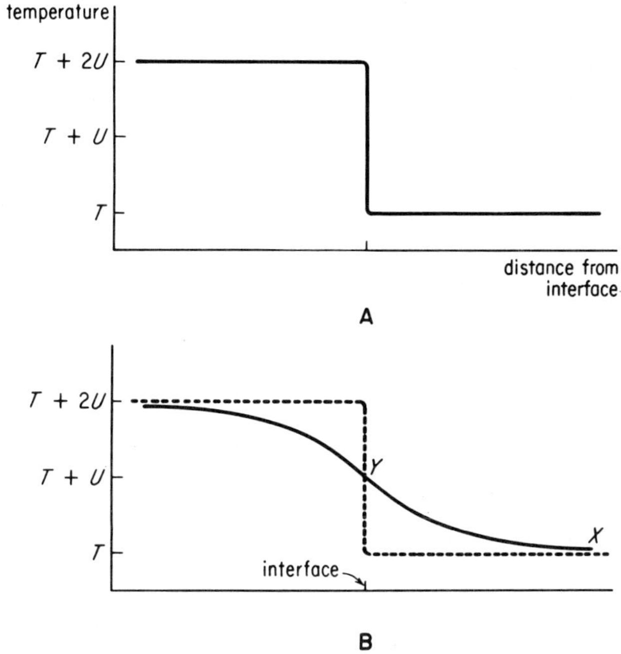

Fig. 11-2 Cooling at a plane interface. (A) Postulated temperature distribution at start of heat-transfer episode; (B) temperature distribution after a lapse of time.

symmetrical about Y. (T = initial temperature of cold body; $2U$ = initial temperature difference at interface.)

3. To begin with, the rock that heats up fastest is very close to the interface; however, soon after the initial moment, the rock close to the interface will be at a temperature close to $T + U$ and thereafter will not heat up much at all; at this time the rock which is heating fastest must be further out. Now, from the law enunciated earlier, the point where the rock is heating fastest with time is the point where the gradient of the temperature-distance curve is changing most with distance, i.e., the point where the curve changes slope most markedly. Thus, the curvature of XY must be small near Y, pass through a maximum and be small again at X.

The process can be summarized in a series of curves, as in Fig. 11-3A.

[The exact form of the curves is found by solving $K \cdot d^2\Theta/dx^2 = S \cdot d\Theta/dt$ for the boundary conditions:

$$\begin{aligned} \Theta &= 1 \text{ for } x = 0 \text{ for all } t, \\ \Theta &\rightarrow 0 \text{ as } x \rightarrow \infty \text{ for all } t; \\ \text{at } t = 0,\ \Theta &= 1 \text{ for } x = 0, \\ \Theta &= 0 \text{ for } x > 0. \end{aligned}$$

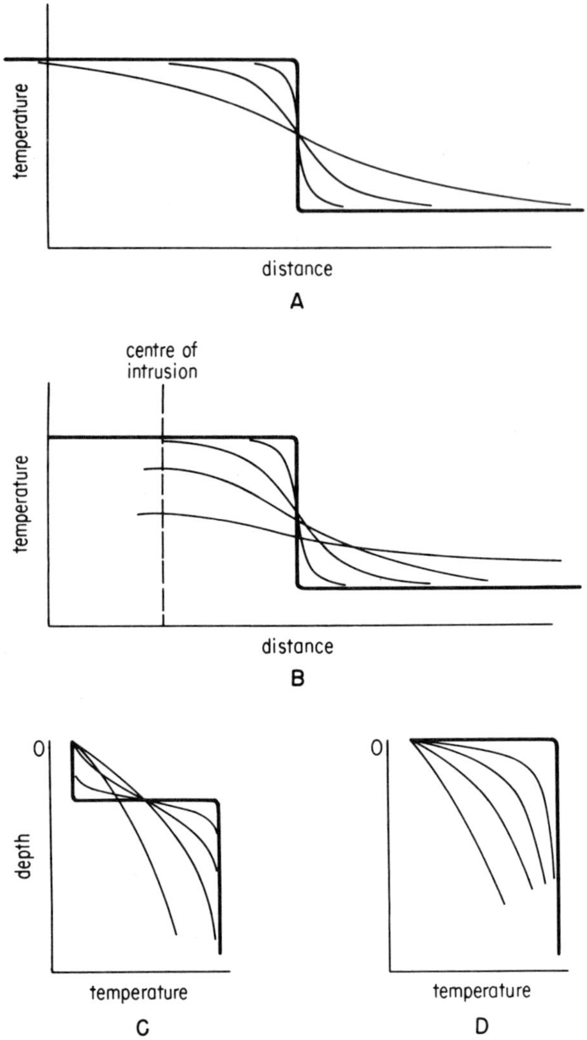

Fig. 11-3 Cooling under geological conditions. (A) A family of curves as in Fig. 11-2B for different periods of heat transfer; (B) modification of (A) where hot body is of finite size; (C) modification of (A) where cold body is of finite size—for example, where a batholith is intruded under a thin cover; (D) modification of (C) where cover is of negligible thickness; the curves have the same shape as in either half of diagram A. (After Larsen, 1945.)

The solution is

$$\Theta = 1 - \operatorname{erf}(x/2h\sqrt{t})$$

where $h = \sqrt{K/S}$ and the function erf (= error function) is defined by $\operatorname{erf} z = \dfrac{2}{\pi}\displaystyle\int_0^z e^{-u^2}\cdot du.$]

Amendments necessary to make the situation more realistic geologically are:

1. In the case of a dike, the hot body is not infinite in extent though the cold body is effectively so;
2. In the case of a batholith roof, the cold body is not infinite in extent;
3. The interface is usually not planar;
4. S and K for the magma do not match S and K for the environment;
5. A latent heat effect may be significant.

Curves for situation (1) are as in Fig. 11-3B, and for situation (2) as in Fig. 11-3C, D. In these, the eventual drop in temperature at the interface, which must occur in nature, begins to appear. The time scale is interesting: for a broad intrusion under 5 km of rock cover, the interface remains within 20 per cent of its maximum temperature for a million years (Larsen, 1945); for a dike or sill 100 m thick, the corresponding figure is 100 years. Amendment (3) mainly affects the duration of the heating episode in the environment—outside a convex magma surface cooling is quicker, and outside a reentrant surface a high temperature is maintained for longer than outside a plane surface—but the highest temperature reached is not likely to be much affected.

Amendments (4) and (5) both have the effect of raising the interface temperature. Firstly, K for the magma may be effectively greater than K for the environment because of turbulence—a body in which temperature gradients are reduced by turbulence behaves as a good conductor. In short, if the cool marginal material is swirled away and replaced by hot magma, the interface temperature will be raised. The value of S for the magma may also be marginally greater than for the environment intrinsically, but a more significant difference in S arises in connection with the latent heat effect. The latent heat of crystallization of silicates may be as much as 60 cal/g*. If crystallization occurs through a cooling interval of 200°C, the heat output is $(200S' + 60)$ rather than just $200S'$. Now $S' \leq 0.3$ cal/g so that the latent heat could more than double the effective specific heat of the magma. This in turn would raise the interface temperature. However, the latent heat is a highly variable quantity, being determined as much by the volatiles in the magma as by the silicates—as the silicates change from magma to crystals, the volatiles change from magma to "gas." The latter change is a kind of evaporation or at least an unmixing, and absorbs heat. Ten g of water evaporating to steam absorbs 5400 cal at 100°C and 1 atm, which is exactly the heat generated by 90 g of accom-

*Up to here, we have used S in cal/cc but the value in cal/g is often more convenient. It is designated S'; thus $S' = S/\text{density}$.

panying silicates crystallizing, at 60 cal/g. Of course, the volatiles do not separate at 100°C and 1 atm, but the comparison shows that a volatile-rich magma may have far smaller effective latent heat than a dry magma. As it happens, the effects discussed under conductivity and under specific heat/latent heat are both smaller in acid magmas than in basic ones; acid magmas, being more viscous, are less turbulent and, having more volatiles, have lower latent heat. So we may conclude that the interface temperature for acid intrusions is probably close to halfway between those of melt and environment, or about 400°C for a melt at 700°C in a shallow environment at 100°C; on the other hand, the interface temperature for a basaltic intrusion may be only 25 per cent below that of the melt (say 1100°C), or 850°C in a similar situation. These figures are in accordance with the metamorphic effects produced.

THE MELTING OF ROCKS

This is a process that cannot be discussed in such simple terms as the cooling. Difficulties are:

1. The heat source is more problematic;

2. There is no clear-cut initial situation from which events proceed in a smooth path toward equilibrium.

Of course, not all cooling processes are as simple as the one described; it approximates the cooling of an intrusion at shallow depth where the interface is very well defined and stable. At greater depth, the temperature of the environment is higher, assimilation of wall rocks is more extensive, and other sources of heat must be considered; in the emplacement of a batholith, for example, melting in one portion can accompany cooling in another.

A crucial question is what makes a rock melt. In the introductory chapter, we noted that at 20 or 30 km depth conditions may be favorable for melting, but still some disturbance in the normal state of affairs has to occur. Possibilities are:

1. Depression of low-melting material ("granitic") to an unusually deep level;

2. Rise of the 700°C isotherm, etc., to an unusual shallow level because of:
(a) hot subcrustal material rising;
(b) hot subcrustal material solidifying;
(c) radioactive material being concentrated;

3. Change of water vapor pressure.

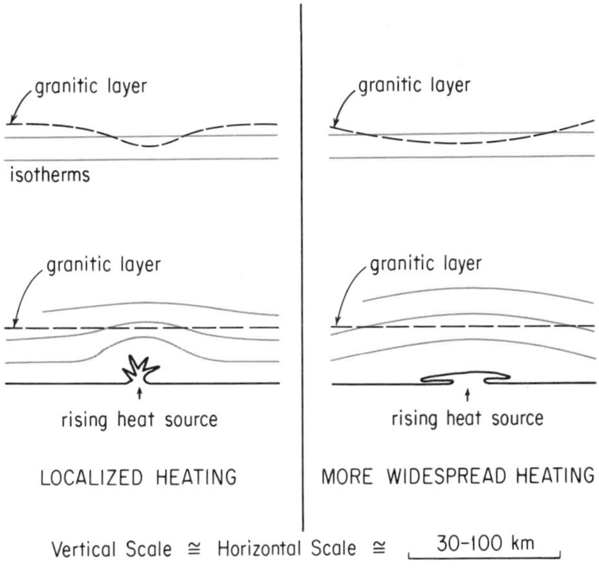

Fig. 11-4 Localized and widespread heating.

For example, suppose a rearrangement in the mantle forces some peridotite up to the 30 km level: the isotherms will rise and, at the same time, water will be driven from the base of the crust to upper levels. If the 700°C isotherm reaches a 20 km level and at the same time the water vapor pressure is uncommonly high, melting may occur. In fact, the *same* heat that permits intrusion of peridotite may later permit the intrusion of granite, and the *same* water that facilitates one may facilitate the other. One heat-and-water pulse can travel up through the crust causing magmatism at different levels, in the same way that one hurricane can travel up the eastern seaboard causing excitement in different cities. We say that heat and volatiles are *lost* on crystallization, but they are not in fact lost—they merely move elsewhere.

We shall, for brevity, speak of a heat pulse reaching a granitic layer, but in doing so we intend all the possibilities above to be implied. It is then possible to envisage a highly-localized or a broadly-spread heat pulse (Fig. 11-4). These will have rather different effects on the chemistry of the resulting melt. Figure 11-5 shows features of the situation that may arise

Fig. 11-5 Partial melting of granitic rocks.

(no attempt at dimensional control). If the heat pulse is localized, temperature within the magma will rise above its first melting point and some of the higher-melting residues will be taken into the melt; on the other hand, if the heat pulse is broad, it is more likely to be dissipated at the margins, with the formation of more magma of the same minimum-melting composition as the first formed. In short, the maximum temperature reached by the magma is determined by conflicting influences—localization of the heat pulse versus spread of heat by conduction, convection and radiation; and the maximum temperature reached fixes the proportion of refractory components taken up, in accordance with the discussion in Chap. 9.

In an embryo batholith 50 or 100 km across, the heat flow will not be the same in all parts. We therefore expect:

1. Various magmas to form by different degrees of fractional melting;
2. Various solids to form from each by different degrees of fractional crystallization;
3. Different degrees of contamination by unmelted solid residues.

The quantitative descriptions of batholith composition that have recently been made show up variations such as could be explained by these mechanisms; but this does not, of course, show that the mechanisms have worked as outlined.

THE EFFECT OF GRAVITY ON MAGMATIC BODIES

A third process that occurs in the crust but obeys the laws of the laboratory is adjustment of density inversions (lower-density material below material of higher density). The standard analog is bubbles rising in water, where the driving force is gravity yet the bubbles move upwards. The bubbles do not defy gravity (in a vacuum, they would fall with the same acceleration as Newton's apple and other famous bodies, until they burst); they merely submit to the greater tendency of the water to flow to the bottom of the container. In other words, the potential energy of the system is reduced by movement of bubbles from a lower to a higher level. In exactly the same way, the potential energy of the crust is reduced if x cc of low-density material rises and x cc of high-density material sinks to take its place.

It is one thing to specify a rearrangement which would result in an energy reduction; it is another and usually more difficult thing to say by what process and at what rate the rearrangement will occur. Fortunately,

Fig. 11-6 Gravity domes. (A) Growth stages of an idealized gravity dome; (B) domes experimentally developed by Ramberg (1963).

demonstration: new honey bottle, invert ... observe from time to time over a few hours — air bubble will migrate to top
Norris 4/70

the process is known by observation in some laboratory situations, and can be extended with confidence to the growth of salt structures of discordant shape (Fig. 11-6): a low-density body, even if more or less uniform and flat-lying, when covered by heavier material will flow horizontally, thinning wherever pressure is locally high and thickening where it is low; if erosion keeps the overlying surface planar, any small irregularity is self-magnifying; at a rate determined by the creep behavior of the rocks involved, a local thickening will grow to a tumor and eventually rise through the overburden, crumpling the rocks as it goes. In the case of a salt horizon, the upward movement continues to the surface and causes salt hills, such as Kuh-i-Namak, Persia; but if granite were to move by this mechanism, either as a viscous melt (density $\sim$ 2.4) or as a mass of crystals (density $\sim$ 2.6), upward movement would cease when the granite reached material as light as itself, such as some sedimentary rocks. It would be expected to spread sideways at this level, to whatever shape the local strengths and weaknesses allowed (Fig. 11-6B, cf. Fig. 5-7).

The hypothesis that granites do move by some such mechanism, driven by gravity and controlled by viscosity, is made attractive by two features:

1. Most granites that have been surveyed are sites of negative gravity anomaly, which suggests that the granites are of lower density than their environment—there is no large granite body that is known to be denser than its environment;
2. As far as they are known, the shapes of granite bodies often seem to be similar to those forecast by this mechanism—flattish and semi-concordant except for a "root" of uncertain depth.

However, the way a structure looks as if it formed is a *very feeble* indication of how it did form; a hydrodynamic calculation is required, involving dimensions, viscosities and times. The nearest approach to this is the work of Ramberg (1963), who made dynamic models to represent earth materials as closely as possible and then modified the results of the model tests by calculation. His conclusions, subject to the many assumptions involved, were:

1. A magma of density 2.4 and viscosity similar to that of obsidian at 800°C, rising through overburden of density 2.8, would take 150,000 years to reach a near-surface location from a source at 30 km depth (the magmatic bodies formed in the model corresponded to dome-shaped granites about 10 km across);
2. For some structures to form, such as the overturned sediments around the Vredefort Dome near Johannesburg, the intrusive body cannot have been as mobile even as obsidian, but must have been almost wholly crystalline, with viscosity approaching that of a solid sediment.

The time required for movement of such a body would be far greater than the 150,000 years for movement as magma, and has not been estimated. However, far greater times are available geologically. Whatever time is favored, 10 million years or 100 million, our knowledge of the viscosities involved is so slight that compatible values could almost certainly be chosen without conflict with known facts.

PHYSICS OF VOLCANIC ACTIVITY

Digression on Scientific Method

In contrast with the two foregoing sections, in which imponderable quantities are rather prominent, this section is like the first—a simplified analog that can be treated with some degree of precision. We should, however, at some point, ask what the *purpose* of such a treatment is. Suppose we assume that a volcanic vent is of uniform cross section and does not change size during an eruption, and on the basis of such an assumption we forecast the velocity with which gas will be ejected. Such a forecast is not the end of the undertaking—if it is velocity of ejection we require, we had much better go to a volcano and make an observation. More usually, the purpose is to compare the forecast with an observation as a check on the initial assumptions. In other words, we assume constant uniform cross section, not because we believe the assumption reasonable, but because this is the assumption we wish to test.

What are the possible conclusions from such a test? Either the observation differs from the forecast by an amount greater than can be attributed to observational error, or it does not. The first result shows that at least one of the assumptions made is wrong; but the second result does not show that all the assumptions made are right—a test cannot prove a hypothesis, it can only fail to disprove it. If a test fails to disprove a hypothesis, it is the absence of any alternative that proves it, if anything does. Most hypotheses, of course, are never proved—they are only shown to be probable in different degrees; so it would be better to say that the probability of a hypothesis is established, not by tests, but by the improbability of its rivals. It is sometimes supposed that a hypothesis that has survived one test becomes more likely if it survives a second test also, but this is not necessarily so; it is so if, and only if, the second test reduces the probability of its rivals.

Petrologists being human, a common situation is that a particular hypothesis is favored and the operator wishes to establish it. If he devises a procedure to test the favored hypothesis, he is proceeding in an unnecessarily roundabout manner; the result he does not want is the only decisive result he can achieve. He will establish the favored hypothesis more effectively if he tests rival hypotheses; in fact, he should perform no test without first asking himself what hypothesis he hopes to disprove. It may be objected that this is an offensively destructive way to proceed, but regrettably there is no other. It may alternatively be objected that the science of petrology deals with rocks, not hypotheses; but I would say that what does not concern hypotheses is neither petrology nor science. It may be some other valuable activity, but science is not the orderly accumulation of facts: it is the orderly accumulation of rejected hypotheses.

The Size of a Volcanic Vent

The particular aspect of vulcanicity to be discussed here is the relation between the distance ejected fragments fly, the weight of ejecta and the size of vent. The assumption we wish to test is that the vent behaves as a tube whose cross section changes neither with depth nor with time. The type of eruption considered is a gas-and-ash discharge, with some larger fragments, of overall rhyolitic composition. Data used are:

1. The range of ejected fragments, which may be five miles in a volcanic outbreak of moderate intensity;

2. The amount of debris delivered in a day, which we take to be a layer 1 m thick over an area 10 km by 10 km, whose average density as deposited is 1 g/cc.

The calculation runs as follows.

In travelling five miles horizontally, fragments probably rise between one and nine miles vertically. Using $mgh = \frac{1}{2}mv^2$, where m = mass of

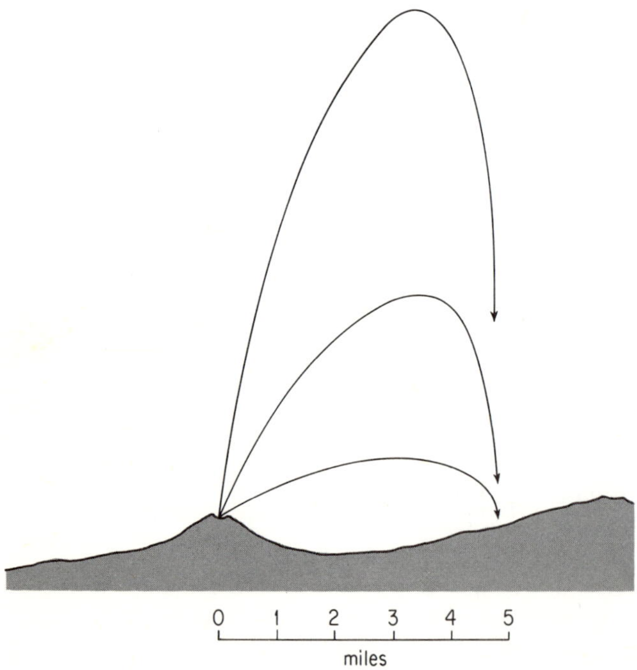

Fig. 11-7 Volcanic ejection paths.

fragment, h = height reached, v = vertical component of ejection velocity and g = acceleration due to gravity, we find that v lies between 550 and 1650 ft/sec (= 1100 ft/sec for an intermediate trajectory reaching four miles high)—see Fig. 11-7.

The initial slopes of the trajectories shown have cosines 0.5, 0.85 and 0.99 respectively; thus possible values for the ejection velocity are 1100, 1300 and 1660 ft/sec, from which we may take 1300 ft/sec as representative.

We now consider 200 g of magma comprising 18 g of water and 182 g of silicates. In a molten state at great depth, where all the water can be dissolved, the volume of this magma (density ~ 3.0) will be 67 cc. At 1100 °C and 1 atm—a condition it may reach somewhere near the orifice—the rock will still occupy 60 cc or so but the water will occupy about 90 liters (4 $\times$ volume of 1 mole at standard temperature and pressure, because of high temperature). That is, 67 cc of magma generates 90,060 cc of emission, an expansion factor of about 1400. In a steady state, the number of grams passing the orifice per second must equal the number of grams passing any other horizontal level in the vent. Thus, *if the vent is uniform*, the velocity of the magma at depth must be 1/1400 of the emission velocity, or about 1 ft/sec or 30 cm/sec. Let the vent cross section be S cm^2; the volume of magma passing any point is 30 S cc/sec and the weight of rock passing any point is $30\ S \times 2.7 \cong 80\ S$ g/sec.

From the data given, the weight of rock ejected in a day is 10^{14} g, or 12×10^8 g/sec. Thus $S = 1.5 \times 10^7$ cm^2 and, assuming a cylindrical vent, the diameter is 40–50 m.

The answer is not absurd, as 50 km or 50 mm would be, since bodies of basalt with circular cross section of about this size have been found in situations where it is reasonable to interpret them as magma conduits. We conclude that the calculation fails to detect absurdity in our assumptions. (More assumptions were made than have been stated—the reader may occupy himself listing unstated assumptions in the calculation.) The next step then is to replace some of the assumptions by different ones, and test again. One may thus be able to set some limits on the range of possible conditions in a vent.

The above calculation is not a serious attempt in this direction; those wishing to see some of the complexities more fully expounded should consult Graton's classic work on the subject (1945). The calculation is presented only as an illustration. It is inadequate because the principles of physics have been applied in an inaccurate, oversimplified way, and not because vulcanicity is beyond the field of application of physical principles. An adequate calculation of this kind can be made with more care, space and time than given here, but we should not expect it to spring up, new-hatched yet complete. Tentative calculations are of value no matter how rough, as long as they are suitable bases for refinement.

CHAPTER TWELVE

METHODS OF STUDY OF MAGMATIC ROCKS

Before leaving the subject of magmatism, it is appropriate to list some of the techniques by which the picture has been built up. The purpose is to show that petrology is a field defined by the subject matter studied, not by the manner of study. The techniques are subdivided as follows:

I. Operations on rocks
   (A) in the field
   (B) in the laboratory
II. Operations on rocklike materials (laboratory only)

Field Studies 1. **Observable Processes**

1. Emplacement. Volcanic rocks only: staff of volcano observatories estimate bulk of emissions, composition, temperature, rate of movement etc.

2. Erosion and sedimentation. While hardly magmatic processes, these are essential parts of the magmatic cycle—at present, ignorance about diagenesis is one of the obstacles to understanding granite formation (Chaps. 9 and 24).

Field Studies 2. **Products**

1. Size and shape of magmatic bodies. Methods are:
a. ground survey
b. air-photo survey
c. geophysical surveys: (i) gravity (the most important); (ii) seismic; (electric, magnetic, radiometric occasionally useful)

2. Magmatic effects. Visible effects such as the following are sought for:
a. chilling at margin
b. baking at margin
c. veining: the order in which veins of different compositions intersect each other is sometimes informative
d. fracturing: the existence of well-defined fragments recognizable as only slightly displaced, as in Fig. 12-1, is rather strong evidence that rock *A* was solid when rock *B* was quite highly mobile (viscosity 10^{10} poises or less)
e. xenoliths: it is helpful if xenoliths of rock types known in the environment, or rock types *not* known in the environment can be recognized (many xenoliths are simply not recognizable); preferred orientation or distribution of xenoliths may rarely be of interest
f. deformation of environment
g. permeation of environment by magmatic components to give gradational margin
h. metamorphism of environment

Laboratory Studies 1. **Geometrical**

1. Whole rock. The texture of the rock is informative, irrespective of mineralogy (Chap. 5); also, observation that mineral *X* habitually encloses mineral *Y* may be useful, although different inferences are possible from such an observation.

2. Individual minerals. The minerals whose individual geometry has been given most attention are *zircon*—the shape of the zircons in one body may be noticeably different from, or similar to, that in another—and *feldspar*—the nature of the twinning is sometimes used to throw light on the history of the body.

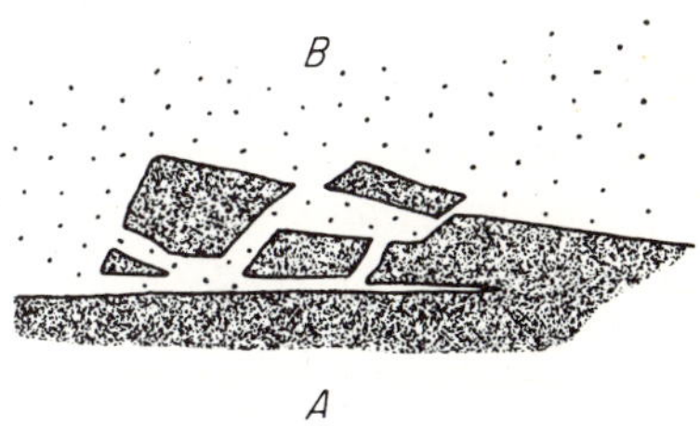

Fig. 12-1 Fragments that can be recognized as only slightly displaced.

Laboratory Studies 2. **Compositional**

1. Whole rock

a. in terms of minerals. The first task is to identify the minerals; for this, methods are: (i) microscope examination of thin sections; (ii) X-ray powder photography; (iii) chemical methods such as selective staining; (iv) differential thermal analysis; (v) infrared spectrography, etc.

The second task is estimation of relative amounts. This is done mainly by point counting on polished surfaces or thin sections, with or without selective staining to help; in some circumstances quantitative estimates can be made by the identification methods mentioned.

b. in terms of ions. Here identification and estimation of quantity are usually done simultaneously. Methods are: wet chemical methods such as weighing, titration, flame photometry, colorimetric methods, etc.; emission methods (where the ions are activated by some process and then detected by the radiation they emit) such as optical, X-ray, neutron activation, etc.

c. in terms of isotopes, by means of the mass spectrograph

2. Individual minerals. After identification as at (1a) above, composition may be estimated as at (1b) or (1c). In addition, some aspects of a mineral's composition can be discovered from its optical properties, measured with a microscope, with or without a universal stage. For minerals it is also possible to perform an X-ray emission analysis activating the ions by an electron beam (electron probe microanalyzer). For methods (1b) and (1c) the mineral has to be isolated. This usually involves powdering the material and separating the mineral of interest by its density or magnetic susceptibility; industrial flotation methods are occasionally adaptable to laboratory purposes.

3. Whatever property is measured on a magmatic rock, it is often of interest either to average its value over a rock unit or else to study its variation from point to point within a rock unit. For such purposes, techniques of sampling and trend analysis have been developed by statisticians. No determination of composition should ever be made until the question has been answered: Is the sample of the right kind to deliver the information sought?

Laboratory Studies 3. **Physical**

More rarely, observation of some physical property of the rock—density, porosity, compressibility etc.—is of interest.

Laboratory Study of Materials other than Rocks

These may be divided into materials related to rocks chemically and materials related to rocks physically.

1. Chemical analogs. Mixtures of pure compounds that contain fewer components than rocks can be studied chemically—results are more reproducible, compositions are more easily controlled, cause of any observed effect is more easily isolated than in experiments on actual rocks. Experiments usually involve heating a sample of known composition, at 1 atm or under pressure, in a controlled atmosphere, for a few hours up to a few months. Chapters 8 and 9 are based almost entirely on the results of experiments of these kinds.

2. Physical analogs. Here the purpose is usually to devise an analog that can be experimented with where the corresponding rock cannot, and yet resembles the rock sufficiently to give information about its behavior. Examples are:

a. gravity models. The size and duration of an experiment with real rock are too great, but models made with putty, clay, wax etc. can simulate real rocks' behavior.

b. flow models. Molten rock is too hot for laboratory experiments on its flow behavior, but the movement of solids in flowing magma and other such problems can be studied by models using cold fluids.

(M. King Hubbert's paper (1937) on scale models is a classic in this field.)

3. Mathematical analogs. It should not be overlooked that a model specified in mathematical terms is also a thing whose behavior can be studied and that can resemble real rocks either closely or not very closely. Most calculations in geology, in fact, do not refer to rocks but to models present, consciously or not, in the operator's mind. As was mentioned before, petrology is very much concerned with idealized concepts, rather than with lumps of rock.

Supplement with:
Krumbein + Sloss. Stratigraphy and Sedimentation.
Freeman.

PART THREE

SEDIMENTARY ROCKS

Except Pyroclastic Sediments

The format of this part matches that of Part Two. The facts we have to explain are summarized first; Chaps. 15 and 16 indicate the lines along which explanations for the facts are to be sought, and Chap. 17 turns aside to list the procedures used to ascertain the facts discussed.

CHAPTER THIRTEEN

OBSERVABLE FEATURES IN SEDIMENT FORMATION

Although the processes of magmatic rock formation are almost wholly inaccessible to direct observation, any child who has played at the edge of a river or lake has observed sedimentation. With slightly more trouble, other sedimentary environments such as deserts, snowfields and the ocean floor can be visited, and experiments can be conducted with natural materials on realistic scales, in contrast to the miniature synthetic analogs of magmatic events. Consequently, it is possible, even in a chapter concerned with observations, to report on processes one by one. In Part Two, processes were necessarily discussed inferentially after the observational chapters, but here a more natural understanding of rocks can be acquired if one follows this program:

1. Observation of processes individually;

2. Observation of interactions between one process and another;

3. Observation of products in situations where the processes can also be observed;

4. Observation of products in situations where the processes cannot be observed.

The last is of course the principal occupation of practicing geologists, but there is no reason for its also being the first activity of beginning students. For the purposes of this introductory text, parts (1) and (3) of the above program are given most attention; parts (2) and (4) are the proper province of those who have already done some groundwork.

MECHANISMS FOR SEDIMENTARY ROCKS

A tiny fraction of the known sediments may be regarded as primary, in that the material has fallen from outer space and has been incorporated into the earth's crust without intervening agency. Meteorites and the constant earthward stream of cosmic dust are in this category. The blazing lumps bring us information about many aspects of planetary physics, and the impalpable dust can be used as a yardstick because its rate of accumulation is a standard of comparison by which accumulation of other materials can be judged. However, neither lumps nor dust is a significant contributor in terms of bulk to the sediments that we see. It can therefore be said that practically all known sediments are derived from material that has been in the earth's interior. Most of the carbon and much of the oxygen now found in limestone probably rose to the surface early in the earth's history, during the formation of the atmosphere, and the other components have been brought to their present level by magmatism or severe metamorphism (see Fig. 2-1). However, beside magmatism and metamorphism, there is a third agent that raises solids and provides matter for erosion; the agent is life, that raises forests and coral reefs. Since life-produced materials differ so strikingly from magmatic rocks and yet function similarly as source material for erosion, we need to consider briefly what kinds of life-produced materials are known.

A moment's thought shows that the boundary between organic and inorganic modes of formation is not sharp and that there are three quite separate ways by which organic formations grade into inorganic:

1. Material of pure organic origin may accumulate simultaneously with inorganic material; an example of this is the deep ocean mud, in which windblown dust from the continents and skeletons of floating marine organisms are mixed;

2. Material of pure organic origin may be modified during transport from its life site to its deposition site, so that its character when deposited is determined as much by the inorganic transport processes—abrasion, sorting etc.—as by its initial characteristics; as an example of this, we might find a shell-fragment beach sand,

composed of calcite of organic origin but now in grains shaped largely by inorganic processes;

3. Material may owe its first development in solid form to a combination of causes, in which organic or inorganic influence may be dominant; an example of this is the calcite precipitated in some lakes—when the sun rises, the water gets warmer for inorganic reasons but also changes acidity because the weeds begin photosynthesis, so that the resulting calcite is of mixed origin; animals' excretions also form deposits of mixed origin and their importance is emphasized, for example, by Twenhofel: "The generalization seems to be valid that few sediments of lakes and seas of slow deposition have escaped passage through the intestinal tracts of organisms . . ." (1950, pp. 193–94).

These three processes may be summarized as contamination, modification and miscegenation. All three play a part in the history of some sediments. Clearly, it is both hopeless and pointless to try to categorize rocks as either organic or inorganic; on the other hand, it may often be interesting to attempt a decision that apportions degrees of influence, such as "The character of the Brig Reef is mainly determined by the species that have built it, but it is modified at the western end by disintegration and in the center by being choked with land-derived detritus."

With the foregoing remarks in mind, we shall begin our survey of sedimentary rocks with those formed purely by organic processes and then proceed to those in which organisms were less important. At the same time, we shall proceed from deposits only slightly affected by transport to those whose character has been much influenced during their transport. The sequence will be as follows:

1. Organic structures with form and texture as in life—forests, swamps, reefs and other animal colonies such as oyster beds.

TERMINOLOGY

Colonial organisms build REEFS, BANKS and MOUNDS. Reefs differ from banks in the degree to which the organisms interlock and support each other, and in thickness—reefs are very robust, and thick from top to bottom, while banks are less robust and thinner. In both, the profile of the unit is fixed by the growth of the organisms, whereas in mounds, the profile is fixed more by detritus that the organisms trap. Where the trapping organism has no durable skeleton, the end product is simply a mound of detritus, but where the trapping organism has a durable skeleton, as crinoids do, the mound comprises both trapped detritus and remains of the trapping agent. Only reefs and banks properly belong here in class 1; mounds belong in class 4 or class 14. Note that as used in this book, a *reef* definitely implies a rock built of organisms whereas in nongeologic contexts, any area of partly submerged rock may be called a reef.

2. Organic remains that have not been much modified by transport, such as some reef detritus, shelly beds, chalk and deep-sea ooze (in fact, most limestone belongs in this class); coal and some chert and oil shale belong here also.

3. Organic excrement, such as guano and some limestones, glauconite, sulfur and sulfides.

4. Organic precipitates and concentrates. In this category are materials that were never *part of* the organism, as in classes 1–3, but would not have formed as sediments without the influence of organisms. The physical influence of organisms is to slow down water movement and so cause suspended solids to settle out (this is the *trapping* action referred to under 1 above), while the chemical influence of organisms causes the separation of some iron minerals, aragonite etc. from solution.

5. Inorganic precipitates such as some aragonite, some iron deposits and some chert.

6. Evaporites.

7. Residues. Chemical residues are, in a sense, the opposite of classes 4 and 5, being formed by selective dissolving rather than by selective exsolving. Physical residues, the result of winnowing, are appropriately mentioned here; they result from nontransport, in contrast to the rocks (see below) that result from a little transport or much transport.

8. Deposits that have fallen with little transport, such as talus, landslides and mudflows (both subaerial and subaqueous).

9. Deposits that owe their character largely to transport, such as river deposits and deposits from turbidity currents.

TERMINOLOGY

DENSITY CURRENT: A density current is any current that flows on account of a density difference between water at a higher level and water at a lower level; among density currents, those where the density difference is due to different amounts of matter in suspension are *turbidity currents.*

10. Deltas, which are deposits that owe their character to the debouching of flowing water into standing water.

11. Deposits that owe their character to surf action, such as some beach sands.

12. Deposits formed by wind action.

13. Deposits formed by ice action.

14. Deposits dependent on more than one process, such as the ocean mud of item 1, p. 130.

In this list, some classes, such as class 1, are very clearly circumscribed, whereas others, such as classes 8 and 9, are not. In fact, classes 8 and 9 grade into each other: they are more in the nature of end members, between which all intermediates are possible, than classes separated by a natural boundary. The first thirteen classes fall into these three groups:

biological processes—classes 1–4
chemical processes—classes 4–7
physical processes—classes 7–13

The grouping by process is much different from a grouping by environment; for example, some of the river deposits and turbidites in class 9 come from widely separated environments, while various rocks from one environment, such as the continental shelf, may be mentioned under different processes (1–5, 7 and 9 in the case of the shelf). In nature, the processes do not act independently, and so statements such as "On the continental shelf, process *A* interferes with process *B* and the observable result is *x*" will eventually be necessary. Here, however, it seems prudent

Fig. 13-1 **Layout for summarizing sedimentology. It is desirable to understand the processes individually before attempting to understand their cumulative effects. (The oversimplification required to draw such a diagram probably vitiates the attempt—it is to the principle involved that attention is directed.)**

	Process *A*	Process *B*	Process *C*	
Environment 1	important	negligible	occurs, to extent limited by *A*	
Environment 2	negligible	. . .	. . .	
Environment 3	. . .	. . .		

to focus attention on one process at a time before attempting to understand the complex of processes that constitutes an environment. Referring to Fig. 13-1, the program amounts to examining the column headings before attempting to read across the rows.

FORMS, TEXTURES AND COMPOSITIONS

In this section, brief descriptions will be given of some sedimentary units *as laid down*. This is an undertaking separate from describing sedimentary units as found in the geologic column, because the former are unconsolidated whereas the latter have undergone all kinds of changes (in dimensions, texture and composition) between being first laid down and being studied as rocks by a geologist. The descriptions that follow are based mainly on observation of sediments in process of being laid down at present; the changes that occur after deposition are discussed in the next section, on diagenesis. The order of treatment matches that of the fourteen-point list in the previous section.

Organic Structures

Very few organic structures are preserved as rocks with their characteristics unchanged. In a way, one may regard a forest as the uppermost stratum of the earth's solid crust, but usually trees are not regarded as within the province of a geologist until after they have fallen and been more closely integrated with other rocks. The same is true of peat and the plants that constitute the "solid" fraction of swamps. By contrast, some plants and animals build such solid skeletons that, while still alive, they show the characteristics that they will later show as rocks. In this category come corals, sponges and some skeleton-building algae.

TERMINOLOGY

ALGAE (singular ALGA): A class of water plant; these may be divided into *soft algae*, such as seaweed, with no durable skeleton, and *hard algae* which have a durable skeleton. The skeleton is usually calcium carbonate but can be silica, as in diatoms, so that *soft algae* and *calcareous algae* are not complementary terms. Of water plants, only the calcareous algae are reef builders; crinoids build mounds and banks.

Although popularly known as coral reefs, modern reefs are principally composed of things other than coral (algae, sponges, molluscs and so on); this mixed character has been a feature of reefs throughout geologic time. The basic process involved is the intake of water carrying Ca^{+2} and HCO_3^- ions, and the conversion of these by the reef builders into usable forms:

$$Ca^{+2} + 2\,HCO_3^- \longrightarrow \underset{\downarrow\atop \text{SKELETON}}{CaCO_3} + CO_2 + H_2O$$

SKELETON — EXHALED OR SYNTHESIZED INTO ORGANISM'S SOFT PARTS

Reefs rarely rise more than 30 m above the contemporary sea floor; however, on a sinking coastal shelf, the colony may grow upwards to maintain an aerated crest, and some colonies have thus kept pace with subsidence to the extent of 900 m total growth. This is a significant fraction of the thickness of the earth's crust, which may be no more than 5000 m under parts of the oceans. Since the Australian Barrier Reef is over 1500 km long, reefs are comparable in dimensions with the largest sill-type intrusions and, from a chemical point of view, they form concentrates of calcium and magnesium to the virtual exclusion of other metals. Texturally, they are unique; organisms are very versatile in the textures and structures they create but are always much different from any inorganic rocks.

Organic Remains

Organic deposits not much modified by lateral transport were listed on p. 132 and will be described sequentially, except for coal and chalk, which have so little in common with other rocks that they will not be mentioned further.

Reef Detritus. Although reef organisms succeed in building, they are subject to two powerful destructive agents, wave action and boring animals. Of the two, the latter is far more powerful, and the waves' principal function is to carry away the fragments that the borers detach; these are of two kinds, the material the borers remove in making burrows and the material between the burrows. The former is fine and may be carried far from the reef; the latter crumbles to coarser fragments and is not carried as far. The product is a *talus apron* (see Fig. 13-2) with surface inclination up to 40°, usually much larger in volume than the parent reef. In texture, the material of the apron is poorly sorted for the most part (the material of the outer rim undergoes sorting during its journey, which may be several kilometers long); but it is usually well bedded, because of variations in the erosive activity. Its composition is, of course, highly sensitive to that of the parent reef. This is particularly true to seaward; on the landward side, the reef fragments may be contaminated with land-derived detritus and lagoon products such as evaporites or plant remains (see below). A striking feature is the abundance of material with grain size below 0.004 mm. Some of this is detrital, but much is probably due to the boring organisms, which

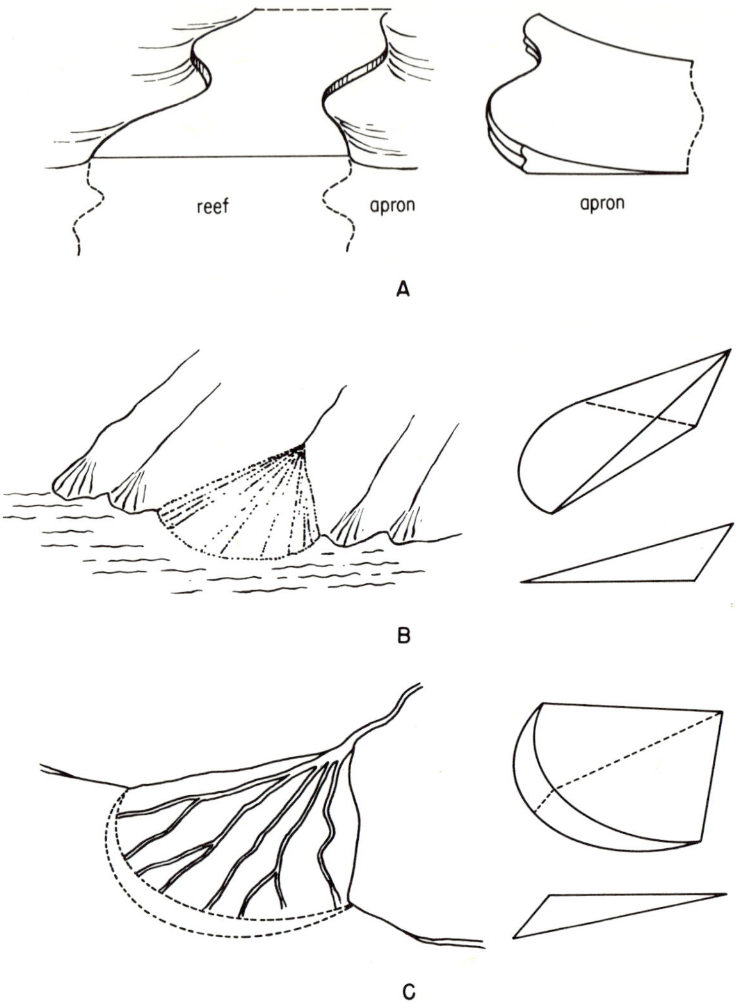

Fig. 13-2 Varieties of wedge-shaped unit (not on a uniform scale). (A) Reef and apron; (B) talus cone; (C) small delta. Large deltas are much broader in comparison with their depth—see Fig. 13-8.

actually eat their way through the rock, passing great quantities of comminuted limestone through their intestines. Borers infest the apron as well as the reef, so that the entire assemblage is dominated by biologic processes and in this respect resembles an ant heap. Reefs were, in fact, the largest structures attributable to activity in the animal kingdom, until the development of the modern city.

Colonies other than Reefs. Some organisms form colonies without forming reefs: examples are found among the brachiopods, molluscs and crinoids. These organisms rarely survive in their life attitudes after death, but they may simply collapse onto the bottom surface on which they lived. In this way, deposits form with high concentrations of a given species, and with many individuals almost completely preserved, although the concentration can be diminished by infilling with detritus after the colony's extinction, or increased by the winnowing-out of detritus that was present during the colony's life. Whether the product is a mound or merely a bank depends on the organism's aptitude for growing on its own dead. For example, crinoids can grow on a substrate of crinoid fragments, but mussels prefer something more solid than other mussels for a foundation. The processes involved are mainly live growth, death and collapse (the trapping function of these colonies is described later in the section on organic concentrates and precipitates). The deposits are of restricted size compared with reefs—under 10 m thick and 100 km in extent, with few exceptions.

Noncolonial Carbonate Remains. The outstanding deposit formed by accumulation of individual organisms is the calcareous ooze of the deep ocean floor. Since late Mesozoic time, foraminifera have proliferated, especially between latitudes 10° N and 10° S. In this zone, accumulation of tests is five or ten times more rapid than to the north or south (Goldberg and Koide, 1962; Arrhenius, 1963). The actual rate, 2 to 10 m per million years, is quite small, but on account of the large area (many times greater than that of the United States), the amount of calcite incorporated in this deposit is greater than that in modern reefs. In fact, as mentioned in Chap. 2, the transfer of calcium from the continents to the ocean floor is at present one-way rather than cyclic, and will lead to a significant change in the bulk chemistry of the continents if continued. Since the present thickness of calcareous sediment is not well known, no one can say how much calcium is already locked up in this way; but, using *possible* figures, 10^8 km^2 by 100 m thick gives a volume of 10^7 km^3, greater than that of the greatest continental basalt floods (Fig. 5-3). (Incidentally, a piece of paper the thickness of this page would have to be extended to five or ten yards across to have the same depth/width ratio; this is a powerful reminder of the thinness of the earth's sedimentary skin.) In texture, foraminiferal ooze is rather fine-grained and monotonous. In composition, it is mainly calcium carbonate, although near the sediment-water interface, the content of organisms' soft parts is appreciable, the soft parts being those of foraminifera and of other animals that feed on their remains.

Smaller accumulations of individual organisms are exceedingly common, especially on the continental shelves. Brachiopods and gastropods, for example, are, in some localities, sufficiently numerous for their remains to constitute a complete rock—the Florida coquina is a modern example. As noted earlier, there is a complete transition from rocks in which each particle is a complete skeleton, through rocks where the skeletons are broken and abraded, to rocks in which inorganic processes of sorting, fracture and abrasion have obliterated the earlier organic characteristics of the material (except for its chemical composition, which is an organic characteristic that no physical process can affect).

Siliceous Remains. The principal silica-secreting organisms are radiolaria, diatoms and sponges. Sponges live in shallow water where carbonates and land-based detritus are also common, so that most sponge fragments are dispersed among other constituents (it is only rarely that sponge fragments by themselves are the main constituent of a rock). By contrast, radiolaria and diatoms can live far from shore, where other sources of sediment may be weak, so that there is a greater chance for organic silica fragments to accumulate uncontaminated. The silica of the organisms is noncrystalline—its internal structure is disordered as in silica glass (see Fig. 5-2A), but may differ from silica glass in the manner of polymerization. The accumulating fragments trap water, some of which is retained when the fragments consolidate; the resulting hydrous silica deposits are one of the forms of chert.

Like foraminiferal ooze, the sediments of this section are bound to be stratiform. In the Pacific Ocean between latitudes 50° and 60° S, diatom accumulations seem to be the main sea-floor sediment at present over an area greater than that of the United States; but no such widespread horizon is known from earlier times. In fact, diatoms are not known at all in pre-Tertiary rocks and recognizable radiolarian cherts are rare and localized. They occur in association with submarine eruptions more frequently than would be expected by random juxtaposition, and the silica released by the eruptions may be the causal link; but no detailed explanation for the association has yet been generally accepted.

Fig. 13-3 Undulations formed by compaction. (A) Before compaction, bedding in the mud is planar; (B) after compaction, bedding in the mud is undulose.

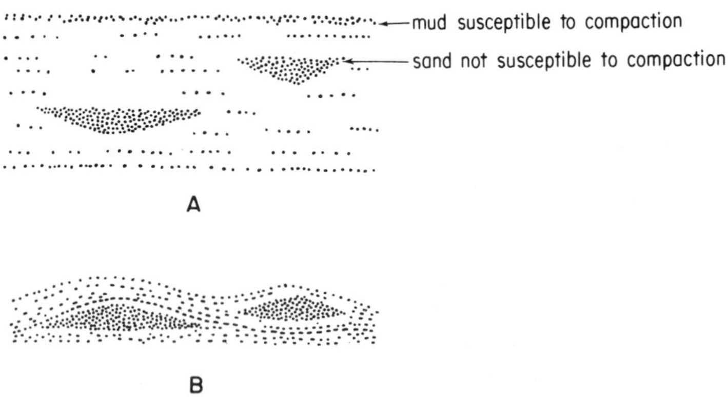

Coal and Oil Shale. In terms of processes, a coal seam resembles a crinoid bank: in one, the preserved plant remains are calcium carbonate and in the other, they are more complicated carbon compounds, but growth, death and collapse is the sequence of events in either case. A swamp, in which many kinds of plants and animals support each other, is another kind of growth that also collapses to sediment. (The fact that reefs are strikingly different in physical robustness should not be allowed to mask the fact that a swamp more closely resembles a reef, as a complex of interacting processes, than any other sediment-forming system.) Rocks derived from swamps range from high-grade coal to slightly bituminous shale, and from rocks in which the organic compounds are preserved (as proteins, waxes, carbohydrates etc.) to those in which degradation to carbon or simple hydrocarbons (oil) is complete. Most rock units of this origin are small—under 10 m thick and 100 km long—like shelly horizons and for similar reasons; they are initially stratiform but because of the extreme porosity of some of the parent materials, undulation due to differential compaction may be strikingly developed (see Fig. 13-3).

Excrement

The quantitative importance of excretion is hard to estimate. It is clear that some small bodies are of excretionary origin and it is possible that some very large bodies are (see p. 131); but, in general, as the size of the unit increases, one's confidence about the part excretion played in its formation diminishes. Fortunately, the seriousness of this uncertainty is reduced for the following reason: the amount of modification a material undergoes in an intestinal passage also diminishes as its bulk increases. For example, sea birds perform a fairly sweeping chemical reorganization between ingesting their food and depositing guano; and guano deposits are highly localized and easily recognized. By contrast, sand worms ingest large quantities of pure quartz and redeposit it unchanged; the only change a sediment suffers on account of sand worms is a screening for whatever tiny percentage of nutrients it contains. Thus, for most human purposes, we do not need to know how much sea sand is excretionary; it is only with more specialized deposits such as guano and some sulfur compounds that the question of biogenesis is of general concern. An exception to the above generalization is provided by some pellet limestones, such as the Shifta formation of Israel, which is clearly excretionary and yet is large in extent.

Regarding guano and other phosphatic deposits, there seems to be little question that phosphorus from seawater has been concentrated biologically, but before being studied by geologists it has usually been redistributed into some mineral form and so is not appropriately discussed at this point. With sulfur and sulfides, the situation is much different: both organic and inorganic processes for extraction of dissolved sulfur (i.e., sulfate and sulfide ions) are known to be possible, but there is rather great uncertainty as to the geologic importance of the organic processes. With regard to the sulfur deposits of Texas and Louisiana, for example, it has been suggested that sulfate was first reduced to sulfide organically and then oxidized to sulfur inorganically (Jones *et al.*, 1956); but where the processes themselves are not known with assurance, it is clearly difficult to assess precisely the importance that organisms have had.

(For a discussion of *glauconite*, whose formation may be biogenic in part, see p. 160).

Organic Concentrates and Precipitates

Concentrate here refers to deposits formed by trapping—the detritus trapped among plants etc. is concentrated on the bottom, in contrast to its former state of dispersal as a suspension. *Precipitate* here refers to solids formed from aqueous solutes without diminution in the quantity of water present, in contrast to evaporites, which take solid form as a consequence of diminution of the water.

The process of trapping has one aspect in common with the process of delta formation: water that was moving with sufficient speed and turbulence to support its contained solids is slowed down and so drops part of its load. If the change in turbulence is great and abrupt, most of the load will be dropped in a nonselective way, whereas if the change in turbulence is gradual, the resulting sediments will be more sorted. Here arises for the first time a point of general importance: the degree to which a sediment is sorted into grains of similar size depends on two factors:

1. The grain sizes available in the material that reaches the deposition site;

2. The power of the deposition process to discriminate between large and small grains.

Clearly, if all material arriving at the site is within a narrow size range, say 0.02–0.04 mm intermediate diameter, no deposition process can generate a sediment with grains spread over a wider range than this. The moderately good sorting of some trapped material is probably due mainly to well sorted supply, rather than to sorting by the trapping process itself.

The area of a trapped deposit is governed by the area over which the trapping colony extends. Its thickness depends on the ability of the trapping organisms to colonize the loose top surface of the detritus they trap. The variety of algae known as blue-green algae, for example, readily do this, yet leave no durable skeleton, so that the geological product of a blue-green algal colony may be a quite thick mound of detritus, with no sign remaining of the organisms that caused its growth.

An example of an organic precipitate is provided by an equation similar to the one that describes the secretion of a calcareous skeleton (see section on organic structures above):

$$Ca^{+2} + 2\ HCO_3^- \longrightarrow CaCO_3 + \underset{\substack{\text{SYNTHESIZED INTO}\\\text{ORGANISM'S}\\\text{SOFT PARTS}}}{CO_2 + H_2O}$$

Even plants that do not build a skeleton need carbon dioxide for photosynthesis. Consumption of carbon dioxide by such plants drives the reaction to the right, and finely-divided aragonite may result. Organisms also affect the acidity of their environment, the concentration of dissolved oxygen and other chemical properties, any one of which may control the deposition of some sediment (cf. Chap. 15). There are two reasons, however, for relegating discussion of the products to the next section. Firstly, it is only very rarely that it can be said of a precipitate: "This would not have formed in the absence of organisms." In some soils, development of aragonite or gibbsite or some other mineral is so localized about each individual root that organic control must have been operative, but on a larger scale the evidence is usually less conclusive. Secondly, no environment is free of organisms and even such a quantity as the chlorine concentration in the sea is to some extent organically controlled. There are therefore not two distinct classes, organic precipitates and inorganic ones, but rather one class in which some precipitates have been influenced greatly and directly by organisms and some have been influenced only in remote and insignificant ways.

Inorganic Precipitates

The precipitation of aragonite is a suitable first example. It was remarked that if plants remove carbon dioxide from water rich in Ca^{+2} and HCO_3^- ions, solid aragonite may form. The same is true if carbon dioxide escapes for any other reason. Two common reasons are lowering of pressure and rise of temperature; it is also possible that placid water will become supersaturated with carbon dioxide, in which case making the water turbulent may provoke loss of the dissolved gas. Any one of these three effects is capable of generating aragonite, but the product will naturally be most abundant where all three effects combine. They do combine where cold, deep water is swept by a current over the crest of a sea-floor rise—on the crest, the water will be at lower pressure and is often warmer and more turbulent; some of the Bahama reefs, for example, seem favorable sites for precipitation from these causes. If the process were to occur, the product would be fine-grained and uniform-textured, but if a fine uniform lime mud is encountered, it is rarely easy to work back and assign degrees of importance to organic and inorganic influences. Only in a situation such as the Dead Sea (an environment effectively insulated from all organic activity) can the question really be settled.

Other deposits in whose history inorganic precipitation seems likely to have been important are iron deposits and some cherts. Some chert is recognizably of organic origin (see discussion of siliceous remains above) and some is recognizably diagenetic, i.e., formed by transformation of a preexisting nonchert sediment, but there is much where, as with lime mud, evidence by which one might reconstruct its genesis has not yet been found.

The more important iron deposits are divisible into two groups, the Clinton type and the Lake Superior type, the former carrying limestone and the latter chert as principal contaminants. Whichever is in mind, however, the favored possibilities are the same as far as transport and deposition of the iron are concerned:

1. The iron is transported as ferrous ions in acid water and deposited when the water reaches an alkaline basin;
2. The iron is transported as ferric oxide adsorbed on organic matter of colloidal size and is deposited when the water reaches a saline basin (the charged ions in the salt water causing the colloidal particles to flocculate and sink).

In the Clinton ores, the contained fossils indicate marine conditions, which are both saline and alkaline with respect to most rivers, so that, as commonly in geology, the problem is to judge the relative importance of two simultaneous processes rather than to make a yes-no decision between them.

The same remark applies to two other aspects of the Clinton problem. Firstly, observation shows that the truth lies somewhere between replacement of limestone by hematite and mere filling of pore space by hematite (some fossils are replaced, while some are overgrown with concentric layers). Secondly, it seems likely that some hematite formed at the sediment-water interface (oolites with alternating layers of silica and hematite suggest this process rather than selective replacement of some layers after formation of the whole oolite), but also some hematite probably formed below the sediment-water interface by agency of interstitial water.

Evaporites

Compared with inorganic precipitates, evaporites are simpler. There are still many uncertainties about the details of their formation, but, like a reef, an evaporite can often be recognized unambiguously by observation.

In form, evaporite deposits are sheets (the reader may be able to think of some small-scale exceptions). Around lakes, they may be annular, solute concentration at the shallow edge being greater than in the center; but large evaporite deposits require a prolonged inflow of solute-bearing water, and the regime by which this is circulated and concentrated can lead to many other patterns of deposition. In dimensions, evaporites are comparable with iron formations: thicknesses range up to 350 m (Permian, Texas) and the extent is that of whatever saltwater body can be isolated sufficiently for evaporites to form. The most striking modern evaporite situation is the Gulf of Karabugaz, adjoining the Caspian Sea. If the waters of this gulf were simply evaporated today without replenishment, a deposit measuring 1 m by 10,000 km^2 could form; but at present, inflow constantly makes up for evaporation and so the thickness the deposits will attain cannot be estimated (Twenhofel, 1950, p. 500).

In composition, evaporites are variable. The most common constituent salts are $CaCO_3$, $CaSO_4$, NaCl and $MgSO_4$; $CaSO_4$ occurs both as anhydrite and as gypsum, the hydrated form, and $MgSO_4$ occurs in three hydrated forms. The order in which these salts form is determined by the ionic abundances in the water, the temperature and the depth. Consequently, evaporite deposits are often stratified, with beds of contrasted composition. Since ordinary detrital sedimentation can be concurrent with evaporation, a "salt" bed may carry 50 per cent or more detrital silicates, with the water-soluble salts as cement. It is supposed that evaporite crystals are small and anhedral when first formed, and that the coarse texture of many evaporites is a diagenetic feature.

A product quite separate from those described above is formed upon evaporation of groundwater. This also is best considered a steady-state process, with a dilute solution constantly percolating from below up into a hot porous surface layer; at whatever level the water table stands, solutes accumulate where the water converts to vapor. They do so only under conditions of exceptional aridity, but the large nitrate deposits in Chile seem to have formed in this way, and similarly indurated sands of less value occur all around the world. These deposits do not attain the purity or the thickness of open-water evaporites; the Chilean nitre beds are only a few meters thick and are valuable only because their chemistry differs from other deposits, not because they are unusual in size.

Residues

A chemical residue results from the selective removal of some part of a rock by solution. For example, when rain falls on a Chilean nitre bed, as it occasionally does, it dissolves out the nitrates and leaves a worthless sand residue. The same kind of effect occurs in the weathering of granite: a little calcium and sodium are extracted, with sufficient decomposition of feldspar to allow conversion of solid granite to a lumpy sand of near-granite composition. By contrast, in some limestone solution cavities, the amount of solute removed may be almost the entire rock, and only a few parts per cent of insoluble residues remain.

A physical residue likewise results from the selective removal of some part of a rock or sediment; the agent is usually water or wind and the part transported is usually the smaller, or the flakier, or the less dense grains. An example is the gravel on the floor of a desert valley: the occasional exceptionally heavy rains cover the valley floor with a poorly sorted mixture of mud, sand and pebbles, but between storms, the wind picks out the dust and fine sand, the remaining sand subsides a little and the pebbles are left conspicuous on the surface. Some river gravels are residues, or *lag deposits*, in the same sense, but some river gravels never had a finer-grained filling and result from nondeposition of mud etc., rather than deposition followed by removal.

The most extensive of present residues are to be found on the continental shelves. The shelves presently receive a certain amount of terrestrial detritus, but over large areas the dominant process now going on is removal rather than sedimentation. In Pleistocene time, the shelves were largely exposed, and the sediments which the sea has transgressed since the Pleistocene are now being washed free of mud. In fact, the process has probably neared its end point in most regions, the finer fraction having been already carried down the continental slope to the ocean floor.

A particularly striking kind of residue is the ilmenite sand found on a few ocean beaches. This is a product of chemical attack, which has dissolved less stable minerals or reduced them to smaller grains, coupled with physical attack which has carried the smaller grains away. The same combination at other localities has produced sands of olivine, magnetite, monazite and other minerals.

Deposits Influenced by Gravity and Water Transport

This bulky and important group (items 8 and 9 in the fourteen-point list, p. 132) is not exactly specified by any short title. Strictly, all subaqueous sediments are influenced by gravity as well as water transport, because it is gravity that makes the sediments settle. The intention here is to recognize two processes:

1. The falling of solids under gravity through a largely static environment (air, water, mud etc.);
2. The transport of solids horizontally by movement of their environment.

Having recognized these as end members, we study here a range of deposits formed by combination of the processes in different ratios—dry talus, landslides, mudflows, slumps, turbidites and river deposits—the characteristics of talus being almost wholly attributable to simple falling and the characteristics of river deposits to fluid transport. Of course, throughout the group, as elsewhere, the characteristics of the end product are greatly influenced by the starting materials, as well as by the transport process.

Under gravity alone, all material would travel merely vertically. An effect important to both the processes mentioned is the deflection of material from a vertical path by the solid surface of the earth. When dry rock falls from a cliff, it bounces outward from the cliff's foot, and a similar effect makes rivers run to the sea. In a sense, therefore, fluid transport as well as simple falling is gravity-driven. The two processes might then be restated as:

1. The direct effect of gravity on the solid particles;
2. The indirect effect of gravity on the particles, exercised through the agency of the environmental fluid (usually water).

Clearly, we must expect the two processes to be intimately interdependent, but for discussion of differences, it is convenient to treat them as distinct end members. The main differences lie in their power to *sort* and in their power to *round*. If angular fragments fall under gavity, their angularity is not noticeably changed, and small fragments fall amongst the large ones; the more they are transported by water, however, the more rounded they become and the better they are sorted. The effects are almost purely physical (the power of organic colloids to keep inorganic material in suspension is the most important chemobiological effect) and speculative aspects are therefore touched upon in Chap. 16. The observational groundwork follows.

There are three important types of slope down which the processes are active:

1. Scarps, both subaerial and subaqueous;
2. Delta foreset slopes (Fig. 13-8);
3. The continental slope, stretching from the outer edge of the shelf to the ocean floor.

The erosion of uplands that proceeds without scarp formation is considerable, but the reader will realize what parts of the sequel are applicable without the erection of a separate fourth category. On delta foreset slopes and continental slopes, the horizontal component of movement far exceeds the vertical, so that in order to begin with products most influenced by end-member (1), direct gravity effects, we have to look at scarp faces.

Subaerial scarp faces are attacked by the weather, fragments fall, and talus accumulates at the foot, in a continuous apron if the face is straight and weathering uniform, or in cones if collapse is episodic or gully-controlled. An idealized talus cone is shown in Fig. 13-2B. Cones are small, as geological features go; they are fed at a rate that is rather constant and eroded at a rate that increases with their surface area, and so a steady state is usually achieved before the height reaches 300 m. Moreover, they form mainly in mountainous terrains and so are not continuous horizontally for more than a few kilometers. At present the world's mountain ranges contain many cubic kilometers of talus, but not many of these will be incorporated in the geologic column intact. In texture the material spans a wide size range, from boulders to dust, and is poorly sorted, if considered in bulk; however, even a talus apron is stratified into wedges that are richer and wedges that are poorer in the smaller grains. (Of course, a dry apron seen from above *reveals* less of its fine fraction than exists, because the small grains are washed or blown down into the crevices between the larger blocks.)

Subaqueous scarps (mainly fault scarps) also accumulate detritus, in part at the time of faulting and in part subsequently; but physical weathering does not occur, so that chemical change is greatly slowed down and, unless the scarp is attacked by boring organisms, growth of aprons between fault movements seems likely to be very slow.

As the material disintegrating is less cohesive and the fragments that move increase in size, talus-falls grade into landslides. At the other extreme, landslides grade into the allochthonous* blocks of gravity tectonics, which may be hundreds of kilometers long. Landslides in the normal sense, i.e., where movement is observable within a single lifetime, are only of local importance in geology.

Mudflows may be regarded as talus falls or landslides that have been provoked and extended by abundant water. Saturation of unconsolidated surface material is the prerequisite; this is the natural condition of subaqueous sediments and may also occur under heavy rain on hillsides of volcanic ash or on desert hillsides, or when the top few feet thaws over ground that is still frozen. In some terrestrial environments, the mechanism is powerful, i.e., great weights of rock are transported each year; but, as with talus cones, the deposits so formed usually suffer further modification before being overlain, and so are not prominent in the geologic column. Historic mudflows on land have measured over 100 km^2, but under 10 m thick. Sorting is negligible and composition, like that of talus, varies with the local bedrock, subject to the restriction that mud-sized particles must be abundant; otherwise the suspension does not become dense enough to carry much rock.

**Allochthonous:* If a rock unit is deformed until one part of it becomes totally detached from the remainder, the detached part may be called allochthonous, in distinction from the remainder which is autochthonous. The term is usually applied to blocks which have become detached by movement on low-angle faults.

Subaqueous slumps or mudflows are probably more extensive and more common than subaerial ones, but of course are less readily observed. There are only a few parts of the sea floor, close to well-staffed research institutes, where observers are present sufficiently often for subaqueous slumping to be observed more or less by routine. Consequently, discussion is largely deferred to Chap. 16. We note here only that the density contrast and the viscosity contrast between a subaqueous slumping mass and its environment are much less than for subaerial slumps. One consequence is that subaqueous slumps occur on far gentler slopes, down to as little as 5° or less. This means that on a delta foreset, slumping and deposition compete: as long as deposition is vigorous, the foreset slope maintains its inclination, but once deposition slackens, slumping soon reduces the inclination of the face, and distributes the material of the delta widely over the surrounding floor. Since the distributary pattern of a delta constantly changes, this kind of dispersal can be a vigorous process at one point on the perimeter even while another point is being actively built outward. The point was made (see p. 144) that many rocks of the present continental shelves are residues—sediments first laid down in Pleistocene or earlier time, now being reworked; the principal exception to this rule is the material now being spread across the shelves by destruction of present-day deltas by slumping.

We now move on to deposits where direct gravity effects are less important and indirect gravity effects in fluid transport are more important. On land, this brings us to flood and river deposits and, under water, to turbidites. In flood water, the solid/liquid ratio is smaller than it is in a mudflow, and in a turbidite the ratio is smaller than in a subaqueous slump. As it happens, the mechanism by which the muddy water (density 1.1 or less) is generated is not the same in the two situations, but, once generated, the manner in which it drops its load in one situation makes an interesting comparison with the manner in the other.

Both a turbidite and a flooding river have a main channel flanked by broad, shallower flow (this is observable in rivers and fairly well substantiated for turbidites), and whereas in the channel either erosion or deposition may be dominant, on the flanking plains deposition is the major effect. Because flooding is a transient state, deposition is from a waning flow, and one might expect a graded deposit to result for the following reason: immediately after the flood is at its height, the flow should drop only sand or silt and should be sufficiently vigorous to retain mud-sized particles in suspension, but as the flood subsides, finer and finer material should be deposited. On river floodplains, this expectation is not wholly borne out: irregularities in the land surface and in the flow pattern lead to ripples and other kinds of nonplanar bedding, and, once formed, the deposits are immediately subject to disturbance by plants and burrowing organisms, so that graded bedding is often absent from floodplain sediments. Under water, the same

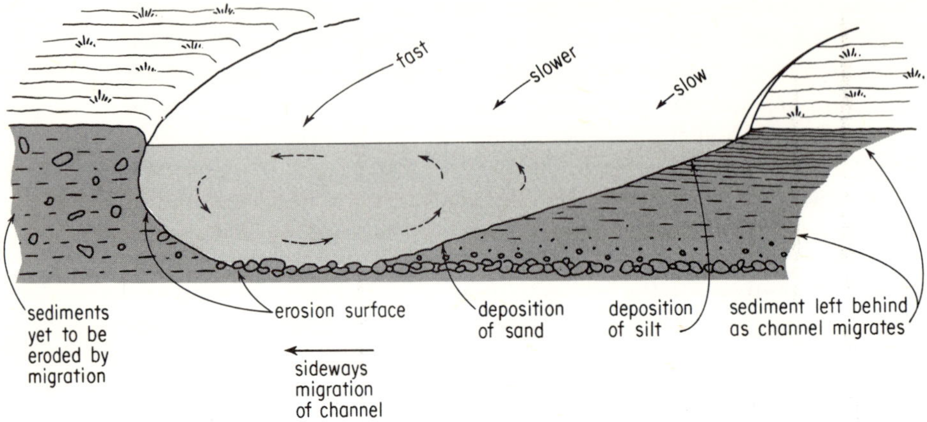

Fig. 13-4 **A simplified river regime. Even in this idealization, particles do not move perpendicularly to the line of the channel: if a sand grain is deposited in the plane of this section, its associated silt grains will be carried further downstream as well as further up the sloping side of the bed before deposition.**

kinds of interference exist but are less powerful, so that graded deposits are found peripherally where delta foreset beds have collapsed to a low inclination, and at the foot of the continental slope. In the latter situation, great areas have been covered; the aprons can be recognized by the extreme flatness of their surfaces (in contrast with the hummocky surface of the true oceanic sediments) and are found to extend over 1000 km out to sea from the west coast of North America (Menard, 1964, p. 195). Close to the foot of the continental slope, an individual bed may be several meters thick, but over the bulk of their area, beds are only a few centimeters or a few millimeters thick.

The dominance of flow processes over simple falling under gravity becomes complete where there is continuous flow in a well established river valley. A river with a stable profile does not fill its own valley very rapidly, except at times of flood; its main effect is to rework old flood deposits by migrating sideways across its valley floor. During such migration, sediments are eroded from one point and deposited at another, with little change in bulk but considerable change in texture (Fig. 13-4). The

Fig. 13-5 **Imbricated particles. The dominance of particles leaning toward the right, as shown, suggests a current flowing from the left.**

eroded bank often contains particles which the river is unable to transport, which merely fall and constitute gravel as at the base of Fig. 13-4; this gravel is therefore a residue or lag deposit, in contrast to the finer layers, whose particles the river has been able to transport.

Lag deposits left by currents of constant direction show a characteristic absent from, for example, continental shelf lag deposits, where the current that has winnowed out the finer fraction is not constant in direction. The characteristic, *imbrication* (see Fig. 13-5), arises from the fact that among platy pebbles of a given size, those leaning downstream are least easily put in motion by the current; a lag deposit may therefore contain a preponderance of pebbles with this orientation. The effect can be seen on some beaches, as well as in river channels, because beaches are also largely lag deposits washed by currents of constant direction (after a wave has surged up the beach, the water washes the beach in retreating seaward).

If a river were to migrate across its valley in an orderly manner, a well stratified deposit as in Fig. 13-4 might result; but of course the great rivers of the world are not orderly—when they were worshipped as deities, they were credited with emotions at least as unpredictable as those of mankind. Besides, only small channels approach the simplicity of Fig. 13-4; in larger channels, sandbanks, islands and backwaters have migratory patterns of their own. Consequently, the valley deposits of a large river contain not only many pockets where the regime illustrated has been in force for a time, but also many irregularities due to floods, droughts, landslides and all kinds of changes in the catchment area.

Before leaving the medley of processes that involve a measure of direct action of gravity and a measure of indirect action via fluid transport, we should notice again the great power of these processes to *sort* solids. If a current existed that was constant in all its characteristics, particles could be divided into those which the current could transport and those which it could not; but natural currents are not constant, and instead of sorting particles into *moved* and *not-moved*, a natural current sorts them into *moved far* and *moved not far*. The mechanics of the action are discussed in Chap. 16; the observable effect, which is our present topic, is to separate small particles from large ones, and if the small particles differ in composition from the large ones, important chemical segregations can also result.

The segregation with the most far-reaching consequences occurs when the rocks subject to erosion and transport are igneous. The principal components of igneous rocks are quartz, feldspar and the ferromagnesian mineral group (Fig. 7-2), and weathering reduces these to particles of different sizes: quartz grains are very little affected, and occur in the weathered product in grains of almost the same size as in the parent rock, but the feldspars and ferromagnesian minerals are altered to sheet minerals, with great decrease in size—in the first place, the change leads

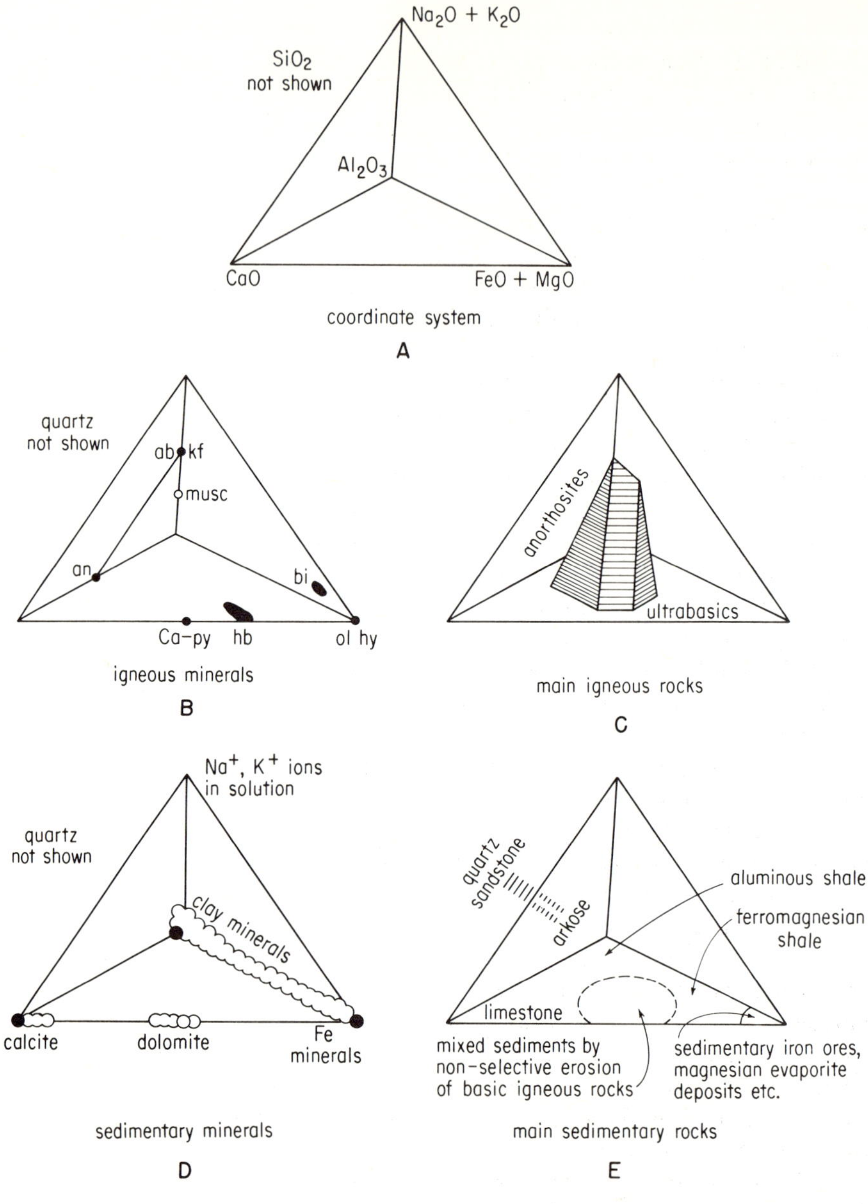

Fig. 13-6 The dispersal of igneous components to form sediments. Using the tetrahedral framework shown (diagram A), any mineral or rock can be given a position according to its richness in lime, alumina, alkali metals or iron and magnesium. Thus, materials especially rich in alumina are represented by points close to the alumina corner; materials equally rich in lime and alumina are represented by points equidistant from these two corners, and so on; materials containing no alumina are represented by points on the triangular face opposite the alumina corner. If this system is used to plot the main igneous rocks (polyhedron in diagram C) and the main sedimentary rocks (diagram E), it is seen that the range of compositions is wider among the sedimentary rocks. **For abbreviations see p. 307.**

to crumbling of the parent crystals and in the second place, the sheet minerals form in very small flakes. At the same time, much calcium, sodium, potassium and iron goes into solution (or into colloidal suspension), and these chemically different fractions are apt to be transported to quite different sites. The process is known as *sedimentary differentiation;* the effect is that the igneous components are dispersed (Fig. 13-6).

The dispersal of igneous components has been emphasized because, if it were not for this effect, sediments would be as restricted in composition as igneous rocks are. There are, however, other important ways of sorting, for not all rivers drain igneous terrains and the material that a turbidity current sorts may be very different from terrestrial source rocks. If a slump occurred, for example, in mixed reef detritus and volcanic ash, the first material deposited would be the larger carbonate fragments, and the quartz would all be in the far-removed fine fraction. This last example emphasizes the point made earlier, that the details and outcome of the sedimentation process are largely controlled by the nature of the input materials.

Deltas

The top surface of a large delta may measure hundreds of kilometers across, and on this top surface—partly subaerial, partly submerged—many of the processes described in other sections of this chapter occur, for example, swamp development. However, swamps do not require discussion here since to a first approximation, what is true of any coastal swamp is also true of a deltaic one. The process discussed here is the one essential and peculiar to delta formation, namely, the flow of running water into standing water with consequent deposition of transported solids. On small deltas, such as those which tend to form when artificial reservoirs are constructed, the change in velocity is abrupt, but on a large delta the situation is much different. It is therefore appropriate to notice some dimensions.

size important re process

The first requirement is to distinguish between a delta and a delta group: when a river has built a delta of some extent, the probability arises that it can find a shorter way to the sea by a change of course (see Fig. 13-7). The life of a single delta is therefore rather well defined and short, in comparison with the life of a delta group. For example, a river with the approximate course of the present Mississippi has been draining North America for at least 30 million years, but the present channel running southeast from New Orleans has been the main outlet for only 500 years or less (Scruton, 1960). A single delta, thus distinguished, is unlikely to exceed 200 km across its front, but there are many rivers in the world at whose mouths deposition is probably active along a front of 100 km or so. When we turn to the distance from the shoreline to the *fringe,* i.e., the line beyond which detritus from the delta-forming river is no longer the

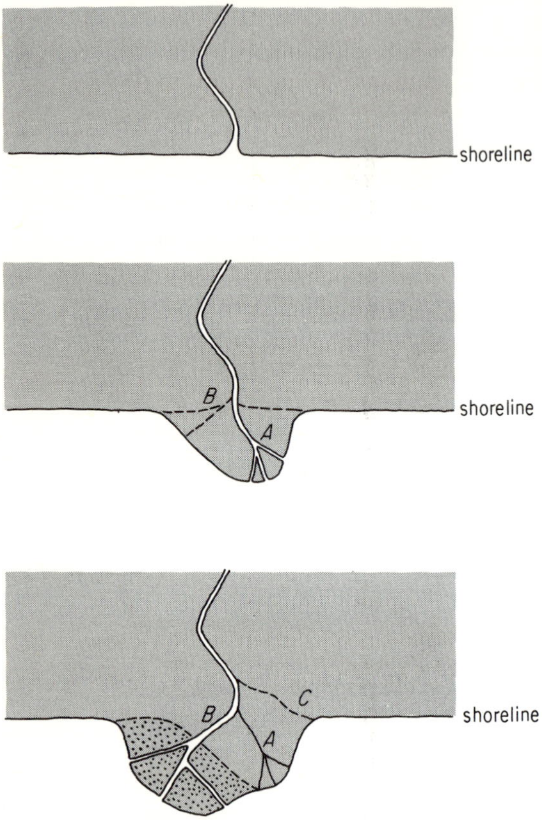

Fig. 13-7 Succession of deltas. If, at a given stage, delta building occurs mainly at the mouths of channel *A*, a time will come when channel *B*, heretofore unimportant, is able to divert the main sediment load. A delta with geometry entirely independent of the one based on channel *A* will then develop (stippled), until it in turn is killed by a new distributary, perhaps at C.

major fraction in the sea-floor sediment, this is found to be 10 or 20 km for the Mississippi, perhaps as much as 50 km for the Orinoco, and 50 km for the Niger (Allen, 1964). The depths of water at the fringe are of the order of 100 m so that the overall gradient is under 1°. At a few localities on the Mississippi delta the slope approaches 3°. While the topset, foreset and bottomset slopes are therefore recognizable (Fig. 13-8), they are all exceedingly gentle and form a smooth curve; the discontinuity between topset and foreset slopes, which exists on very small deltas, as it does on sand dunes, is not detectable on the important deltas of the geologic record.

The petrological significance of these figures lies in the fact that they show the drop in velocity at a distributary mouth to be a rather gentle phenomenon. There is, in general, no reason to expect the silt deposited on the face of a large delta to differ from silt deposited by any other flow whose vigor declines with distance but remains constant with time—there is simply a smooth increase in the proportion of clay particles to silt in samples successively further out to sea. As was mentioned at the beginning

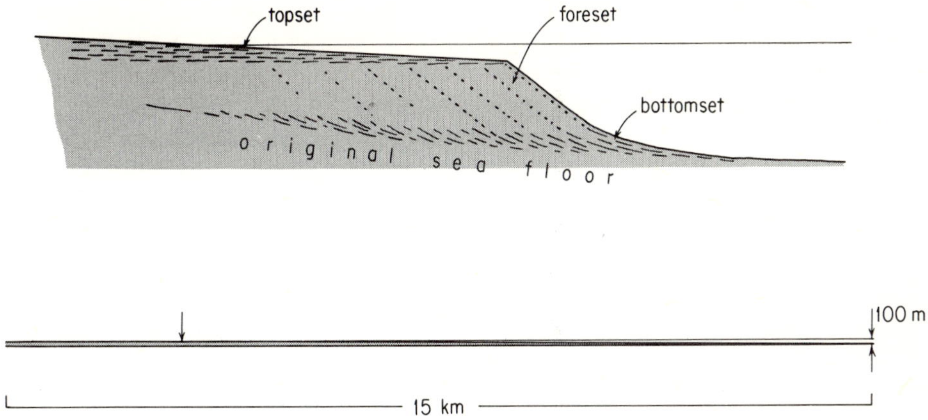

Fig. 13-8 **Deltas in profile. The three parts of a delta, and a large delta in profile with no vertical exaggeration; the arrow marks a typical position for the shore line.**

of this section, other processes affect delta sediments (even on such gentle slopes, slumping, for example, can occur) but the delta process itself, as specified above, leads to deposits whose texture is in no way peculiar.

Independently of the physical aspects, the chemical effect of salt water on the fresh-water clay suspension of a river deserves mention here. Much of the material carried by a great river is of colloidal size (0.1 to 0.001 micron) and, when unflocculated, such material settles at an extremely slow rate. However, charged particles such as sodium ions, which are present in seawater in vastly greater abundance than in river water, cause these colloidal particles to cluster and precipitate more rapidly. If it were not for this effect, deltas would be even thinner and broader than they are. The flocculation effect is, of course, not confined to deltas, but is far less important away from them in terms of quantity of sediment. In fact, it is the combination of flocculation with fall in velocity that makes marine delta groups the mighty accumulations that they are. Murray (1960) has estimated that a group of overlapping deltas (as in Fig. 13-7) may reach a total thickness approaching 1 km, and that in Mesozoic and later times, a total of 15 km of sediment has been loaded onto the Gulf of Mexico sea floor, in large part by deltaic processes.

Surf-modified Deposits

Most marine deposits and many terrestrial ones are subject to some kind of current flow, with attendant sorting, but the group now to be treated is the special class that owes its petrologic character to *ceaseless milling* by inshore waves. Such deposits are found in the littoral zone and below it, to a distance that varies with the nature of the coast and of the incoming waves—where storm-raised ocean rollers come in unimpeded, the zone of wave action is wider than on protected coasts of gentle climate. To maritime people, the relentless grinding of the sea is one of the most impressive aspects of natural power, but its geological products are rather insignificant in bulk; even an exposed reef disintegrates more by action of boring organisms than under the pounding of the sea. Surf, in fact, winnows preexisting deposits more than it forms new ones, so that most surf-modified deposits are residues; there is a lower size limit to the particles found, below which any particles once present have been washed away. This is a distinction from, for example, windlain deposits, where there is no lower size limit but there is an upper one, above which particles are too big to be carried by the wind (Friedman, 1961; this matter is further discussed in Chap. 16).

In addition to sorting, surf is also a powerful agent for rounding, and consequently beaches carry some of the roundest particles in the natural world. In fact, a surf-worn shell sand is a good example of a deposit with mixed organic and inorganic character; the composition is organically controlled but the texture may be conferred largely by inorganic processes.

Deposits Formed by Wind Transport

Observing the present land surface, we see deposits of windblown sand measuring 1500 km by 500 km, like the basalt floods (but thinner). The desert area of North Africa and Saudi Arabia is twenty times greater even than this, but only about one quarter is covered with loose sand, in drifts and patches; much of the remainder is sufficiently stable to support shrubs, and some is bare rock. On the whole, however, the Sahara is the closest that any land formation comes to matching in size the foraminiferal blankets of the great oceans. Of pre-Pleistocene formations, the Torridonian of Scotland, the New Red Sandstone of England and the Navajo Sandstone of Colorado are examples that are believed to be windlain, at least in part, although this idea has not been definitely established for any of them.

Windblown sand units have no characteristic external shape, such as talus cones, beaches or deltas have, but internally their steep foreset cross-beds are a prominent feature. Judging by present dunes, the foreset slopes can be as high as 30 m, but since the top is often removed before formation of higher beds, heights in preserved windlain sands are rarely over 5 m. Beside foreset cross-beds, windlain sands seen in section can show various kinds of trough and ridge, as well as small-scale ripple marks (heights measured in centimeters). Windlain sands are moderately to very well sorted, but the distribution of grain sizes differs from that in a well sorted beach sand, for, as noted above in the discussion of surf-modified deposits, beach sands have a lower size limit whereas the limit on a windblown sand is at the upper end of the size range. In windlain sand, as in marine sand, quartz is the most abundant constituent, taking a world view (the reason for other minerals tending to generate smaller particles was given in the discussion on p. 149). Of course, windlain sand is subject to the same restriction as all other sediments—the product at any one locality is governed by the source material available.

Deposits of particles greater than sand size are not formed by wind, but windlain deposits of particles in the silt and clay size fractions are widespread on the present earth's surface. This is to some extent in spite of two factors that hinder the formation of windlain dust deposits:

1. Exposed clay and silt land surfaces are more easily colonized by plants than sand surfaces, and so are protected from being put in motion by the wind;
2. Once raised into suspension, windblown dust can be carried greater distances than sand; much is dispersed into neighboring sedimentary environments, including the oceans, instead of being deposited in localized concentrations.

Much dust comes from regions where too much salt or too much agriculture has made the land barren, or where the climate is too cold and dry—the Pleistocene glaciation may be the cause of most of the Pleistocene dust deposits now exposed. Much windlain dust has a curious texture, highly porous yet coherent; windpacked snow has the same properties, which may be related to the angularity of the particles, but the matter is not well understood. Windlain dust is difficult to recognize in older formations; perhaps compaction masks its character, or perhaps the deposits are mostly reworked into other kinds of sediment before being overlain and preserved.

Deposits Formed by Ice Transport

It is necessary to notice the difference between the following groups:

1. Rocks formed in a glacial climate;
2. Rocks whose distribution is controlled by glaciers;
3. Rocks actually deposited by ice.

The third group comprises simply the different kinds of moraine. The second group includes moraines and also outwash gravels, eskers and kames which, while controlled by ice, are laid under water. The first group includes all the foregoing and also such deposits as the windblown dust off cold deserts and some varved sediments in lakes. It is only the moraines whose petrology is characteristic of ice; the other glacier-influenced deposits are discussed elsewhere, in the sections appropriate to their process of deposition.

Rocks formed in a glacial climate are very extensive at present because of the recent Ice Ages, but moraines are rather restricted. By the nature of the glaciation process, they are rather localized in the first place and many, for example those over Canada and Northern Eurasia, have been somewhat modified during the years since their formation. Live moraines are now more extensive in Greenland than anywhere else (unless we include the ground moraines which presumably exist beneath much of the Antarctic ice cap); thickness rarely exceeds 100 m and deposits are probably not continuous over areas greater than 100 km^2.

The conspicuous textural feature of ice-laid rocks is their poor sorting. Having been subject to neither water nor wind, they are thus comparable with fault breccias (in fact, the ice undersurface is a thrust sole of a special kind). All particle sizes are likely to be present, from boulders down to clay; in this respect ice deposits differ from, for example, talus that has been filled in with sand or mud. In composition, an ice deposit is likely to be rather close to the average of all rocks the ice has traversed, because ice is a less selective eroder than water and wind. For the same reason, neighboring pebbles are often widely contrasted in composition (the finer grains should not be so widely contrasted, however, because the pulverizing process is selective, and reduces some minerals faster than others).

Mixtures

We have considered thirteen processes, some with subdivisions, all going on now and many going on together within a very small space. It would be natural to expect that most rocks would be affected by two or three processes at once, but two factors operate to reduce such multiplicity:

1. Some of the processes are mutually exclusive; for example, evaporites and oil shales are not mixed by being formed *simultaneously* in the same spot—the salinity that permits evaporite formation excludes the oil-generating organisms; one sediment may succeed the other closely, but this association does not constitute mixing;

2. Some of the processes go on at vastly different rates; the separation of deep-ocean sands from the oozes is an example—most of the time, ooze accumulates in the presence of no sand whatever and then, in a few hours, a sand unit may be thrown down by a turbidity current with no ooze whatever (at least, since 1 mm of ooze takes about a thousand years to accumulate, the contribution during a few hours can be neglected); here, one sediment does not exclude the other but merely swamps it.

However, some processes can go on together at comparable rates. Suspended particles and evaporite crystals can settle together (see above under discussion of evaporites) and in parts of the deep ocean, windblown dust and organic ooze are intimately mixed. Of course, a mixture of material from different sources deposited by a single process is a very common phenomenon, but this is not the same as a deposit to which two processes have contributed simultaneously.

DIAGENESIS

In accordance with a precept enunciated earlier, before describing diagenesis we shall define it, as follows. After deposition, it is in theory possible to define a volume of rock by means of an envelope that passes outside particles *a*, *b*, *c*, . . . but inside particles *p*, *q*, *r*, ... Such an envelope will remain well defined under compaction and deformation, and also under removal or addition of material as long as the amount removed or added is small in comparison with the amount initially contained. Usually, in fact, the envelope will remain well defined until its contents either disintegrate (under erosion, solution or tectonic crushing) or melt. There is therefore a class of events that occurs between the time the envelope becomes definable (the end of deposition) and the time it becomes unrecognizable (during disintegration or melting). These events include diagenesis, metamorphism and weathering, and, in brief, what is not weathering or metamorphism is diagenesis. The boundaries between these three are far less easy to agree upon than the boundary that embraces them, but fortunately there are few purposes that are hindered by disagreement in detail. Some geologists refer to changes in the range 100 to 200°C and 10 to 1000 bars as diagenesis and some as metamorphism, but such differences do not materially hinder our attempts either to understand rock formation or to put natural resources to good use.

Having partly defined diagenesis, the next matter is to describe it, and this will be done by means of examples; but the examples are prefaced by the question, "In what sense is diagenesis observable?" The purpose of this chapter has been to review the facts about sedimentary rocks that we should like to be able to explain. However, unlike sedimentation, most diagenetic changes take too long to be watched. How, then, are *facts* about diagenesis ascertained? The answer is, of course, that in some situations the nature of the parent material is more or less apparent, and diagenesis is "observable" only when this condition is met.

The clearest examples come from diagenesis of fossils. When a mineral such as chert, pyrite or hematite has surface or internal geometry that shows that it has replaced an animal's shell or filled a mold, diagenesis must be involved (where it fills space that was a cavity or soft part of the animal during life, the argument is not so absolute). Similarly, if a calcite fossil of a species whose live members build aragonite is found, or if a fossil consists of coarse crystals where the live organism is known to have had fine structure, there must have been diagenesis. The preferential solution of a shell to leave a mold is also a kind of diagenesis. Molds and pseudomorphs after crystals are in the same category as those after fossils.

The filling of pores in the original deposit is a second kind of diagenesis that is sometimes easily recognizable; the Chilean nitrate-cemented sands are one instance, and the filling of the voids in a talus cone by fine sediments is another. Calcite-cemented sandstones are not quite so simple, because conceivably calcite mud and quartz sand could be thrown down together, for example, if the sand were blown out to sea by wind. Suppose then that rounded quartz grains are seen to be cemented by coarsely-crystalline calcite: one can conclude that diagenesis has occurred, but cannot choose among the three possibilities that exist:

1. The calcite was of original deposition and only recrystallization is diagenetic;

2. The calcite is a subsequent introduction but has not changed texture since emplacement;

3. Both emplacement of calcite and recrystallization to a coarser texture occurred diagenetically.

Layers made of interlocking quartz polyhedra also deserve attention. If such a layer has, for example, gradational margins, the possibility that it is a vein is excluded; there are also arguments against the direct crystallization of large quartz units from surface waters. Some remaining possibilities are that it is modified inorganic detritus (probably) or modified chert (possibly). On the detrital hypothesis, one cannot tell whether the original porosity has been filled by overgrowths of quartz from an external source or by cannibalization of smaller grains, so that again one can only be sure that diagenesis has occurred without being sure exactly in what manner.

A third situation in which diagenetic effects are observable is portrayed in Fig. 13-9. (If composition contrasts occur with irregular distribution and no recognizable control, diagenesis is possible without being certain.) Here the necessary picture of what the rock was like before diagenesis occurred is provided by the material distant from the fractures in A or the shells in B. An analogous situation exists on an emergent or submergent coastline where changes consequent upon subaerial exposure or marine immersion respectively can be studied by comparison with material not yet affected by the change. It turns out, for example, that if marine limestones

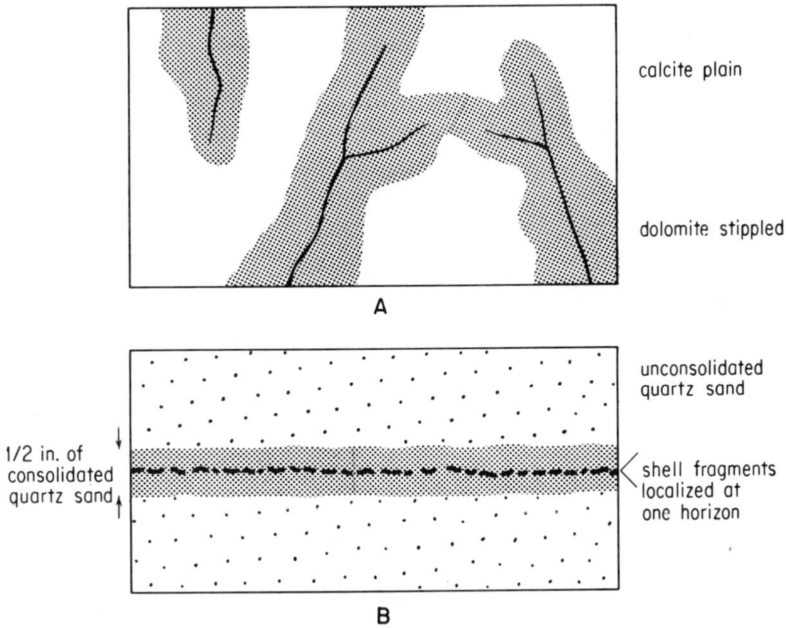

Fig. 13-9 Localized diagenesis. In both cases the localization is a clue to the diagenetic process.

are exposed to sun and rain, aragonite dissolves and some calcite forms in its place—but the replacement is not always immediate, so that cavities can be formed.

A fourth, rather specialized, situation exists in the case of dolomite, $Ca_{1+x}(Mg, Fe)_{1-x}(CO_3)_2$. The ratio of oxygen isotopes in this mineral is believed to serve as an indicator of whether it formed simultaneously with its associated calcite or diagenetically from it; so far, measurements indicate that even recently-formed dolomite, such as that found in the bottom sediments of some present-day lakes, is an alteration of a preexisting carbonate rather than primary crystals (Degens, 1965, p. 119). This finding seems to mean that all dolomite is diagenetic, although for some dolomite rocks we may be unable to decide at what stage in their history they were dolomitized.

Lastly, the diagenesis of clays deserves mention, although uncertainty as to exactly what changes occur is probably greater here than in any of the foregoing examples. Some formulas of clay minerals are given in Table A1-3. The principal clay minerals presently being carried by rivers into the sea are illite, montmorillonite, kaolinite, vermiculite and chlorite; in addition to these, glauconite is found abundantly on the sea floor. However, shales carry pyrophyllite and muscovite as well as the minerals listed, so

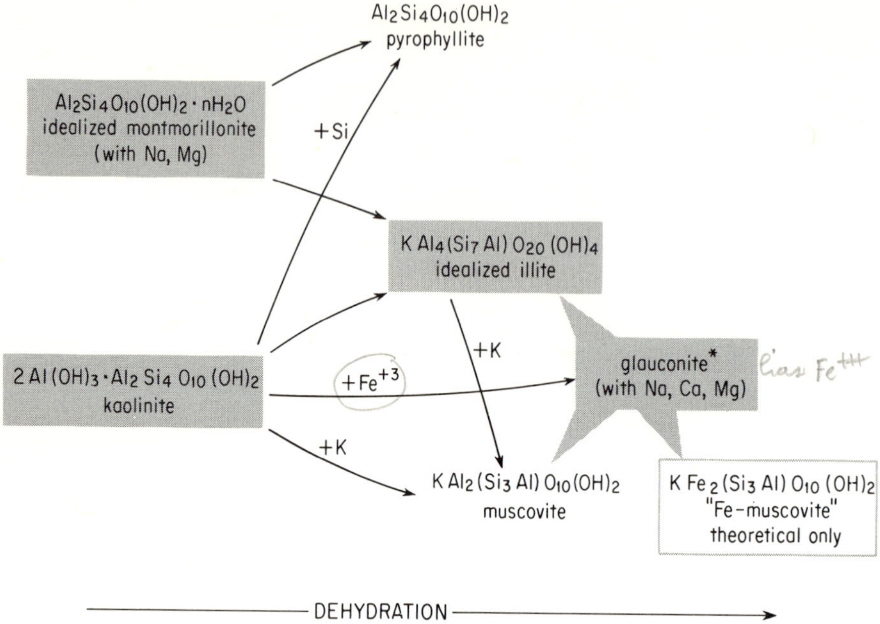

Fig. 13-10 Possible dehydrations among the aluminous clay minerals. ****Glauconite* **is chemically intermediate between the three molecules indicated.**

that changes must occur among the aluminous minerals, to some extent as shown in Fig. 13-10. What happens among the ferromagnesian minerals is more uncertain: eventually, under metamorphism, biotite appears and presumably takes up the constituents originally present in chlorite, vermiculite and glauconite, but the intervening stages are not at all well known.

Another observable fact is that much of the glauconite on the sea floor, and some of the other clay minerals, occur in pellets 0.02 to 0.04 mm in diameter. The uniformity in the size of these pellets suggests that some animal may be involved in their formation, but whether the ferric iron which glauconite contains is also conferred on it biologically is not known. Like dolomite, glauconite is found in recent sediments and unlike dolomite, it is very rarely found in situations that suggest subsurface formation (as in Fig. 13-9), so that it is currently believed to be almost always formed by marine diagenesis of loose clay.

In contrast to what may be called the "constructive" diagenesis of clay minerals, discussed above, there is also the possibility that the "destructive" trend from feldspar to clay will be extended, with further disruption of the silicon tetrahedral structure, thus:

$$\underset{\text{FELDSPAR}}{2\,KAlSi_3O_8} \longrightarrow \underset{\text{KAOLINITE (+ SOLUTES)}}{Al_2Si_2O_5(OH)_4 + 4\,Si(OH)_4 + 2\,KOH} \longrightarrow \underset{\text{GIBBSITE (+ SOLUTES)}}{2\,Al(OH)_3 + 2\,Si(OH)_4}$$

The powerful agents which perform this disruption convert the ferromagnesian minerals simultaneously to $Fe(OH)_3$, variably dehydrated to hematite, goethite etc. The resulting concentrations of sesquioxides are called *laterite* or, if aluminum is sufficiently preponderant, *bauxite*. Observation leads to the conclusions:

> **1.** Laterite can be formed from almost any kind of rock, given favorable circumstances;
>
> **2.** Favorable circumstances are: tropical climate, alternately wet and dry; low relief; neighboring rocks that increase the acidity of the groundwater.

As with other diagenetic changes, it is easier to establish what has occurred than why or how, but some of the relevant considerations are touched on in Chap. 15.

To complete this descriptive gloss, some of the physical changes consequent upon diagenesis are listed. The outstanding physical change is loss of water, with reduction in porosity. Other effects include change of color, density, permeability, roundness, coherence, rigidity and fracture behavior. To a large extent, diagenesis affects only the smaller textural features, so that gravel, for instance, remains recognizable after transformation to conglomerate. However, the contrast between one bed and the next, or between a lump and its matrix, can be drastically changed; for instance, a gravel texture might be made quite inconspicuous or might be dramatically emphasized. Exceptions to the statement that diagenesis affects only small features exist; for example, concretions can grow to a length of several feet.

CHAPTER FOURTEEN

RECOGNITION, CLASSIFICATION AND NOMENCLATURE OF SEDIMENTARY ROCKS

Whereas the preceding chapter was about rocks, the present chapter is about people. Specifically, it is about a mental activity—the decision-making process engendered when people look at rocks. Naturally, the discussion has points in common with the corresponding sections of Part Two (Chaps. 5, 6); the reader is invited to refer back to those sections in order to perceive the common ground.

RECOGNITION

There are two separate aspects to the recognition problem:

1. The recognition of a rock as sedimentary or not-sedimentary, as in Chap. 4;

2. The recognition of the kind of sedimentary rock that is under study, e.g., marine sand, glacial clay etc.,

but the second can be considered independently of the first. For example, a rock may be so fine-grained, dark and tough that the question of whether it is an igneous flow or a mudstone may not be immediately resolvable; but while the possibility remains that it may be igneous, much can be said about the kind of sediment it is if it turns out to be sedimentary, e.g., it is poorly sorted, probably subaqueous, from current-free environment and so on.

With regard to aspect (1), the four-way division in Chap. 4 (p. 25) is recalled. Observations that put a rock in the upper right quarter of the diagram include:

> Abundant fossils;
>
> Predominance of well rounded grains or oolites;
>
> Abundance of hematite;
>
> Abundance of halite, gypsum and other evaporite minerals;
>
> Abundance of rounded lumps that are loose in their matrix, except for products of spheroidal weathering.

Observations that put a rock in the upper half of the diagram, without excluding the possibility that it is pyroclastic, include:

> Layering, especially with cross-bedding, grading, mud cracks, ripple marks etc., which exclude the possibility of parallel magma sheets;
>
> Abundance of fragments that are loose in their matrix, except for fault breccias in igneous rocks;
>
> Abundance of clay, except for clay produced by subsurface alteration of granite etc. (the possibility of bentonites prevents this criterion from being put in the previous list).

Observations that put a rock in the right-hand half of the diagram, without excluding the possibility that it is a low-temperature vein or alteration product such as an ore body, include:

> Abundance of quartz;
>
> Abundance of calcite, except for the very rare carbonatites (Chap. 7).

The number of exceptions and qualifications that are needed in any attempt to make such criteria rigorous is a reminder about geological practice: it is far quicker and easier to reach a conclusion by forgetting about all the improbable possibilities than by systematically finding evidence to exclude them. Conclusions so reached will *usually* be correct, and statements that embellish the most probable conclusion with expressions of uncertainty and all kinds of less probable possibilities are often of no more *use* for the purpose in hand. The level of intellectual rigor attempted will therefore vary from one study to another. For example, if the question is *where* to drill for oil, the optimum site is all that needs be selected—whether the probability of its proving barren is 20 per cent or 50 per cent does not need to be established; but if the question is *whether* to drill, it is desirable to think of all the unfavorable possibilities and to estimate the sum of their probabilities.

Aspect (2) of the recognition problem, concerning the kind of sediment, is rather different, in that it is never finished. If a rock consists to ninety per cent of rounded quartz grains, it simply *is* sedimentary and no further attention to aspect (1) is needed; but after some questions about its transport history, source area and so on have been answered, there are always more that can be attacked: field study can be made more extensive or intensive, and laboratory techniques can ramify without limit. We stop because we have achieved some purpose, not because no more can be discovered.

How do we achieve a purpose involving recognition of genesis? (Even where one's basic question is forward-looking—for example, "Will bridge piers built of this rock last eighty years?"—it may be best answered by looking backward into the rock's past.) The main thing to note is that nothing that is said in Chap. 15 is of any direct help. The statements of Chap. 15 are cast in the form, "Process x leads to features y and z" whereas for recognition we need, "Feature p can arise only from processes s or t." Statements of the latter kind are rather few; an example is, "Parallel incised striae in coarse rocks form only under ice." Even this is merely a conjecture, whereas "Ice can form parallel incised striae" is observable, at some cost in time and trouble. The conjectural nature is exemplified by ripple marks: it was widely believed that these formed only in shallow water, until deep-sea cameras recorded them on the ocean floor. A compilation of such conjectures in detail is not appropriate here; it will merely be remarked that, even more so than in igneous rocks, single definitive criteria are rarely used. It is even more important in sedimentary studies to establish the shape of a unit, the degree of internal homogeneity and its relations with its neighbors. Conclusions are more often reached by observing what different kinds of units are associated than by observing a single unit in isolation. Some of the techniques available for use are mentioned in Chap. 17.

CLASSIFICATION AND NOMENCLATURE

As with magmatic rocks, the naming of sediments has two purposes: (1) to permit references to classes of rock, e.g., "limestones are important as building stones," and (2) to permit reference to individual rocks, e.g., "the rock at Pin Point is a limestone." A term that is good for one purpose is not necessarily good for the other; particularly with purpose (2), any term that exactly expresses the operator's observation or opinion is preferred to a term from a system of nomenclature that means more or less than is intended. Both purposes are best served by terms that are exactly defined, but sedimentary rocks, possibly even more so than magmatic rocks, are burdened with names that have become embedded in current usage and now defy exact definition. Here a minimum of terms will be introduced, partly for use later in the text and partly to emphasize and advocate the current practice of distinguishing rigorously between texture, composition and origin.

Textural Terms

The most basic textural terms are the grain-size terms (Table 14-1). Using these, any unmetamorphosed sediment can be described by grain size—if the bulk of the rock is not wholly in one size fraction, a composite term such as *sandy silt* or *silty sand* can be used. The terms imply nothing about origin except that the rock is sedimentary, nothing about composition and nothing about texture except for the properties in parentheses.

Note 1: If you speak of a sandstone without qualification, your hearer will think you mean a quartz sandstone, but there is no inconsistency in the terms *olivine sand*, *magnetite sand* etc.

Note 2: Breccia and *rubble* do not even imply sedimentary origin.

TABLE 14-1 GRAIN-SIZE TERMS[a]

Intermediate Diameter	*Single Particle*	*Unconsolidated Aggregate*	*Consolidated Aggregate*		*Metamorphosed Product*
	boulder	—	—	rudite	psephite
256 mm (c. 10 in.)					
	cobble	gravel (rounded)	conglomerate (rounded)		
64 mm (c. 2 in.)		OR	OR		
	pebble	rubble (angular)	breccia (angular)		
2 mm					
	sand grain	sand	sandstone OR arenite		psammite
$\frac{1}{16}$ mm[b] (ϕ4)[d]					
	silt grain	silt	siltstone OR siltite		pelite
$\frac{1}{256}$ mm[c] (ϕ8)[d]					
	clay particle	dust (dry) mud, clay (wet)	mudstone[e] OR lutite		

[a]after Lane *et al.*, 1947

[b]single particle readily visible against contrasting background; feels gritty between fingers

[c]single particle invisible to naked eye; no gritty feeling between fingers

[d]the phi scale of grain size: a measurement of n on the phi scale is equal to $\frac{1}{2}^{n}$ mm. Although the divisions between sand, silt and clay were originally established in fractions of a millimeter, it is now customary to use phi units or decimal equivalents

[e]because most mudstones are fissile, the term *shale* (= fissile mudstone) is in much more common use.

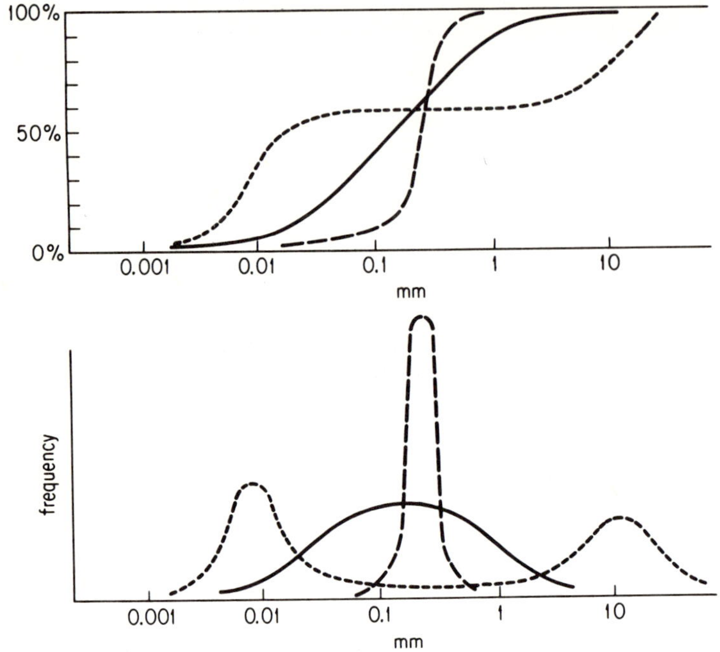

Fig. 14-1 Cumulative curves and frequency curves for particle size. Each curve completely represents a single sample. In the upper (cumulative) diagram, any point on a curve shows the percentage of the sample that consists of grains smaller than the size corresponding to the point. In the lower diagram, any point on a curve shows the relative frequency in the sample of grains whose diameters lie within x of the size corresponding to the point, where x is small in comparison with the range of sizes in the sample. For small x and large sample, the shape of the curve is not affected by the particular value of x used, except in vertical scale.

The metamorphic terms are mentioned here to make the distinction clear. They do not indicate the grain size of the rock as seen; they indicate the grain size of the hypothetical sediment before metamorphism; a pelite may contain biotite flakes 5 mm across. Thus "the rock is a pelite" states a hypothesis about its history; "the rock is a lutite" states a property of the rock which anyone can confirm by present observation. Apart from this difference, the equivalence is close—*pelite* was originally no more a composition term than *lutite*, although the reader should realize that it is now rather widely used to indicate aluminous composition.

Although not exactly terminology, the graphical representation of grain size is conveniently mentioned at this point (Fig. 14-1). Where accurate data or a specific hypothesis is to be presented, it is a waste of time to use words since a diagram is quicker as well as more informative. Where precision is desired, it is necessary to state whether percentages quoted are by weight or by volume, and also exactly what diameter is to be measured or estimated—the width of a particle as seen down a microscope is not at all simply related to the width of a square hole that will just permit its passage, as in sieving. Fortunately, when referring to a specimen as sandstone, we can often be confident that, however the percentages and diameters were to be established, the majority would fall within the limits implied, so that the intricate question of exactly what we mean by "diameter" need not be faced.

Other textural properties which every sediment must exhibit to a greater or lesser degree are sorting, rounding, porosity and fissility. No special terms are involved, however, except *shale* = fissile lutite (which deserves mention as constituting over half the world's sediments, according to some systems of estimation).

Compositional Terms

Many compositional terms are simply mineral or metal adjectives:

aluminous	dolomitic
calcareous (calc-, calci-)	feldspathic
ferruginous	micaceous
magnesian	quartz, quartzose
siliceous	

These naturally tend to be combined with a grain-size term, thus: *calcareous siltstone*. Very few composition terms are nouns in their own right like *limestone*, which has no textural implication whatever. Several combinations of composition and grain-size terms have one-word equivalents, however, such as:

argillite:* aluminous lutite (adj. argillaceous)

arkose: feldspar-quartz arenite

chert: siliceous lutite

marl: calcareous argillite

*More so than most terms, *argillite* and *argillaceous* suffer from having been used in the past in many different senses. A geologist who uses either term needs to decide what meaning he intends by it and inform his readers on this point. The meaning given above has these advantages:

1. The relation between the terms *argillite* and *lutite* is made clear: all argillites are lutites; some lutites are argillites;
2. The proper relation of noun and adjective is maintained between *argillite* and *argillaceous*, as between *arenite* and *arenaceous*.

The reader should be aware that many geologists do not follow the usage here advocated.

greywacke: sediment with appreciable weight fraction in each size range from mud through sand and in each composition group—quartz, feldspar, sheet silicates and rock fragments

The minimum amounts of the different components necessary to conform to the definition vary with different authorities, but the general relation of greywacke compositions to other sediments is shown in Fig. 14-2.

All the terms introduced so far have been observational; not one has implied anything about origin.

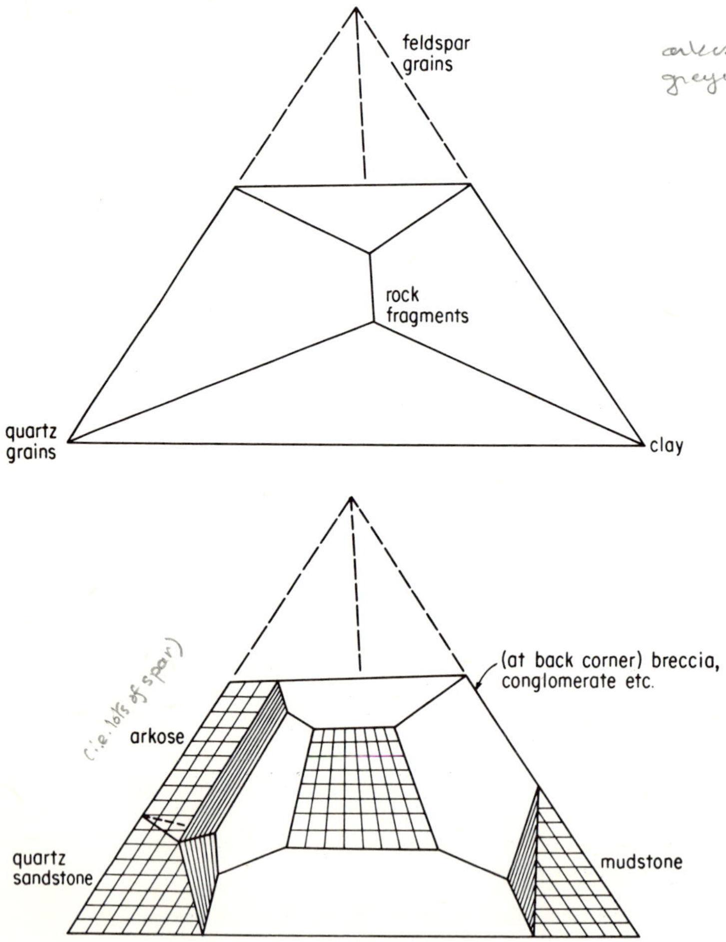

Fig. 14-2 General relation of greywackes (in central cavity) to other detrital rocks. The upper diagram shows the mineral components considered and the feldspar-rich field not represented among rocks. The lower diagram shows the same tetrahedron and the approximate composition fields of some common rocks.

Genetic Terms

All sediment material is *detrital* or *nondetrital*. Detrital grains are grains that have remained solid throughout the sedimentary cycle under discussion; nondetrital material is material contributed to the sediment from solution, during the cycle under discussion, by organic or inorganic means. (Nondetrital material such as a reef limestone may become detrital material in the next erosion cycle, and all detritus has, ultimately, a nondetrital origin.) Deposits are called detrital or nondetrital according to the nature of the greater bulk of contained material. The clay minerals pose a slight problem because we do not know how far the lattice is disrupted when a feldspar or ferromagnesian mineral weathers to clay; certainly a clay particle has not remained solid throughout the cycle in the same way that a quartz particle may, because at the beginning of the cycle probably very little clay was present. In a sense then, a clay particle is as much a product of chemical transformation as a marine hematite deposit. However, because the clay particles form early in weathering and thenceforth behave as inert solid grains under transport, in the same way that quartz does, the clay particles in a sediment are commonly considered part of the detrital fraction.

Similarly, all sediments are *subaqueous* or *subaerial*, according to the location in which the majority of the components were assembled (though a subaerial tuff, for example, may be lithified subaqueously, or vice versa). Usually, glacial deposits are separated as a third category, and understood to include all deposits that owe their peculiarities to ice action, even though they accumulated in air or under water. Subdivisions of these classes can be made by locality of deposition—*piedmont, deltaic, lagoonal, pelagic* etc., and by mechanism—*fluvial* or *residual deposits, turbidites* etc. Thus one sediment sample could be described as a *fluvial piedmont quartz arenite* and another as a *chemically-precipitated lacustrine calci-lutite*. For many purposes it is not necessary to specify all four aspects—mechanism, situation, composition and texture—but, whatever the purpose, it is necessary to decide which of these aspects one wishes to specify by the name.

Detail 1: An evaporite would not be called a silt even if the grain size were in the silt range; the terms in Table 14-1 are applicable mainly to detrital particles, including, by convention, "detrital clay." Fine calcite sediments, however, are called calcilutite even if they are believed to have been precipitated chemically.

Detail 2: Strictly, "residual" is the opposite of "transported" and, if wind is the agent in question, an eolian-residual sediment forms in one locality while an eolian-transported sediment forms in another. However, for most geological purposes, "residual" is sufficient, whatever the agent, and "eolian" is understood to mean deposited by the wind.

A few terms combine genetic and observational meanings: thus *tillite* = glacial deposit with wide range of grain size; *fanglomerate* = rudite deposited at the change of slope at the foot of a steep gully or cliff. If textural, compositional and genetic meanings of terms are clearly recognized as separate, much difficulty will be avoided in devising terms for particular occasions that make the reader understand what is intended. Terminology, however, is not standardized—what one geologist takes to be rigorously implied by a term may be merely a usual connotation, admitting exceptions, to another. When two people use the same term in discussion, it is generally unsafe to assume that they mean the same thing by it.

CHAPTER FIFTEEN

CHEMICAL ASPECTS OF SEDIMENTARY ROCK GENESIS

A complete sedimentary episode comprises *disintegration*, *transport* and *reintegration*. In this comprehensive view, chemical processes during sedimentation perform a complete conversion from a mineral assemblage stable under one set of conditions to a mineral assemblage stable under another (or, even where the starting material or the end product is metastable, this is the direction in which the chemical processes tend). It follows that questions about the chemical processes can be summarized thus:

1. What is the stable end product under the new conditions?
2. By what mechanisms does the change-over proceed?

It is rather probable that the new conditions will be in the range 0–100°C, 1–1000 atm and in presence of water. It is true that sediments form under ice below 0°C and in deserts where water may be "absent"; but in both these circumstances, chemical change is very slight—as with igneous rocks, as long as water is present changes proceed, but when the last water disappears chemical standstill supervenes. Much sedimentation goes on within an even more narrow range and in fact very close to the standard chemical condition, 25°C and 1 atm. The approach, therefore, is first to discuss the processes of change as they would occur at this condition, and subsequently to adapt the conclusions to take account of different temperatures and pressures. A third necessary distinction is between the stability of a phase in bulk and the stability of the same phase on a surface. An example is provided by water: under 1 atm, water becomes unstable with respect to steam at 100°C if in bulk; but if the water is adsorbed on a glass surface in a sufficiently thin layer, it will remain stable up to 150°C or more. In summary then, the order of treatment will be:

1. Stability relations between bulk phases at 25°C and 1 atm;
2. Modifications for surface effects at 25°C and 1 atm;
3. Modifications for other temperatures and pressures;
4. Mechanisms for approaching stability.

STABILITY OF BULK PHASES

The stability of a mineral in a sediment is indirectly related to its stability in a magma, so that the topic can be introduced by reference to Fig. 8-6, the quartz–albite system. The figure shows that if a quartz-albite mixture is at temperature T_2, crystals of quartz will or will not be present according as the composition of the mixture lies to the right or to the left of B. In other words, the stability of quartz as a solid phase is determined by the concentration of quartz in the fluid phase. But it was remarked in Chap. 3 that the stability of a phase is determined by its free energy. There must therefore be a link between silica's free energy balance and its concentration in the solution or melt. If we now turn to the sedimentary environment at 25°C and 1 atm, the concentration of a solute becomes the *principal* influence on its free energy. It is therefore desirable to explore the relation between these two. The ideas developed will be used in the sections that follow on carbonates, feldspars and iron minerals, and are of importance again in connection with metamorphism.

The influence of concentration on free energy is studied with this reaction as an example:

$$\underset{\text{SOLID SCAPOLITE (MEIONITE)}}{(3\ NaAlSi_3O_8)\cdot NaCl} \rightleftharpoons \underset{\text{SOLID ALBITE}}{3\ NaAlSi_3O_8} + \underset{\text{IONS IN SOLUTION}}{Na^+ + Cl^-}$$

As always, the side with the lower free energy is the stable side of the reaction (see Chap. 3); but the energy balance is affected by the concentration of the solution—the more a solution is diluted, the lower the molar free energy of the solute drops. At 25°C and 1 atm, the molar free energies of meionite and albite are fixed quantities, that of meionite being greater than three times that of albite by a fixed amount, A. Suppose a salt solution could be made of such strength, S, that the free energies of one mole of sodium ions and one mole of chlorine ions exactly totalled A; then on immersing albite and meionite in this solution an equilibrium situation would exist. (A mole of sodium ions is x g of ions, where x = weight of a sodium ion/$\frac{1}{16}$ × weight of an oxygen atom, i.e., the normal definition of a mole is extended to species which are not molecular.) If the solution were then diluted to twice its bulk, the free energy of the sodium ions contained would drop, and so would that of the chlorine ions, and the right hand assemblage would have an energy advantage over the left hand assemblage (meionite). The reaction would run to the right, with enrichment of the solution in Na^+ and Cl^-, until the strength returned exactly to S (the time taken would be infinite, but this is no stumbling block in the idealized world under discussion). Thus, for a given temperature and pressure, there is an equilibrium concentration of solutes, toward which the system will tend.

For ideal solutions, if the concentration is halved, the free energy of any species of solute is diminished by $RT \ln 2$ cal/mole (R = the gas constant, T = absolute temperature; at 25°C, $RT \ln 2$ = 410 cal/mole). Highly dilute solutions approach this ideal behavior, but in stronger solutions the relation breaks down. If one were able to increase the concentration of a solution by jumps and wished to increase the free energy of a mole of solute by steps of 410 cal, *to begin with* jumps that doubled the concentration would serve exactly; but when the solution became stronger, each jump would have to more than double the concentration. A solution would thus be formed which actually had a concentration c but behaved in energy respects like a solution of concentration a (less than c). The value of a, the *apparent* or *effective* or *thermodynamic concentration* is called the "activity" of the solute species, and will be shown in this text by brackets; thus $[x]$ = activity of species x, regardless of whether x is an ion or a molecule.

Up to this point, the conditions of the hypothetical experiment have been such that the only source of sodium and chlorine ions has been meionite, and they have necessarily been formed in equal numbers. We next remove this restriction and imagine the solution to be diluted to twice its bulk with caustic soda. The effect per mole of chlorine ions would be as before—their free energy would drop by 410 cal; but if the caustic soda was suitably strong, the net effect might be to increase the overall sodium concentration and so increase the free energy per mole of sodium ions. Conceivably, in fact, the molar free energy of sodium could be increased by exactly 410 cal. In this case, the equilibrium would be exactly preserved—diluting the chlorine would tend to make the reaction run to the right, but increasing the sodium concentration would tend to make it run to the left, with net effect nil. The equilibrium situation would not be changed, and there is another thing that would also not be changed—the product

$[Na^+] \cdot [Cl^-]$. This example introduces the generalization: as long as the product $[Na^+] \cdot [Cl^-]$ is maintained at the appropriate value, equilibrium will be maintained however much the individual activities change.

A second reaction of interest is:

$$\underset{\text{SOLID}}{CaSO_4} + \underset{\text{IONS IN SOLUTION}}{CO_3^{-2}} \rightleftharpoons \underset{\text{SOLID}}{CaCO_3} + \underset{\text{IONS IN SOLUTION}}{SO_4^{-2}}$$

Again, suppose that an equilibrium condition exists and the solution, containing carbonate and sulfate ions, is diluted to twice its bulk. The molar free energies of carbonate and sulfate will both change, but since they both change by the same amount and in this case occur on opposite sides of the reaction, the equilibrium condition is not upset. In the first example the equilibrium condition was:

$$F_{Na^+} + F_{Cl^-} = \text{constant} \quad \text{or} \quad [Na^+] \cdot [Cl^-] = \text{constant}$$

The corresponding conditions in this second example are:

$$F_{CO_3^{-2}} - F_{SO_4^{-2}} = \text{constant} \quad \text{or} \quad \frac{[CO_3^{-2}]}{[SO_4^{-2}]} = \text{constant}$$

The general nature of these relations is perhaps now apparent without further examples: an equilibrium is a situation where by adding and subtracting a number of free energy terms, a particular numerical result is attained; but because of the proportionality between a change in free energy and the *logarithm* of a change in activity, adding and subtracting free energies implies multiplying and dividing activities. Thus in a complicated hypothetical reaction:

$$\underset{\text{SOLID}}{AB} + \underset{\text{SOLUTES}}{C^+ + D^- + E^{+2} + \ldots} \rightleftharpoons \underset{\text{SOLID}}{UV} + \underset{\text{SOLUTES}}{W^+ + X^- + Y^{+2} + \ldots}$$

the equilibrium condition would be

$$F_C + F_D + F_E + \ldots - (F_W + F_X + F_Y + \ldots) = F_{UV} - F_{AB} = \text{constant}$$

or

$$\frac{[C] \cdot [D] \cdot [E] \ldots}{[W] \cdot [X] \cdot [Y] \ldots} = \text{constant}$$

These ideas will now be applied to some more realistic sedimentological situations. The first and outstandingly important reaction is

$$\underset{\text{MOLECULES}}{H_2O} \rightleftharpoons \underset{\text{IONS}}{H^+ + OH^-}$$

All that has been said above about solute ions applies to solute molecules, so that at 25°C and 1 atm, pure water ionizes to an extent fixed by

$$F_{H^+} + F_{OH^-} - F_{H_2O} = \text{constant (in this case, zero)}$$

or $$\frac{[H^+]\cdot[OH^-]}{[H_2O]} = \text{constant}$$

The term $[H_2O]$ is the "effective concentration of water molecules in water," a slightly awkward concept. It must be realized that the equation in activities is applicable even if all species are present in, say, 90 per cent alcohol; here, we might find $[H^+] = 1\%$, $[OH^-] = 1\%$, and $[H_2O] = 8\%$, a perfectly intelligible value (where $x\% = x$ particles per hundred particles). This is a reminder that concentration may be taken as the ratio

$$\frac{\text{number of items of species } x}{\text{total number of items of all species}}$$

(or it may be taken as a ratio by weight or by volume rather than by number, or it may be taken as the number of moles of species x in a given weight or volume of water—a form which will be used later; any of these forms may be used as convenient, as long as in a given calculation, one form is used consistently). Using the first form, we should write

$$\text{concentration of } H_2O \text{ molecules} = \frac{\text{number of } H_2O \text{ molecules}}{\text{total number of particles including ions}}$$

As a matter of fact, in distilled water only one molecule dissociates into ions for about five hundred million that do not, so that the ratio is 500,000,000/500,000,002, whose departure from 1 may be neglected. At such high concentrations (as well as at very dilute ones) the activity of a species more or less equals its concentration, so that $[H_2O] \cong 1$. Even in seawater, $[H_2O]$ is 0.98 or more, so that for geological purposes the departure from 1 is almost always negligible and we assume

$$[H^+]\cdot[OH^-] = \text{constant}$$

If $[H^+]$ and $[OH^-]$ are expressed in moles per 1000 g of water, the value of the constant is 10^{-14}.

By virtue of this relation, it is not necessary to specify both $[H^+]$ and $[OH^-]$ for an aqueous geological environment. Conventionally, $[H^+]$ is stated in the form $[H^+] = 10^{-p}$, e.g. 10^{-6} or, more simply, just "p = 6" or, more unambiguously, "pH = 6." Note that a *low* value for pH means a *high* effective concentration of hydrogen ions, or what is commonly called an acid solution. It follows, for example, that in an acid solution the reaction

$$\underset{\text{SOLID ANALCITE}}{NaAlSi_2O_6\cdot H_2O} \rightleftharpoons \underset{\text{SOLID PYROPHYLLITE}}{AlSi_2O_5(OH)} + \underset{\text{IONS IN SOLUTION}}{Na^+ + OH^-}$$

would tend to run to the right, while alkalinity would tend to make it run to the left.

In the remaining three parts of this section on bulk phases, the foregoing concepts are applied to some more geological situations.

Carbonate-Water Relations

After water itself, the most conspicuous geological material concerned in solute reactions is $CaCO_3$. The following discussion is based largely on treatment by Garrels (Garrels and Christ, 1965), whose work should be consulted for a more scholarly and far-reaching presentation. We shall study the effect of rainwater on limestone, with two preliminaries:

1. Effect of atmospheric CO_2 on rainwater;
2. Effect of distilled water on limestone in absence of atmospheric CO_2.

The effect we wish to study, of rainwater running on a limestone surface in equilibrium at once with CO_2 in the air above and $CaCO_3$ in the ground below, is a kind of combination of these two effects. There are many interdependent equations that can be made relevant, but the following data items are an essential minimum:

1. $\underset{\text{MOLECULES IN AIR}}{CO_2} \rightleftharpoons \underset{\text{MOLECULES IN WATER*}}{CO_2} \qquad \dfrac{[CO_2]_{aq}}{[CO_2]_{air}} = 10^{-1.5}$

2. $\underset{\text{SOLID}}{CaCO_3} \rightleftharpoons \underset{\text{IONS IN SOLUTION}}{Ca^{+2} + CO_3^{-2}} \qquad [Ca^{+2}]\cdot[CO_3^{-2}] = 10^{-8.3}$

3. $\underset{\text{MOLECULES IN WATER}}{CO_2 + H_2O} \rightleftharpoons \underset{\text{IONS IN SOLUTION}}{H^+ + HCO_3^-} \qquad \dfrac{[H^+]\cdot[HCO_3^-]}{[CO_2]_{aq}} = 10^{-6.4}$

4. $HCO_3^- \rightleftharpoons H^+ + CO_3^{-2} \qquad \dfrac{[H^+]\cdot[CO_3^{-2}]}{[HCO_3^-]} = 10^{-10.3}$

5. $H_2O \rightleftharpoons H^+ + OH^- \qquad [H^+]\cdot[OH^-] = 10^{-14.0}$

(a) Equilibrium between Atmospheric CO_2 and Rainwater. The relation of items (1), (3) and (4) to other possible equilibria is shown thus:

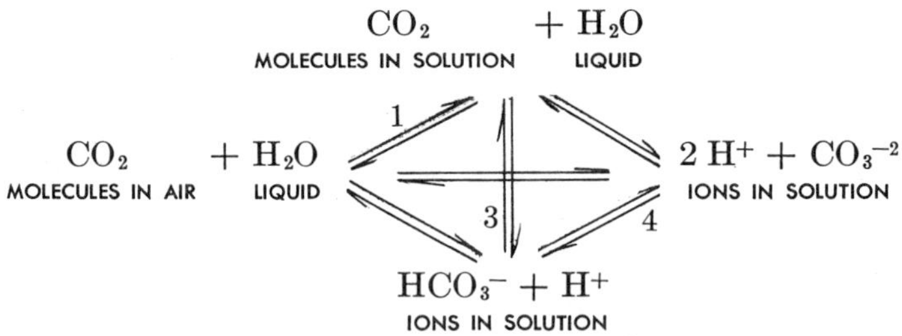

*Whether these exist as independent molecules or associated with n molecules of water makes no difference to their number; and since n is not known, no increase in precision is gained by writing H_2CO_3 or similar combinations.

From data items (1) and (3), if equilibrium is attained, then

$$[H^+]\cdot[HCO_3^-] = 10^{-7.9}\,[CO_2]_{air} = 10^{-11.4}$$

for normal atmosphere. We see at once that:

1. The less the atmospheric CO_2, the less acid the rain;

2. For a given atmospheric CO_2 content, if rain is made acid from some other source, e.g., by lightning or industrial fumes, it will take up less CO_2 as HCO_3^- ions;

3. If the rain gains bicarbonate from some other source, its acidity will fall.

Effect (3) is studied quantitatively later. Here we remark that:

1. By data item (4), if a carbonic solution is at all acid—say $[H^+] > 10^{-6}$—$[CO_3^{-2}]/[HCO_3^-] < 10^{-4.3}$. This means that $[CO_3^{-2}]$ is negligible in comparison with $[HCO_3^-]$ and that reaction (4) proceeds very little toward the right;

2. By data item (5), if $[H^+] > 10^{-6}$, $[OH^-]/[H^+] < 10^{-2}$. This means that $[OH^-]$ is negligible in comparison with $[H^+]$ and that reaction (5) proceeds very little toward the right. Consequently the hydrogen ions must be largely from some other source.

These indicate that *if* rainwater becomes more acid than pH = 6, its acidity will be mainly attributable to reaction (3); now every hydrogen ion generated by reaction (3) is accompanied by a bicarbonate ion, so that their concentrations are equal and, with sufficient precision (since rainwater is dilute), $[H^+] = [HCO_3^-]$. From above, $[H^+]\cdot[CO_3^-] = 10^{-11.4}$, giving $[H^+] = 10^{-5.7}$, i.e., rainwater *is* perceptibly acid.

(The procedure of using approximations based on an assumption in a calculation that shows the assumption to be valid is a rigorous means of finding an acceptable solution, but fails to show that it is the only possible solution; but the existence of a second acceptable solution seems improbable —no one has ever distinguished two kinds of rain; moreover, the calculation can be made without using the above approximations and still leads to only one answer.)

The use of this calculation is not restricted to atmospheric equilibria. In the same way that water in a bottom sediment differs from seawater, so the "air" in a porous soil differs from normal air. Plant life, for example, may raise $[CO_2]_{air}$ in soil pores from $10^{-3.5}$ to 10^{-2} or more. If the approximation $[H^+] = [HCO_3^-]$ were maintained, this would generate $[H^+] > 10^{-5}$, even more acid. As a matter of fact, in a soil all kinds of other effects interfere and the result is as often an alkalizing one; but some of the extremely acid waters known from soils may be in part due to the bicarbonate effect.

(b) Equilibrium between Pure Water and Limestone. The equilibria relevant to this situation differ from those of section (a) only in the primary "source" of CO_2; in section (a) this was the reservoir of atmospheric CO_2 and here it is the limestone. Accordingly the condition

$$[H^+]\cdot[HCO_3^-] = 10^{-7.9}\,[CO_2]_{air}$$

is replaced by a different condition:

number of Ca^{+2} in solution	=	number of C atoms in solution	=	total number of items HCO_3^-, CO_3^{-2} and CO_2 molecules

Again, the diluteness of the solution justifies rewriting this as

$$[Ca^{+2}] = [HCO_3^-] + [CO_3^{-2}] + [CO_2]_{aq}$$

From data item (4),

$$[CO_3^{-2}] = \frac{10^{-10.3}[HCO_3^-]}{[H^+]}$$

and from data item (3),

$$[CO_2]_{aq} = [H^+]\cdot[HCO_3^-] \times 10^{6.4}$$

In the special circumstance $10^{-9.3} < [H^+] < 10^{-7.4}$, both $[CO_3^{-2}]$ and $[CO_2]_{aq}$ are less than $\frac{1}{10}$ of $[HCO_3^-]$, so that the main process of calcite solution is

$$CaCO_3 + H_2O \rightleftharpoons HCO_3^- + Ca^{+2} + OH^-$$

and $[Ca^{+2}] \cong [HCO_3^-]$. For $[H^+]$ in the stated range, $[OH^-] > 10^{-6.5}$ and therefore it is also true that calcite solution provides most of the OH^- ions (in comparison with only a few from dissociation of water); that is, $[OH^-] \cong [HCO_3^-]$ also. Forming the product of data items (2) $\times$ (5)/(4) gives

$$[Ca^{+2}]\cdot[HCO_3^-]\cdot[OH^-] = 10^{-12.0}$$

so that if $[Ca^{+2}] \cong [HCO_3^-] \cong [OH^-]$, then $[OH^-] = 10^{-4.0}$, $[H^+] = 10^{-10.0}$. In this case, the result obtained does not agree with the assumption made; in particular, from data item (4), $[CO_3^{-2}] = \frac{1}{2} \times [HCO_3^-]$ so that $[CO_3^{-2}]$ is not truly negligible and $[Ca^{+2}]$ is closer to $1.5 \times [HCO_3^-]$. The difference this makes to the final result is very small; the outstanding conclusion is that mixing calcite with pure water generates marked alkalinity. What happens, then, when limestone and rainwater meet?

(c) Equilibrium between Limestone, Water and the Atmosphere. Where rainwater falls on limestone, bicarbonate ions are derived from both the limestone and the air; hence neither of the approximations already used will serve ($[HCO_3^-] \cong [H^+]$ in the discussion of rainwater; $[HCO_3^-] \cong$

$[Ca^{+2}]$ in the discussion of limestone). However we can remain assured that the total of positive charges will equal the total of negative charges. Writing n_{Ca} for the number of calcium ions, and so on, gives:

$$\text{total positive charge: } 2n_{Ca^{+2}} + n_{H^+}$$

$$\text{total negative charge: } n_{HCO_3^-} + 2n_{CO_3^{-2}} + n_{OH^-}$$

If we continue to take activities as equal to concentrations, they will be proportional to the n's in these expressions, whence

$$2\,[Ca^{+2}] + [H^+] = [HCO_3^-] + 2\,[CO_3^{-2}] + [OH^-]$$

Data item (2) gives $[Ca^{+2}]$ in terms of $[CO_3^{-2}]$; item (4) gives $[CO_3^{-2}]$ in terms of $[HCO_3^-]$ and $[H^+]$; and items (1) and (3) give $[HCO_3^-]$ in terms of $[H^+]$. The solution of the resulting equation is $[H^+] = 10^{-8.4}$ whence $[Ca^{+2}] = 10^{-3.4}$.

Geological implications of this result are:

1. Even though water be acid when it falls, it becomes alkaline if it reacts with limestone. This is merely one instance of a common phenomenon; although contact with sulfides and with various kinds of plant tends to make water more acid, contact with common rocks tends to make it more alkalinc (see also Fig. 15-4).

2. The calcium content calculated above amounts to 0.016 g/liter, enough to form 0.015 cc of calcite, or one part in 65,000 by volume. If two years of tropical climate provided 650 cycles of wetting and drying, any pore space that filled with rain and emptied by evaporation would gain one part in a hundred by volume of calcite; the effectiveness of such a quantity in cementing uniform spheres together is indicated in Fig. 15-1.

The second of these implications is to be compared with observation. Many tropical exposures of sandy-textured carbonate deposits such as shell sands are known, which are loose or friable when freshly exposed, but consolidate soon after exposure to rain. Where the time taken for surface hardening is a year or two, the mechanism indicated above seems adequate. Where hardening is quicker, one may postulate that the calcite which is dissolved

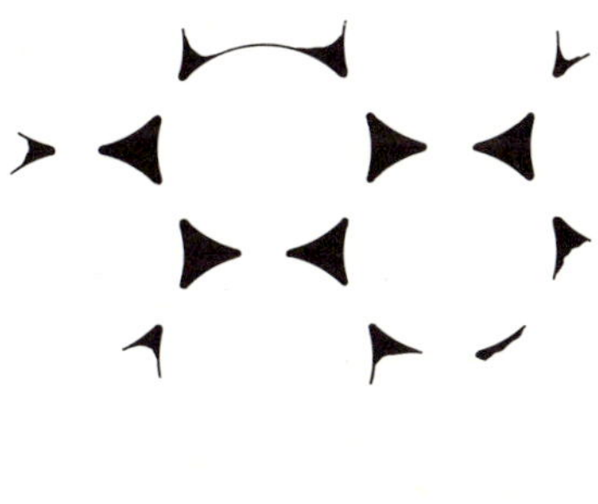

Fig. 15-1 Cemented spheres. The spheres shown are in contact, with one hundredth of the initial void space filled by cement.

over a large volume is deposited over a small volume, i.e., is concentrated near the surface, like the Chilean nitrates (see chap. 13). If the carbonate mineral dissolved is aragonite, the effect will be enhanced, since aragonite is more soluble in rainwater than calcite. Queries may of course be raised as to the attainment of equilibrium at each successive drenching; however, the calculated result seems at least of the right order of magnitude to explain the observed surface hardening. Another rather clear example was illustrated in Fig. 13-9B, although there the time scale is unknown.

The analysis up to this point has dealt only with the simplest possible equilibria. Interactions between limestone and subsurface waters of different properties or seawater are much more complex. The purpose has been simply to show the way in which testable predictions about sedimentary processes can be made on the basis of physical chemistry.

Feldspars and Clays

The dominant process in the erosion of the earth's crust is the conversion of feldspars to clays. Some clay minerals and their interrelations were shown in Fig. 13-10, from which the following is repeated:

$Al_2Si_4O_{10}(OH)_2 \cdot n\ H_2O$
IDEALIZED MONTMORILLONITE
≡ pyrophyllite + water
(often has Na, Mg for Al)

$Al_4Si_4O_{10}(OH)_8$
KAOLINITE
which can be thought of as a combination of pyrophyllite and gibbsite and can alternatively be written $2\ Al(OH)_3 \cdot Al_2Si_4O_{10}(OH)_2$

$KAl_5Si_7O_{20}(OH)_4$
IDEALIZED ILLITE
≡ pyrophyllite + muscovite
(often has H_3O^+ and Mg for Al)

The fact that the formulas given are simplifications and that clay minerals are highly variable in composition must not be forgotten. With these simplified forms, the following may be written:

1. albite + water $\rightleftharpoons$ kaolinite + $\underbrace{Na^+, Si(OH)_4, OH^-}_{\text{IN SOLUTION}}$
2. albite + water $\rightleftharpoons$ montmorillonite + $\underbrace{Na^+, Si(OH)_4, OH^-}_{\text{IN SOLUTION}}$

3–4. potash feldspar + water $\rightleftharpoons$ as above, substituting K for Na

5. potash feldspar + water $\rightleftharpoons$ illite + $\underbrace{K^+, Si(OH)_4, OH^-}_{\text{IN SOLUTION}}$
6. anorthite + water $\rightleftharpoons$ kaolinite + $\underbrace{Ca^{+2}, OH^-}_{\text{IN SOLUTION}}$

The reader may quantify these at will; for example, (5) becomes

$$5\ KAlSi_3O_8 + 20\ H_2O \rightleftharpoons KAl_5Si_7O_{20}(OH)_4 + 4\ K^+ + 8\ Si(OH)_4 + 4\ (OH)^-$$

The similarity between these possibilities is obvious; they also resemble the main limestone-water reaction of the previous section, in that hydroxyl ions are formed, so that water often becomes alkaline through prolonged contact with rock. The silica content is written in the hydrous form $Si(OH)_4 = H_4SiO_4 = SiO_2 \cdot 2\ H_2O$ in the belief that a Si^{+4} ion in water does form bonds with several nearby oxygen atoms, but no opinion is offered as to the strength of these bonds or the rigidity or permanence of the resulting groups; they may in fact be no better organized than the embryonic associations imaginatively represented in Fig. 5-2A.

It must be emphasized that although the above equations are real in an arithmetic sense, there is no evidence to show that one or another actually occurs as written. We know that feldspathic rocks crumble and rivers drop mud, so that the overall effects must be somewhat as written, but ignorance is rather profound about the details of the sequence. It is not known, for example, whether the silicon tetrahedra of the feldspar are all separated, broken up and rebuilt, or whether whole sheets of tetrahedra peel off the feldspar structure and retain coherence while reshaping themselves into sheets of mica type; but at least it can be said that most of the tetrahedra carrying aluminum in place of silicon have to break, because the clays carry most of their aluminum in octahedral sites; for example, if (6) occurred as written, at least half the tetrahedra would have to change their central cation and would probably be totally disrupted in the process. Another thing that is not known is the extent to which one clay mineral changes to another during diagenesis. One might suppose that in a given situation, albite would change either to kaolinite or to montmorillonite, whichever was thermodynamically preferred; but in many analyzed mud samples, some of each product is found. Does this mean that albite weathers in two ways simultaneously (this is possible) or that albite always changes to kaolinite first and, almost always, some kaolinite goes to montmorillonite after? (There are also, of course, other possibilities.) The fact is that the study of clays is so difficult in comparison with the study of carbonates that answers to these questions are only beginning to be constructed. It is to be hoped that the increased use of clay minerals in ceramics, ion exchangers and other industrial applications will require and therefore promote better understanding; but enough is known already about their internal constitution to show that their behavior will always be difficult to discuss in precise terms.

Discussion up to this point has concentrated on the aluminous clay minerals; magnesian clay minerals also require brief mention. The most abundant species are vermiculite and chlorite; these correspond in chemistry fairly closely to montmorillonite and kaolinite respectively, with most of the aluminum replaced by magnesium and iron (see Table A1-3). They form from the ferromagnesian minerals of the parent rocks by processes as obscure as those resulting in aluminous clays. There is further resemblance in that while much aluminous clay under metamorphism turns to muscovite, the magnesian clays often turn to biotite; however, no magnesian intermediate stage corresponding to illite has yet been generally recognized.

Iron Minerals and Oxidation Potentials

In the same way that bicarbonate solutions were used as an example of the link between water acidity and sedimentary changes, so here an attempt will be made to show, while discussing iron minerals, the importance of the oxidation potential of a solution in determining its geological effects.

It has been mentioned that aqueous solutions contain, beside water *molecules*, a number of dissociated ions, including hydrogen ions. A hydrogen ion is a comparatively elementary particle (a proton), with no orbital electrons and hence no appreciable ionic radius. Accepting that these exist in aqueous solutions, we turn to the possibility of electrons occurring unattached with a finite concentration. Of course, no one can see electrons and evidence indicates that if they are unattached at one moment they will be attached at another. What can be observed is that solutions behave *as if* there were a population of unattached electrons, of density more or less constant with time. The density of the population is very small—measured values of the *effective concentration*, [e], have been of the order of 10^{-12} mol/1000 g; but customarily [e] is not measured absolutely. Instead, advantage is taken of the fact that if two solutions *a* and *b* of different [e] are connected by a wire and voltmeter, the voltmeter reading will be *proportional* to the ratio $[e]_a/[e]_b$. This means that if any solution of interest is compared, even indirectly, with a standard solution, its relative [e] can be established without ever finding its absolute [e]. The standard in use is a hydrogen electrode. Potential differences measured with respect to it are symbolized Eh, and all we need to know is that a potential difference of 0.059 volts corresponds to a ratio $[e]_a/[e]_b = 10$.

Coming closer now to the matter of iron minerals, we notice a very simple reaction:

$$Fe^{+3} + e \rightleftharpoons Fe^{+2}$$

If the effective concentration of electrons is low, this will tend to run to the left and if [e] is high, to the right. (It is because such changes as the conversion of ferrous compounds to ferric have been known for a long time as "oxidations" that the potential differences here discussed are known as *oxidation potentials;* a high oxidation potential means a low concentration of electrons). Many reactions can legitimately be written similarly in terms of an increase or decrease in the number of unattached electrons, and whenever this is done, a change in the concentration of unattached electrons will affect the reaction in the same way as a change in the concentration of any other solute involved, as on p. 173. As an example, the relation of magnetite to hematite is discussed:

$$2\ Fe_3O_4 + H_2O \rightleftharpoons 3\ Fe_2O_3 + 2\ H^+ + 2\ e$$

This reaction is analogous to the scapolite reaction, and a similar argument suggests that the equilibrium condition will be $[H^+]\cdot[e] = k$, an unknown constant. This implies

$$pH + \frac{Eh}{0.059} = \text{constant}$$

Check: If pH rises by 1, Eh needs to fall by 0.059 for constancy; these imply that $[H^+]$ diminishes tenfold, by definition of pH, and [e] rises tenfold as mentioned in the previous paragraph; net effect on $[H^+]\cdot[e]$ is nil.

If pH and Eh are used as coordinates, this equilibrium condition can be plotted as a line separating conditions where magnetite and water are stable from conditions where hematite and water are stable (Fig. 15-2). The

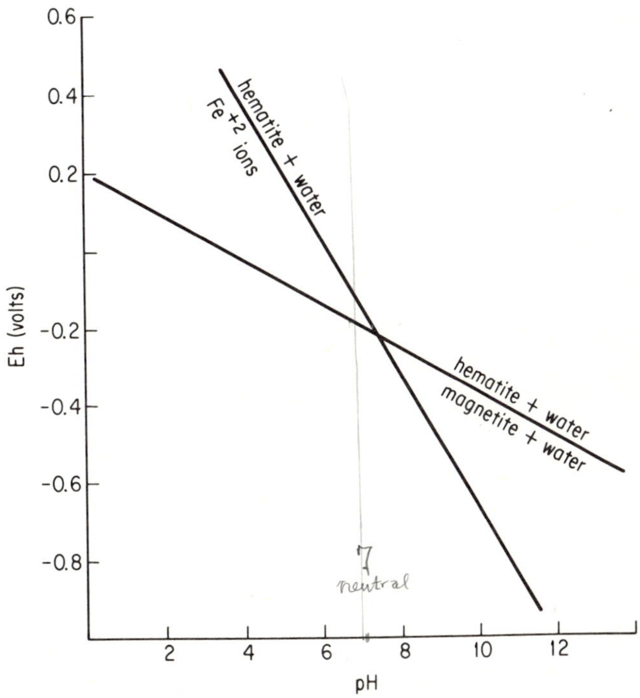

Fig. 15-2 Equilibrium conditions for two oxidation reactions. (After Garrels and Christ, 1965, Fig. 7.6.)

height of the intercept is fixed by the size of the constant k which, as before, is fixed in turn by a sum or difference of a number of free energy terms.

Another possibility of interest is the precipitation of hematite from ferrous solutions:

$$2\ Fe^{+2} + 3\ H_2O \rightleftharpoons Fe_2O_3 + 6\ H^+ + 2\ e$$

Equilibrium here will be determined by

$$\begin{aligned} 6\ F_{H^+} + 2\ F_e &= 2\ F_{Fe^{+2}} + 3\ F_{H_2O} - F_{Fe_2O_3} \\ &= \text{constant at given temperature} \\ & \text{and } Fe^{+2} \text{ concentration} \end{aligned}$$

or $\quad 3\ F_{H^+} + F_e = \text{constant}$

As on p. 174, this implies $[H^+]^3 \cdot [e] = \text{constant}$, or

$$3\ pH + \frac{Eh}{0.059} = \text{constant}$$

A line of this kind is also shown on Fig. 15-2; to plot this line it is necessary not only to work out a free energy sum but also to select an arbitrary value for Fe^{+2}. Lines for other values of Fe^{+2} will be parallel to the first and close to it but not coincident.

Parts of this diagram are rather unrealistic; for example, ferrous ions will not oxidize to hematite in a region where magnetite is more stable—

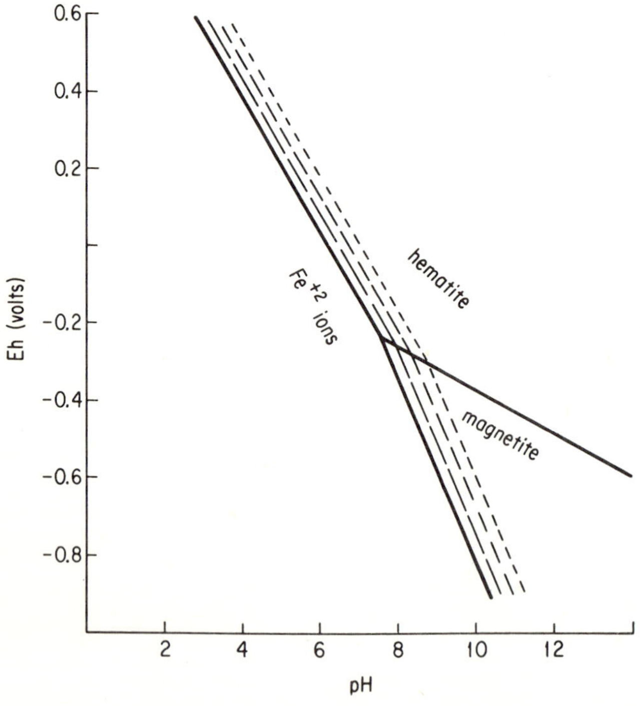

Fig. 15-3 Stability fields of three forms of iron in presence of water. The boundary of the field for ferrous ions is drawn with a full line for an arbitrary effective concentration and with broken lines for lower concentrations. The fact that water itself is not stable at the bottom edge of the diagram has been neglected for purposes of the drawing—see Fig. 15-4. (After Garrels and Christ, 1965, Fig. 7.6.)

a third possibility must exist:

$$3\ Fe^{+2} + 4\ H_2O \rightleftharpoons Fe_3O_4 + 8\ H^+ + 2\ e$$

The equilibrium boundary for this reaction must slope even more steeply, being given by

$$4\ pH + \frac{Eh}{0.059} = \text{constant}$$

The mutual relations of these three equilibria are shown in Fig. 15-3. The fact that *as a solute species is diluted, its stability field is enlarged* will be found of importance again in studies of metamorphism.

The three reactions considered all turn out to be sensitive to the acidity of their environment as well as to its oxidation potential. Can we

Fig. 15-4 Range of Eh and pH in natural waters. The sloping lines show boundaries outside which water decomposes, to oxygen above and hydrogen below. The dashed lines are *very approximate* limits to the stability of some minerals, inserted to show the potential use of the diagram. (After Baas Becking, Kaplan and Moore, 1960.)

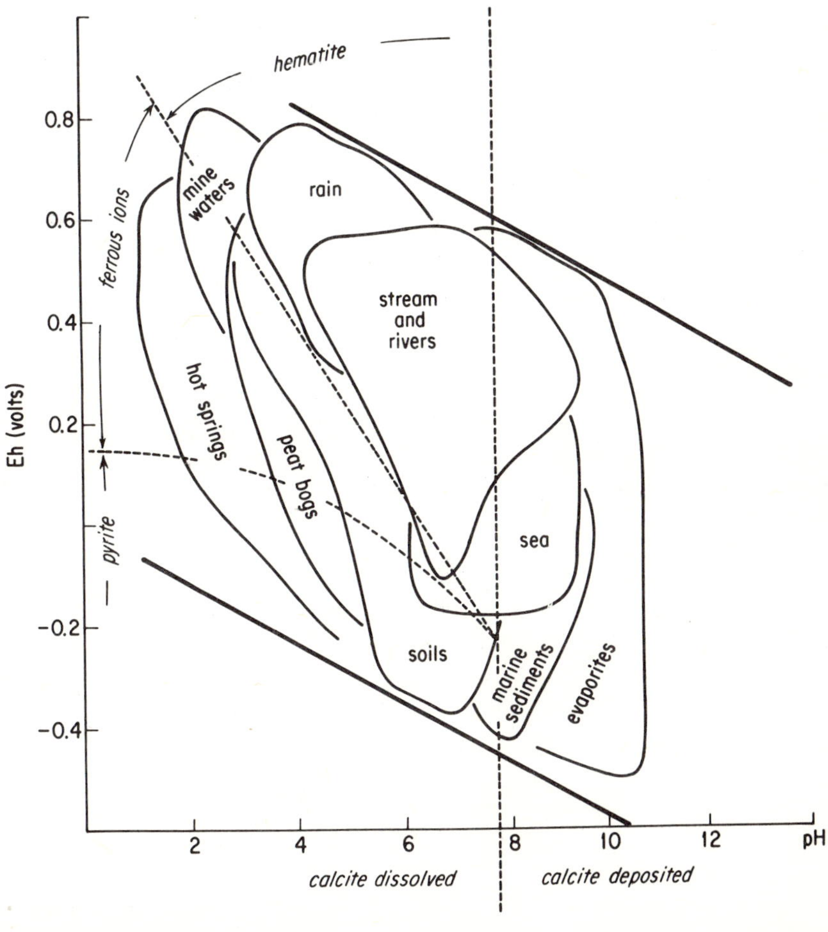

find any reaction that is controlled by oxidation potential alone? One has already been mentioned—$Fe^{+3} + e \rightleftharpoons Fe^{+2}$—but ferric ions in nature are exceedingly rare, being stable only in very acid and oxidizing waters. A second reaction in this class

$$\underset{\text{PYRITE}}{FeS_2} \rightleftharpoons Fe^{+2} + \underset{\text{NATIVE SULFUR}}{2\,S} + 2\,e$$

also requires exceptionally acid conditions.

It must be apparent that both Eh and pH of a natural solution are powerful controls over the minerals that coexist stably with them. Figure 15-4 gives an idea of the range of properties of natural waters and *approximate* boundaries for the stability fields of three minerals. For further discussion of such topics, the reader should refer to the work of Garrels and Christ (1965).

SURFACE EFFECTS

It is hard to emphasize sufficiently the important difference between events that occur in a homogeneous medium and events that occur on surfaces. The fact that water within a few ångstroms of a solid surface has different properties from water in bulk was mentioned at the start of this chapter. A further instance of the difference is its freezing behavior: if a bucket is cooled below 0°C, *all* the water in it freezes quite quickly; but if a wet siltstone is cooled below 0°C, a measurable fraction of the water does not freeze, even at −30°C. A detailed explanation cannot be given but it certainly seems that the change in behavior is related to the *orderliness* of the water molecules close to solid surfaces (in a siltstone, with small pores, much of the water is "close to" a solid surface in this sense, whereas in a bucket the proportion close to the walls is negligible). Support for this idea of orderliness comes from the swelling behavior of some clay minerals on wetting: the swelling, attributed to entry of water molecules between the silicate sheets, proceeds stepwise (Bradley, Grim and Clark,

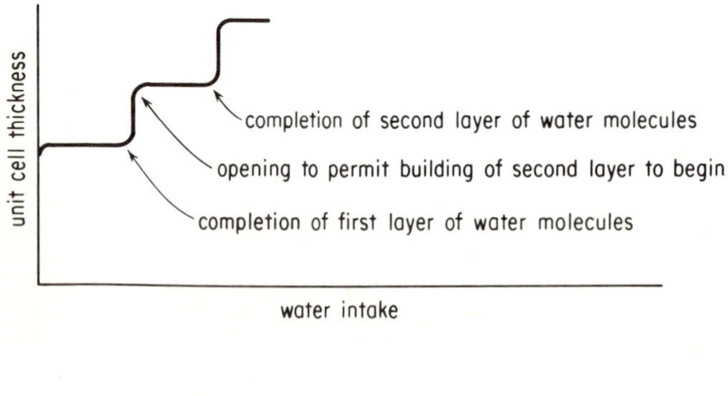

Fig. 15-5 Hydration of montmorillorite. (Based on Bradley, Grim and Clark, 1937.)

1937; see Fig. 15-5). It seems that once the sheets part to admit one molecule, many creep in and build a monolayer, and that this first layer more or less completes itself before the building on of a second layer begins. Water adsorbed on a free surface is not expected to be as orderly as water confined between two surfaces, but it is clear that we should entertain a picture of a few layers of highly ordered molecules close to the surface and successive layers outward of gradually increasing disorder until such a distance is reached that the influence of the solid surface is no longer felt.

Some numbers are instructive. The silicate tetrahedral layers in montmorillonite are about 10 Å thick, and if the number of monomolecular layers of water between them is four, as it may often be, these will total about 10 Å also (Fig. 15-6A). Typical dimensions for a montmorillonite "crystal" are 30 Å by 300 Å as in Fig. 15-6B, and an attempt to portray such a crystal with adsorbed surface water as well as an internal layer is also made in Fig. 15-6B. The reason for putting commas around "crystal" is apparent; with such small particles, there are too few repetitions for the orderliness which is a crystal lattice's outstanding characteristic to be manifest. It can be imagined that in a suspension of particles such as these, none of the water will be as mobile as bulk water—all will be within the sphere of influence of one or more flakes, and so show some tendency toward an ordered structure. The peculiar jelly-like behavior of such suspensions can readily be observed in a beaker, where as little as 10 per cent by weight of montmorillonite is enough to completely take away the water's mobility.

If the bonding of the water molecules is thus changed, it is inescapable that all aspects of solute behavior are changed also. Solubilities will be different, with the corollary that a solute will be unevenly partitioned between bulk water and surface-adsorbed water (this is in addition to the tendency, which may or may not be strong, for the solute particles themselves to form bonds to the solid surface). An example is furnished by the following: if lumps of feldspar are immersed in distilled water, the pH will remain in the neighborhood of 7; but if sufficient feldspar is ground up in the water, the pH rises to 9 or 10, i.e., the water becomes alkaline. This is quite a different effect from that discussed on p. 180—the feldspar does not react or dissolve in a normal sense, but does upset the H–OH balance in the water by some physical surface effect. (In the surface region, "reaction" and "solution" are hard to define, and in fact even "solid" and "liquid" are not as comprehensive and mutually exclusive a pair as usual—the degree of disorder and the ratio of water components to other components change rather continuously across the region of interest—see Fig. 22-2B.)

An effect of greater importance is the uneven partitioning of sodium and potassium between bulk water and surface-adsorbed water. The importance lies in the influence the partition has on the bonding of sodium and potassium into sea-floor sediments; the connection between this process

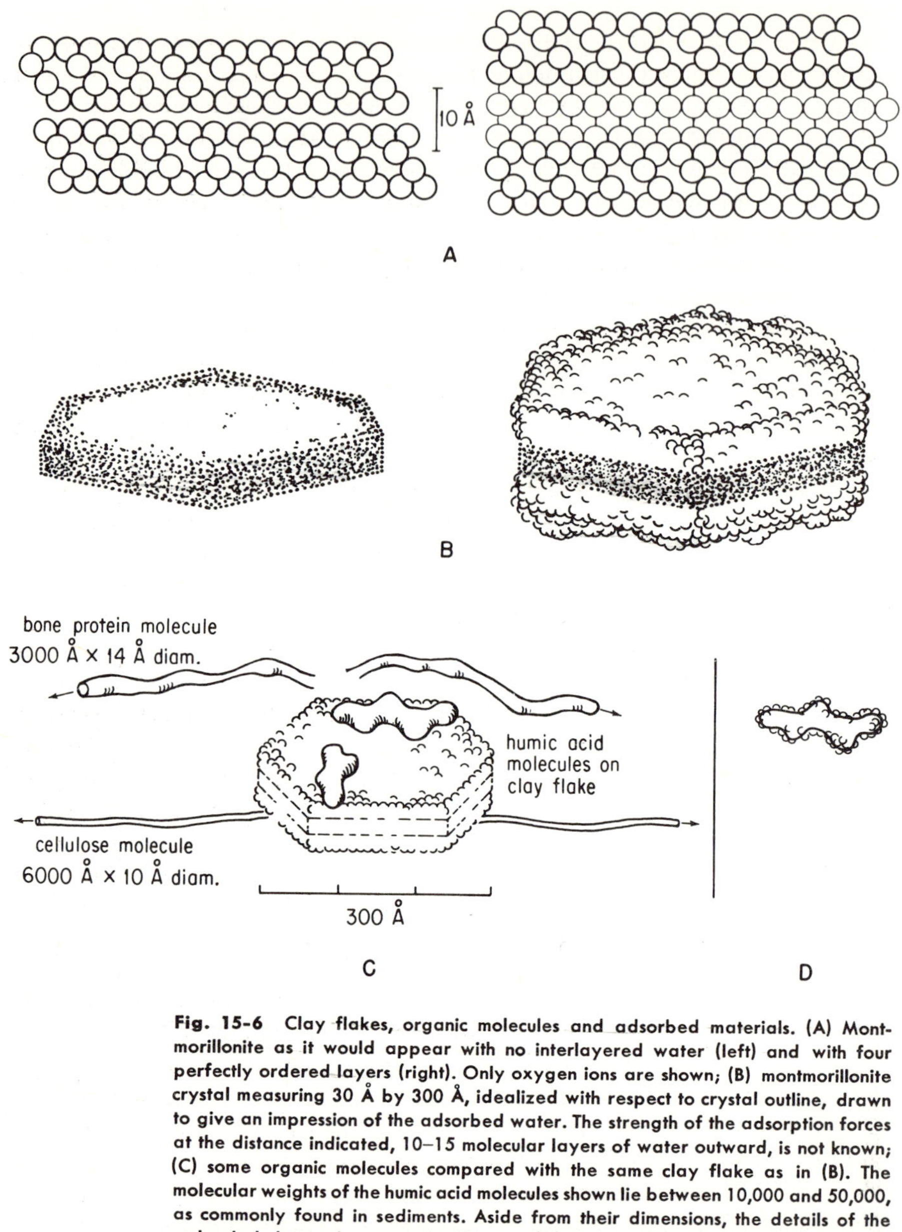

Fig. 15-6 Clay flakes, organic molecules and adsorbed materials. (A) Montmorillonite as it would appear with no interlayered water (left) and with four perfectly ordered layers (right). Only oxygen ions are shown; (B) montmorillonite crystal measuring 30 Å by 300 Å, idealized with respect to crystal outline, drawn to give an impression of the adsorbed water. The strength of the adsorption forces at the distance indicated, 10–15 molecular layers of water outward, is not known; (C) some organic molecules compared with the same clay flake as in (B). The molecular weights of the humic acid molecules shown lie between 10,000 and 50,000, as commonly found in sediments. Aside from their dimensions, the details of the molecules' shapes have no significance; (D) the same humic acid molecule as in (C), as it might appear with twice its weight of hematite adsorbed. This could be accomodated in a monomolecular layer of oxygen over most of its surface, with Fe^{+3}, of negligible volume, in interstitial sites. The actual manner of adsorption in nature is not known.

(Dimensions from Grim, 1953, and Degens, 1965.)

and the possible fractional melting of sediments to granitic magma has already been mentioned (Chap. 9). The bonding seems to occur during the conversion of mud to shale; reactions (1) through (5) on p. 180, and especially (1) and (2), appear to run to the left. Of the necessary reagents, solid silica may be available in the mud, but where does the sodium come from? Ultimately from the sea, it must be supposed; but how far below the water-sediment interface do the reactions occur? How much sodium is buried with the mud in its pore spaces, and how much migrates down to feed the reaction as it occurs? The answers to these questions depend on the surface-adsorbed water, but techniques for sampling bottom sediments are only beginning to be adequate to provide the information needed.

Among surface effects, it would be wrong to ignore organic influence. When organic matter decays, a welter of molecules results which defies individual specification. In general, the greater the decomposition, the lower the molecular weight. The abundant material known as "humic acid" seems to consist of molecules with weights from 1000 up to 50,000, fundamentally elongated but rich in aromatic sub-units. Figure 15-6C permits a comparison between some organic and some inorganic dimensions (apart from the overall dimensions, the details of shape are purely imaginary).

Parts of most humic acid molecules tend to form weak bonds with water. If the molecules also carry units that form bonds to clay, the coherence of a fine sediment will be further enhanced, and its tendency to retain water increased. Here lies the probable explanation for the stability and high water content of organic-rich ooze. The capacity of organic material to prevent hematite or goethite from precipitating is also made intelligible: organic matter can keep in suspension about twice its own weight of ferric oxide or hydroxide, which seems a lot; but this amount has only half the volume of its organic carrier, so that it can actually be accommodated without difficulty (Fig. 15-6D).

EFFECTS OF TEMPERATURE CHANGE

The effect of a change in temperature on any chemical reaction can be divided into two parts:

1. Effect on rate of reaction;
2. Effect on the end point toward which the reaction proceeds.

A change in temperature may change a reaction's end point by changing the stability fields of possible products. For example, if the free energy of possibility *B* falls more rapidly with rise of temperature than the free

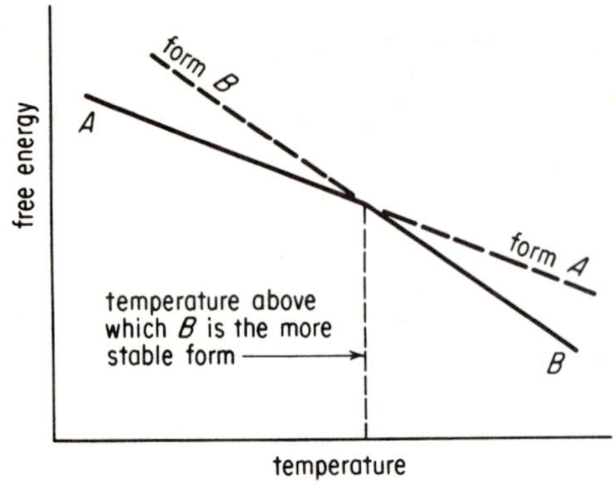

Fig. 15-7 Change of free energy with temperature. The positions of the lines, but not their slopes, are subject to change with pressure and concentration.

energy of possibility *A*, *B* will become the more stable product at some sufficiently high temperature (Fig. 15-7). A particular instance of this generalization is the change of a minerals' solubility with temperature. The solubility of most minerals increases with temperature but an important exception is calcite, whose solubility is related to atmospheric CO_2 by means of the equations given earlier in this chapter. A rise in temperature tends to drive dissolved CO_2 molecules out of the water into the air; this in turn diminishes the concentration of carbonate and bicarbonate ions that can be held, and so encourages the precipitation of aragonite or calcite. Since lowering of pressure has the same effects, the flow of water from cold, high-pressure depths to warmer, shallower levels is particularly favorable for carbonate precipitation, and this mechanism *may* account for much carbonate on tropic banks; but the difficulty of isolating this effect from that of organisms has already been emphasized (Chap. 13).

A comparison between the effect of a change in temperature and the effect of a change in the concentration of a solute is as follows: doubling the effective concentration of a solute changes the free energy balance of a reaction in which the solute is involved by 410 cal/mol, as stated earlier; for many reactions in the surface environment, this effect can be matched by a temperature change of from 5 to 50°C. (For environments on the land surface and the continental shelves, departures of pressure from 1 atm have negligible effect on the reactions discussed, but at depths of 1000 m or more, the pressure is sufficient to make stability relations perceptibly different from those at the surface.)

RATES AND MECHANISMS

Up to this point, the chapter has been concerned with the question: in such-and-such circumstances, what minerals will be most stable? It remains to face the fact that, even if this question were answered, we should still not know what mineral to expect to find. Metastable minerals are abundant, and it seems as if the closer one comes to the surface, the more abundant they are. One reason is that organisms often synthesize compounds in metastable form—much aragonite occurs as shells. Another reason is as follows: suppose that a solution *A* on reaching the sea can precipitate either *B* or *C*, two forms of the same chemistry but different atomic structure. Whichever has the lower free energy will ultimately predominate, but whichever *forms faster* is apt to be abundant in early stages. It is often true that the form with more order in its atomic structure has the lower free energy, but the form with more disorder takes less time to develop—what is true of packing a suitcase is also true of packing a structure. An outstanding example is the formation of dolomite, an exceedingly common mineral in sedimentary rocks yet one which long defied synthesis under sedimentary conditions. The dolomite structure contains sheets of carbonate radicals separated by sheets of metal ions (Ca or Mg), with the peculiarity that any one sheet contains only calcium or else only magnesium ions. When a carbonate crystal grows on the sea floor from the disorderly population of ions in solution, this degree of order is not easy to attain. Figure 15-8 shows the possibilities, in the usual situation where more calcium is being precipitated than magnesium. It should be noticed that even if process (3) in Fig. 15-8 is far slower than process (2), which seems probable, most dolomite will form via process (3) because of the ease with which the ions pass through process (1); also, the slower process (3), the greater the abundance of (*B*) relative to (*C*) in the geologic column. A few estimates indicate that dolomite crystals can actually form from disordered ones in a few thousand years—a shorter time, probably, than that taken for the primary growth of some igneous crystals. (The evidence that dolomite crystals do not form by process (2) is a comparatively small amount of data on their oxygen isotope contents, as mentioned in Chap. 13.) Nothing in this section indicates that *all* dolomite forms so quickly or soon after de-

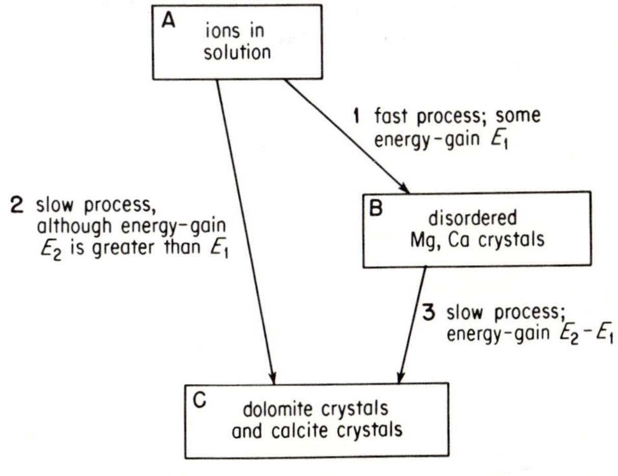

Fig. 15-8 Possible processes of dolomite formation.

position; in fact, evidence of the kind shown in Fig. 13-9 shows that much does not.

Another important example of this kind is provided by silica, which can form from solution in either amorphous or crystalline form. The crystalline form is quartz; the amorphous form differs from quartz in being less dense and more hydrous, and in having no regular atomic structure. In this respect it resembles silica glass, and in fact it may be questioned whether any discontinuities exist in the transition from "silica gel" (highly hydrous) through "opaline silica" (a few per cent of water by weight) to silica glass (anhydrous). Because of the difference from quartz, water kept over amorphous silica will dissolve 120 parts per million of SiO_2 at 25°C, whereas water kept over quartz will only take up 6–10 ppm. This means that most surface fresh water (10–100 ppm.) is undersaturated with respect to amorphous silica, while much is oversaturated with respect to quartz; yet very little quartz forms in comparison with cubic miles of chert. Part of the explanation of this paradox lies in organic activity, which produces radiolarian chert (the radiolaria build amorphous rather than crystalline shells), and part probably lies in the comparatively great speed with which amorphous silica can be thrown down, if the solute content ever reaches the supersaturation range—chert can grow rapidly enough to entirely outweigh the more continuous but exceptionally slow growth of crystalline quartz. (Growth of crystalline quartz from *interstitial* water is not necessarily so slow, and in fact passes continuously into the phenomenon of quartz veining, which probably occurs at temperatures over 100°C and may be quite quick; seawater is a third category, being undersaturated even with respect to quartz for the most part).

A third situation in which growth rate may be a stronger influence than fundamental stability is in the conversion of feldspar to clay. Although in theory albite should convert either to kaolinite or to montmorillonite according to the circumstances, it is conceivable that both form quite readily and that the less stable one takes a very long time to change over to the more stable one.

CHAPTER SIXTEEN

PHYSICAL ASPECTS OF SEDIMENTARY ROCK GENESIS

It is fair to warn the reader that in studying the physics of sedimentary rock formation, the closer one gets to one's objective, the more difficult and uncertain one's progress. The processes under study may be divided into

disintegration
uptake
transport
deposition
diagenesis

Of these, probably the transport process is best understood; deposition is understood less well, especially where flocculation and other effects characteristic of small particles are involved; and diagenesis, which is most intimately connected with the rock we actually see, is hardly understood at all in terms of physics. It is regrettable that an understanding of the mechanics of clay-flake transport should help so little in understanding shales, but it is divided from them by uncertainty as to what controls deposition and by extreme uncertainty about how the particles are affected after that. Part of the difficulty is that the physical processes get more complicated—wet porous aggregates are more difficult to describe by equations than single particles widely dispersed in water; and part is that in diagenesis, chemical effects interfere with the physical effects. For example, in the squeezing of water from a clay, solute ions tend to remain behind and lead to changes in the viscosity of the pore-fluid and the properties of the clay-water surface layers.

Notwithstanding these remarks, transport processes clearly have a large influence on sediment characteristics, and, for the sake of a simple beginning, we discuss them *as if* the particles were chemically inert. Although, of course, in nature uptake precedes transport, for present purposes it is discussed later.

UPTAKE, TRANSPORT AND DEPOSITION

It is convenient to make two distinctions at the outset: one is between *laminar flow* and *turbulent flow* and the other between *bed load* and *suspension*. Anyone who has turned off a faucet has observed the first distinction: fast water tends to be turbulent but as the water flows more slowly, a steadier flow supervenes. The distinction is strictly one of degree; that is to say, particles in turbulent motion merely follow more complicated paths than particles in laminar flow; but the change from very simple flow paths to very complicated ones occurs over a small change in the external conditions, and is so unmistakable that one is rarely in doubt whether to call an observed flow laminar or turbulent. The characteristics of suspension and bed load require more extended discussion.

Suspensions

It was mentioned in Chap. 3 that the stable condition of the earth's atmosphere is when the density decreases continuously upward in such a way that a balance exists between the tendency of gravity to pull more particles down and the tendency of the lower particles to escape from their high-density situation (upward) to regions of lower density. The same balance is achieved in muddy water and in true solutions, only the scales being different. In a true solution, diffusion is rather effective and corrects for all but a tiny tendency for the heavier ions to concentrate low down; in calm muddy water, the particles have to be rather small (of the order of 100 Å or 0.01 micron) for a significant fraction to maintain itself indefinitely above the bottom; but in turbulent water, the turbulence may be regarded as increasing the diffusion coefficient by many factors of ten, so that the surface water of a turbulent river is almost as muddy as its depths. In all these situations, the basic concepts are as follows. We begin by considering the suspended particles to be identical in all respects except position and velocity.

1. As with diffusion of heat, so with diffusion of particles:

quantity that diffuses across a
boundary per unit area per unit time = k × gradient

In a heat situation, it is the temperature gradient that is involved (Chap. 11); in diffusion of particles, it is the concentration gradient.* In the special case of standing water where concentrations change only with height and diminish upward, diffusion alone would lead to particles passing upward across a given level in numbers proportional to the concentration gradient at that level. (The constant of proportionality, or diffusion coefficient, is taken, for the present, to be the same at all levels in the water. In some important geological situations, this assumption is unjustified; discussion of such situations comes later.)

2. Because the particles are identical, they will all have the same terminal velocity for falling, so that gravity alone would lead to particles passing downward across a given level in numbers proportional to the concentration at that level.

3. If a stable suspension exists, it must be true for any level that the number of particles crossing upward per unit time equals the number crossing downward; that is to say,

k_1 × concentration gradient = k_2 × concentration
(number crossing upward) (number crossing downward)

Here both k_1 and k_2 are constants fixed by the physical properties of the particles and fluid, so that nothing is lost by writing

concentration gradient = k_3 × concentration

Behavior that meets this requirement is as follows: if the concentration at a height H is C, then for every unit of height above H, the concentration drops by a *factor f*†—if at h cm above H, the concentration is $\frac{1}{2}C$ then at $2h$ cm above H, the concentration will be $\frac{1}{2} \times \frac{1}{2}C$ (Fig. 16-1).

*Concentration gradient at a point P: if a volume V of fluid around a point Q contains n particles, the concentration around Q is n/V (If V is small, so that there are no particles within V for say three-quarters of the time, and only one particle within V during the remaining quarter of the time, it is permissible to calculate on a basis of $n = \frac{1}{4}$ particles, in spite of the fact that the particles are indivisible). If we make V smaller and smaller, the concentration *around* Q approaches a value which we may call the concentration *at* Q. Now consider points Q_1 and Q_2 a distance x apart on opposite sides of P, i.e., both a distance $x/2$ from P. If the concentrations at Q_1 and Q_2 are C_1 and C_2, the gradient *around* P is $(C_1 - C_2)/x$ and again, as x is made smaller, this ratio approaches a value which we may call the concentration gradient *at* P.

†If $dc/dh = k_3c$, $dh = dc/k_3c$ and $h = (\ln c)/k_3 + k_4$ or $\ln c = k_3h + k_5$ where the different k's are all unspecified constants. If h increases by 1, $\ln c$ increases by k_3 and c changes by a factor e^{k_3}. In the natural situation, k_3 is negative so that this represents a lowering of concentration.

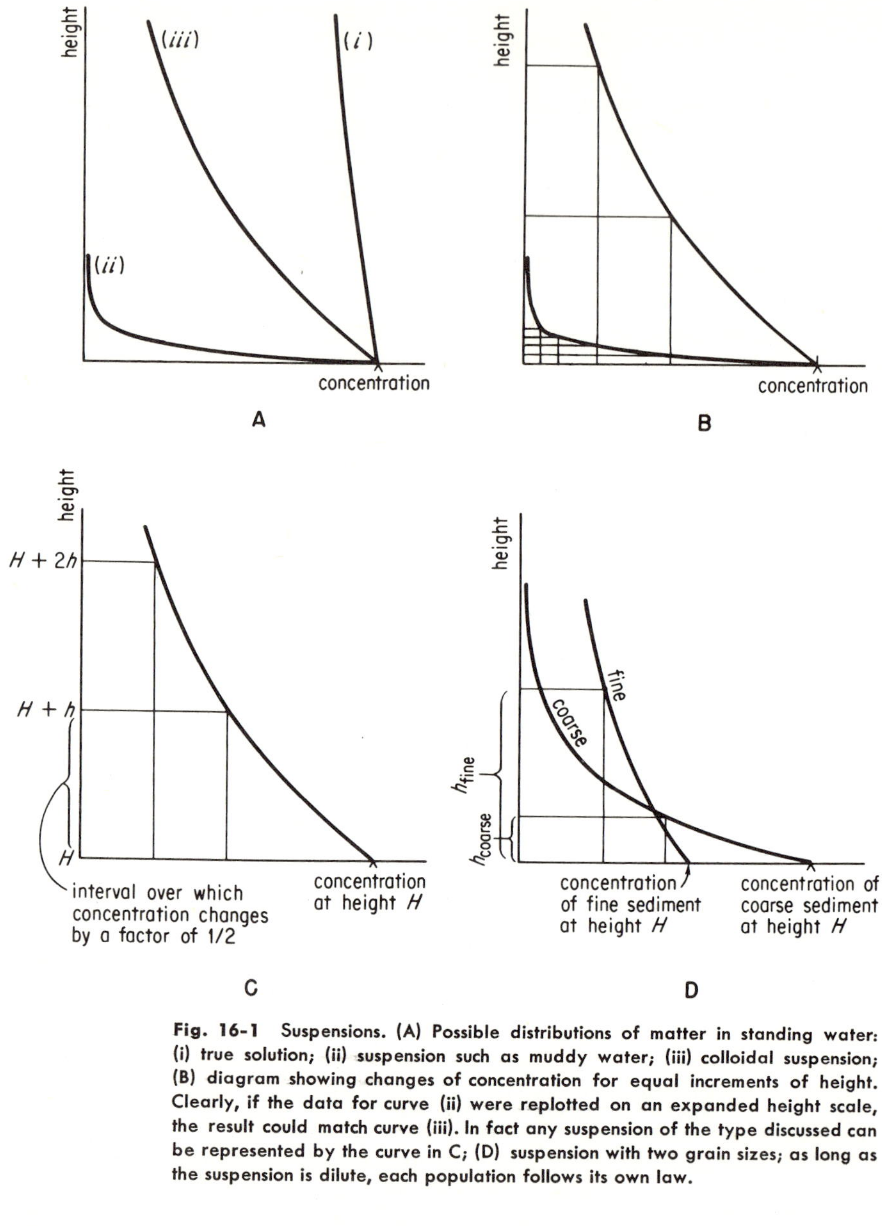

Fig. 16-1 Suspensions. (A) Possible distributions of matter in standing water: (i) true solution; (ii) suspension such as muddy water; (iii) colloidal suspension; (B) diagram showing changes of concentration for equal increments of height. Clearly, if the data for curve (ii) were replotted on an expanded height scale, the result could match curve (iii). In fact any suspension of the type discussed can be represented by the curve in C; (D) suspension with two grain sizes; as long as the suspension is dilute, each population follows its own law.

These basic concepts mean that any suspension of identical particles should be completely specified by stating two quantities:

1. The concentration at a specified height;
2. The factor by which the concentration changes per centimeter

(or meter or kilometer), *or* the height change over which the concentration changes by a factor of $\frac{1}{2}$.

We now ask whether natural suspensions behave in such an orderly way. Conspicuous differences are:

1. Natural suspensions do not have identical particles;

2. Natural suspensions are not static; especially in turbulent water, the situation at one moment differs greatly from that at another moment;

3. In many of the suspensions of geologic interest, the fluid host (air or water) has a mean horizontal velocity as well as the eddy velocities due to turbulence.

4. Partly in consequence of item (3), the degree of turbulence is different at different levels in the flow, and this leads to the apparent diffusion coefficient having different values at different levels.

The first difference can be allowed for as follows: we remark that to every particle, some height h can be assigned—the height over which the concentration would halve if a suspension of many such particles were formed; then all the particles in a mixed natural suspension can be grouped into those that have been assigned h_1, those that have been assigned h_2 etc. Even though the particles that have been assigned h_1 differ in size, shape and density, still, for purposes of "self-support," they will behave as a population of identical particles and distribute themselves according to a curve as in Fig. 16-1. Similarly, the particles that have been assigned h_2 will distribute themselves according to a different curve of the same family (see Fig. 16-1D). As long as the particles are sufficiently few for particle–particle effects to be neglected, a complete description of a natural suspension could in theory be built up in this way. (To make plain a limitation on these remarks, consider three grains, a small magnetite grain and two quartz grains of different sizes. In one set of conditions, the magnetite grain might behave hydraulically in the same way as the smaller quartz grain, but under different conditions this connection would break down and the magnetite grain's behavior might resemble more the larger quartz grain. It may be possible, in fact, to work out the conditions under which a sediment formed by noting what sizes of magnetite and quartz coexist. This means that a grouping such as has just been postulated would be valid under only one set of conditions; but this does not weaken the principle involved.)

The second difference, that natural turbulence fluctuates from moment to moment, is in some circumstances unimportant, because of the way turbulence effects average out over a time interval; for example, in turbulent water, the pressure at a point varies in a highly irregular manner, but if an average pressure is formed over a period of time, its value is close to the pressure that would exist at that point if the water were not turbulent. In the same way, *after time-averaging*, the concentration in a turbulent suspension will probably be found to change in a simple way with height, whereas at any instant the change with height would be very erratic. (These remarks must not be taken to imply that "turbulence makes no difference." Turbulence makes an enormous difference to a particle's effective diffusion coefficient—in almost obliterating the change of concentration with height, turbulence has the same effect that a great increase of diffusion coefficient would have. It was mentioned in Chap. 11 that turbulent magma behaves thermally as magma with a very high conductivity; the two effects are of exactly the same kind.)

The third difference, that natural suspensions are often flowing horizontally, need not in itself affect the properties of the suspension. Of course, usually it is horizontal velocity that causes the turbulence that maintains the suspension, but if the same distribution of turbulence could be maintained, for example by mechanical stirrers, the distribution of matter with height in a tank could be made much the same as that in a channel flow of constant characteristics.

The fourth difference, that turbulence changes with height, leads to actual distributions of sediment departing from the behavior shown in Fig. 16-1. Examples of such departures will be discussed after the effect of the rigid bed.

Bed Load

The bed of a stream influences the transport process in four ways:

1. It supports particles while they slide or roll;
2. It permits bouncing;
3. It influences the degree of turbulence in the water flowing over the bed;
4. It influences the concentration of particles in the water just above the bed.

Where a stream bed is solid rock, rolling and sliding may be important, but in the more common case where the stream bed is formed of separate grains or pebbles, rolling and sliding are not important. It must be expected that the stationary particles of a bed will be mostly larger than the moving particles and in that case more distance is travelled by bouncing than by rolling or sliding.

We next ask whether bouncing particles constitute a part of a stream's load that is independent of and additional to the suspended load. It seems that they are not; as soon as a particle has bounced, it becomes a particle *in* the fluid, whose future is determined by its position and velocity, and such a particle is in no way different from a particle that has acquired a similar position and velocity by the random processes of suspension. Of course the presence of a bed influences the concentration of such particles. Therefore, recalling that a suspension can be specified by stating (1) the concentration at some reference level, and (2) the manner in which the concentration changes with height, we may say that processes involving the bed fix the concentration at a reference level just above the bed and that all concentrations at higher levels are related to the reference level by the laws governing suspensions. In short, all particles except those rolling and sliding are suspended, but all concentrations in a suspended load are influenced by processes at the bed. In this formulation, the *bed load* contains only particles that are sliding or rolling—a very small fraction of the total load.

To explore further the influence that a bed has on the whole suspended load, it is necessary to consider how turbulence may change with height. In Fig. 16-1, all curves are drawn as if the diffusion coefficient were invariant with height, whereas in natural flows the amount of turbulence, and so the effective diffusion coefficient, may vary greatly from one height to another. To begin with a simple example, Fig. 16-2A shows how particles of a given size would be distributed if a flow could exist in two parts, one below with a high but uniform degree of turbulence and one above with a low but uniform degree of turbulence; in this figure, both parts of the profile are exponential curves like those in Fig. 16-1. A related but more plausible situation is illustrated by Fig. 16-2B; this differs from Fig. 16-2A only in that the change from high turbulence to low turbulence has been made continuous over a small interval of height, so that the sharp corner in the concentration profile disappears. Figure 16-2C shows the kind of profile that results if the flow has maximum turbulence at some height above the base. The convex part of the curve above the zone of high turbulence is a consequence of the flow becoming less turbulent with height; a similar convexity occurs in Fig. 16-2B for the same reason.

It is now possible to summarize the influence that the bed has on the suspended load:

1. The bed helps to determine the manner in which the flow's turbulence changes with height and so influences the shape of the concentration profile;

2. The bed helps to determine the concentration of particles in the fluid just above the bed, and this numerical value fixes all other numerical values up the concentration profile.

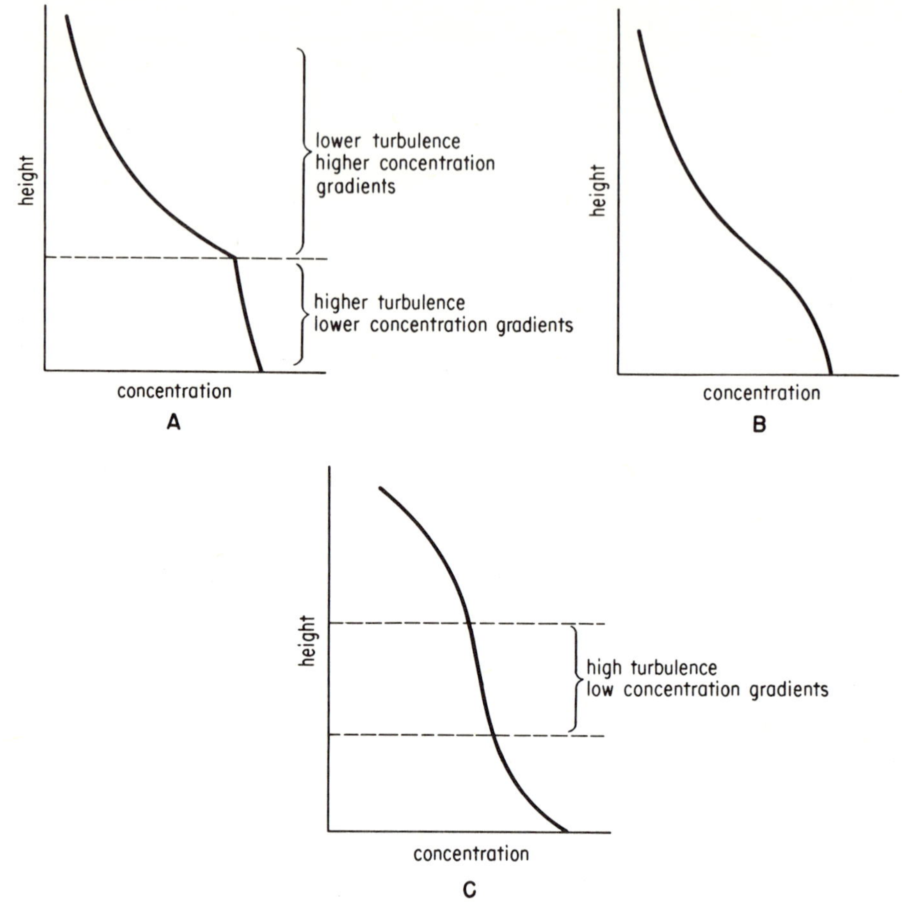

Fig. 16-2 Nonuniform turbulence. The diagrams show changes of concentration with depth for particles of a given size, (A) if turbulence were to diminish upward in a discontinuous way; (B) if turbulence were to diminish upward continuously; (C) if turbulence effects were to have maximum intensity at some distance above the bed. In natural flows, there is an infinite variety of possible profiles.

Sorting and Energy Levels

In this section, we give attention to the fact that for a population of particles of a given size in a flow situation, there will usually be some at rest on the bed as well as some in suspension. The ratio of particles suspended to total particles can be nearly zero, as for boulders in a sluggish river, or nearly one, as for clay flakes in a torrent, but in many geological situations the ratio is at neither extreme. We must then expect an equilibrium condition to be approached in which the frequency with which particles come to rest on the floor equals the frequency with which particles are picked up; there will then be an equilibrium ratio of suspended particles to particles at rest. If a situation arises where the ratio of suspended particles to particles at rest is above the equilibrium ratio, deposition will occur and if below, uptake will occur. Corollaries are:

1. The same flow that would deposit particles on a clean surface could take up particles from a thickly covered surface;

2. A flow can deposit particles of one size while taking up particles of another size.

It is an oversimplification to suppose that all flows are either overloaded with sediment and therefore depositors or underloaded with sediment and therefore scourers.

The idea just mentioned of a population of items being partitioned between one possible situation and another is very useful and widespread. In Chap. 15, the partitioning of a solute between bulk water and surface-adsorbed water was discussed. To take another example, if alcohol, water and benzene are shaken together, the products are a solution of alcohol in water and a solution of alcohol in benzene; the alcohol is partitioned between the two immiscible liquids. The significant feature is that such partitions usually change if the energy content of the system changes. In the alcohol example, warming changes the energy content and also causes a change in the partitioning of the alcohol; in the sedimentation example, changing the velocity or turbulence of the flow changes the energy content and also causes a change in the partitioning of the particles—the greater the energy, the greater the proportion in suspension.

The power of a current to *sort* the grains on its bed is a consequence of such partitions. If the bed of a flow carries pebbles, sand and silt and the flow is gentle, even for silt particles the majority may be at rest. If the energy of the flow increases by an increase in velocity or turbulence, the partitioning will change, with more particles of all sizes being suspended. However, while the proportion of pebbles suspended is still small, the proportion of silt particles suspended may become very large. In simple terms, we usually say that the flow becomes powerful enough to transport silt while not powerful enough to transport pebbles; but the occurrence of a few pebbles that move and a few silt grains that stand still needs to be borne in mind.

A practical illustration of the foregoing ideas follows. For a given flow, we shall oversimplify the situation and speak of a critical size, S, below which there is transport and above which there is no transport. If such a flow were to winnow a sediment whose initial grain size frequency distribution was as in Fig. 16-3A, and then deposit the transported grains in a new location, the ideal end condition would be as in Fig. 16-3B. The

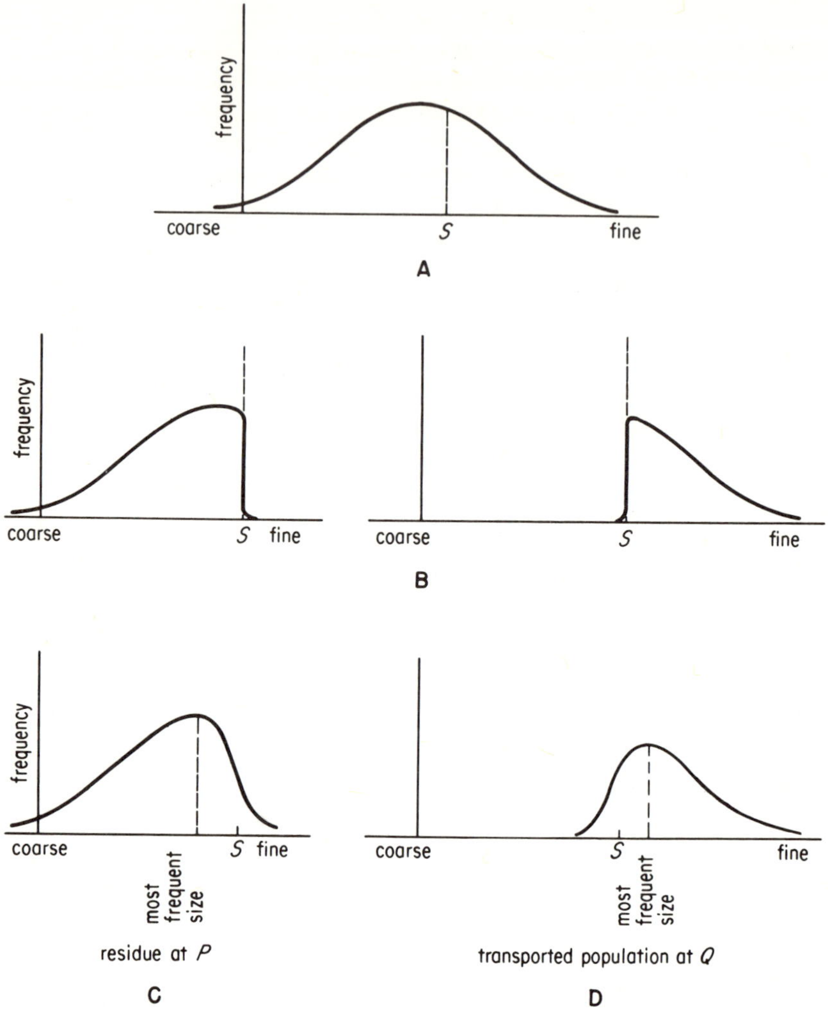

Fig. 16-3 Grain-size frequency distributions under winnowing. (A) Postulated original grain-size frequency distribution at location *P* (*S* = size of largest particle moved by the flow to be imposed); (B) idealized situation: on left, grain-size frequency distribution of residue at location *P*; on right, grain-size frequency distribution at location Q, where scour from *P* is laid down; (C) natural situation: grains at location *P* include many larger than the most frequent size and few smaller; at location Q, the opposite is true.

reader will think of abundant reasons for natural distributions having slopes as in Fig. 16-3C, rather than vertical cut-offs as in Fig. 16-3B, but it remains true that residues should, in general, have a long "tail" of particles larger than the most frequent particle size, while material deposited nonselectively after a transport episode should have a long "tail" of particles smaller than the most frequent. These expectations are fulfilled by recent beach sands (residues, well washed by receding waves) and inland dune sands (transported), as reported by Friedman (1961). If deposition is selective—for example, if the flow wanes but does not die completely—the fine particles which could form the long "tail" of the frequency distribution may be carried beyond Q; in this case, the grains deposited at Q, although transported, might not have a particularly asymmetric frequency distribution. Another possibility is that the initial population may be asymmetric. For example, if wind attacks a beach sand and builds a dune inshore from the beach, this dune will contain few large grains, because the wind is unable to transport them, but it will also contain few small grains, because the beach sand has been washed free of these by the surf. It is thus to be expected that dunes behind beaches will carry some of the best-sorted sands in the world.

Another aspect of the sorting process, conveniently mentioned at this point, is sorting by shape. The best known effects are due directly to horizontal flow (as opposed to turbulence eddies): along the base of the flow, where particles roll and bounce, round particles move most readily and disc-like particles leaning downstream move least readily (Fig. 13-5). The consequent imbrication of residues, as illustrated, is comparatively easily recognized, both on beaches and in rivers. Preferential transport of round particles is less easy to demonstrate in nature but is believed to be characteristic of some dunes behind beaches (Shepard and Young, 1961). Certainly, a difference in behavior between spherical and angular grains was observed in the wind-tunnel experiments of Williams (1964, especially p. 284): movement of angular grains was confined very closely to the tunnel floor, whereas 20 per cent of the round grains in motion in some experiments were found more than 15 cm above the floor.

Grading

Although in theory a sediment could be graded in composition while being uniform in grain size, most gradations are size gradations and the following discussion is confined entirely to size effects. Grading by size can be seen in vertical section; it is also true that the surface of a sediment may be graded laterally—this is sometimes referred to as horizontal grading although in fact the surface or time plane referred to is unlikely to be horizontal; grading in directions other than these two does not require separate discussion. Graded *residues* exist: for example, the lag gravel in a river bed is apt to be coarse near the head and to grade laterally, becoming finer downstream; also, winnowing affects the surface of a sediment strongly but only affects more mildly the grains one or two diameters below, so that there is a marked vertical gradation on this score—presumably common, for example, in desert valley gravels (see Chap. 13, the section on residues). Geologists' attention, however, has been confined mainly to graded *deposits*.

It is important to distinguish at the outset two kinds of flow:

1. Flows whose properties at a given point change with time;
2. Flows whose properties at a given point remain constant with time but whose properties vary from point to point.

The latter are more simple and will be discussed first. Some rivers approximate this condition, and currents in tidal water, such as the flow off marine deltas and some offshore currents, may have constant properties except for the tidal fluctuations; for some geological purposes, these also behave as constant flows. Where flow becomes more vigorous downstream, deposition is usually negligible. The important situation is where flow diminishes in vigor downstream; this happens in rivers because of lower gradients, and on delta fronts because the area of cross section of the flow expands away from the apex of the delta. Lower velocities downstream lead to deposition.

It will be remembered from the section on suspensions that, for particles of a given size, the number in suspension above any point on the floor is fixed by:

1. The depth of water;
2. The concentration maintained immediately above the floor by interaction between the water and the floor;
3. The way in which this concentration decreases with height above the floor.

A decrease in velocity affects item 2 directly and also affects item 3—the velocity affects the turbulence and turbulence affects the concentration profile. Thus a drop in velocity has a double effect on the suspension: firstly, the suspension's base concentration is reduced, and secondly, the decrease of concentration with height becomes more marked. These effects outweigh any effect due to increase in depth and result in the deposition first of large particles and later of small ones.

It is interesting to note that lateral gradation of other kinds can be described in the same terms. When a suspension flows over a patch of sea grass, the velocity of the water is reduced and sediment is deposited, but the effect occurs mainly because the drop in velocity leads to a drop in turbulence.

Flows whose properties at a given point change with time include those mentioned above, in the following sense. A river meanders; a delta advances seaward; a patch of sea grass changes in area; for any of these reasons, what was a point of low velocity may become a point of higher velocity or vice versa. Well known consequences are the upward increase of grain size in delta profiles and the upward decrease of grain size in river sediments.

The other category of flows whose properties at a point change with time are transient flows such as floods, slumps and most turbidity currents. These also lead to grading both vertically and horizontally. Vertical grading is caused by the change, at a given point, from fast flow to slow flow as the current subsides. In horizontal grading, two effects combine:

1. At higher levels in the flow where the velocity is great, the ratio of small particles to large is also great; this means that if the flow began at one instant and suddenly came to a complete standstill at another, the particles that had travelled farthest in the flow episode would mostly be small ones;
2. In addition to this effect, natural flows do not come to a complete standstill at one instant; instead, the interval during which large particles are carried is followed by intervals during which only smaller particles are carried—small particles not only move further in a given time compared with large particles but also, the transport of small particles goes on longer. For both reasons, a transient flow makes a horizontally graded deposit.

Quantities and Dimensions

The discussion of uptake, transport and deposition has been carried to this point in an entirely nonnumerical way; we have assumed that the fluid has a density, a velocity, a viscosity etc. without stopping to inquire what they might be. In fact, most of what has been said is equally applicable to transport by water and by wind, in spite of the great differences in the properties of the two media. In more sophisticated theories, differences between the media become important but, as far as actual behavior is concerned, a large measure of similarity remains—see Fig. 16-4. In the two experiments illustrated, there is also similarity in the rate of transfer of material: although the aerial suspension is less dense by a factor of 10, it moves faster by a factor of 6, so that for a given cross section, the wind carries nearly as much matter per unit time as the water. When it is recalled that water flows in channels of limited width, while wind sweeps the earth's surface in flows hundreds of miles wide, it will be realized what a powerful agent wind is, at least in arid climates where surface particles are loose.

When we speculate about velocities that have been effective in producing rocks of the geologic record, observation of present rivers, tides and so on is for the most part an adequate guide; but an important omission from many people's direct experience concerns turbidity currents on the continental slope and ocean floor. The following estimates, based on a

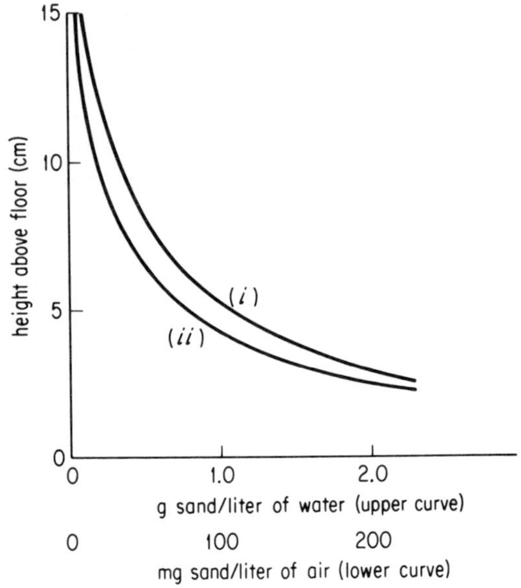

Fig. 16-4 Concentration profiles in water and air. Profile (i) is for waterborne sand, mean size 0.16 mm, mean water velocity over measured profile 2.4 mph (Vanoni, 1946, run 14). Profile (ii) is for windborne sand, mean size 0.31 mm, mean wind velocity over measured profile 14.2 mph (Williams, 1964, run on symmetrical sand, intermediate wind, p. 276).

compilation by Menard (1964, Chap. 9) are intended to give at least an order of magnitude; necessarily, they refer to major currents, because minor currents do not force themselves as effectively upon observers' attention.

Turbidity Currents. Figure 16-5 shows some of the dimensions involved. It appears that flow within the channel follows a very long narrow path (though natural channels are not straight as pictured), but overflow out of the channels is also appreciable. We must therefore keep in mind two simultaneous processes—the linear progress of the main current and the sideways spreading of the overflow to cover a broad area rather than a long line. Taking an estimate by Menard of the rock displaced by a slump at the head of the Magdalena River Canyon, 3×10^8 m^3, we note that this could cover an area 200 km by 75 km with sediment to a depth of 2 cm (see stippled area in Fig. 16-5B). It seems reasonable to suppose that this volume of rock mixes with at least its own volume of water and travels down the *canyon* as a slurry; such a slurry would approximately fill the canyon *XY* in Fig. 16-5A, 20 km long. For travel in the *channel*, a suspension of perhaps 1 g of rock in 20 cc of water is more probable; thus dispersed, the rock could still be spread through a length of 20 km if the channel were half the size of those illustrated—the dense slurry would turn to a thin suspension mostly by sideways and upward spreading. Of course, this is not a sudden change, and much of what travels down the canyon is deposited near the start of the channel, but the picture of a "plug" of muddy water progressing down the channel is probably realistic. During progress, the

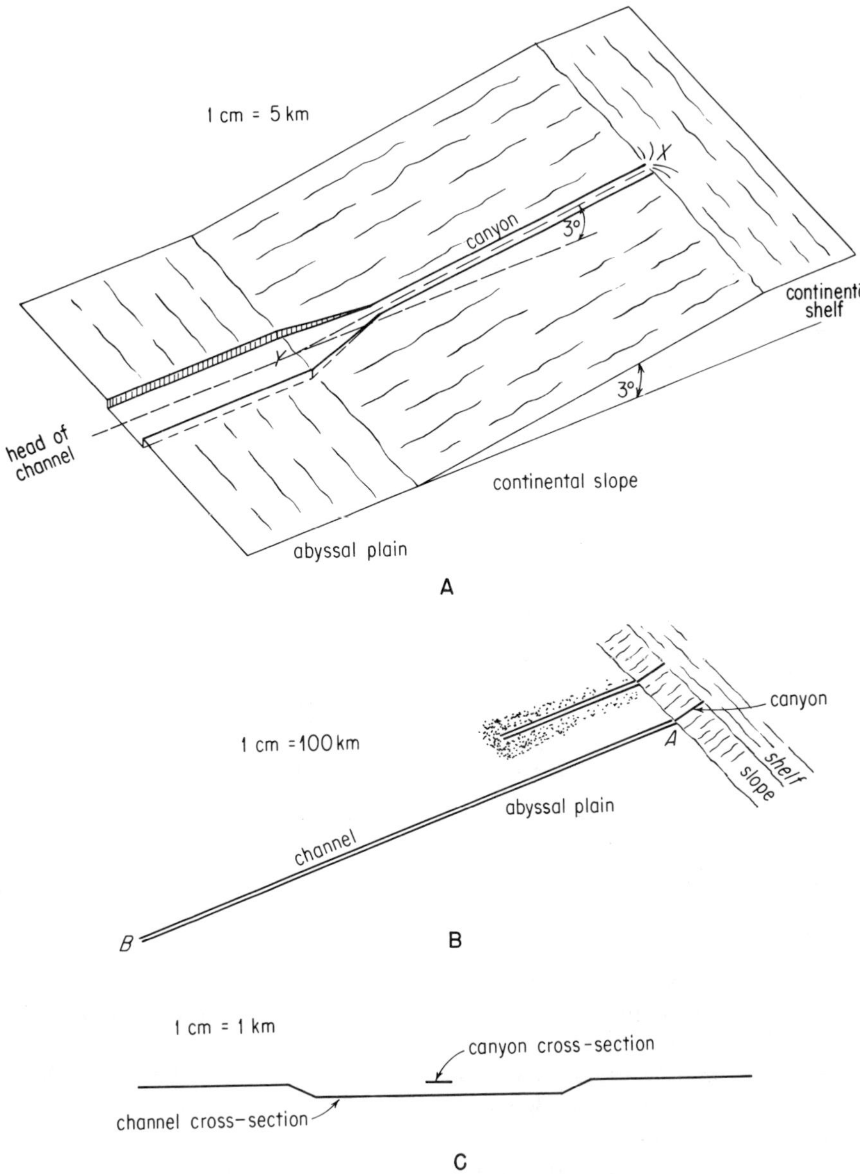

Fig. 16-5 Submarine canyons and channels. The dimensions shown come within the range of those exhibited by known canyons and channels; no attempt has been made to show the slight meanders that natural channels show. (A) A canyon and channel-head. The canyon illustrated measures 20 km by 400 m by 60 m (proportions 50:1:1/7). Inclination is exaggerated—the actual drop should be only 2 mm (= 1000 m) between *X* and *Y*; (B) two complete channels. The shorter of the channels illustrated measures 200 km by 4 km by 400 m. The dotted area on the abyssal plain could reasonably be covered with sediment by a single flow; (C) cross sections.

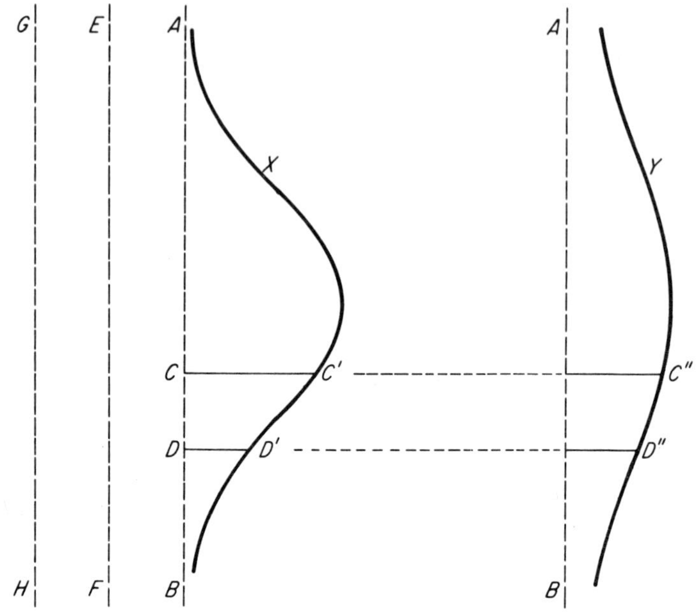

Fig. 16-6 Change of velocity profile with time. Suppose a fluid flows in such a way that at some instant all particles on a line *AB* are moving along parallel paths perpendicular to *AB* (for example, perpendicular to the page). Then a curve *X* can be drawn such that each distance CC' is proportional to the fluid's velocity at C; such a curve is called a velocity profile. If at one instant the velocity contrast between C and *D* is great (steep slope on $C'D'$), at a later instant we expect the contrast to be less (more gentle slope $C''D''$). Compare these curves with those in Fig. 11-3B.

suspension becomes less dense both by deposition and by mixing with clear water; the channel also becomes shallower out to sea, so that overflow sideways is possible all along its length.

One might ask what keeps the plug moving for hundreds of kilometers, but it is just as pertinent to ask what there is to bring it to a standstill. In fact, we have here a point relevant to all questions of deep ocean currents: a slump such as the one envisaged delivers a large amount of energy to the sea at 3000 m depth or more. Where does this energy go? It is dissipated as heat wherever sluggish water drags a current to a standstill, but as a brake, this is a very inefficient process. It appears that the slurry may move down the canyon at a speed of the order of 50 km per hour (such speeds have been deduced for several currents from the times of breaking of successive submarine cables). A slug of mud $\frac{1}{2}$ km wide and 20 km long moving at such a speed is a phenomenon incomparably greater than any known terrestrial avalanche or landslide; it is thus reasonable to suppose that gentle currents, of a few centimeters per second, could be set up at a very great distance from the center of disturbance.

It is interesting to note that the curves in Figs. 11-2 and 11-3 are to some extent relevant: it has been remarked that diffusion of particles is analogous to conduction of heat, and the same is true of "diffusion of velocity." If at one moment there is a high velocity contrast between water at one point and water at another, then at a later moment the fast-moving water will have slowed and the slow-moving water will have accelerated (see Fig. 16-6). Under suitable conditions (all particles' paths to remain parallel; velocity profiles on all lines such as AB, EF and GH identical; fluid incompressible, etc.) curve X will be modified to curve Y with passage of time exactly as one curve in Fig. 11-3B is succeeded by another. Unfortunately, when the fast-moving water in question is a bottom current, it does not have slow-moving water below as well as above, but has a stationary ocean floor instead. This introduces difficulties analogous to the difficulty of combining Fig. 11-3B with Fig. 11-3D; but the point remains valid that the time needed for a flow to decline from 2 m/sec to 1 m/sec is very long even where the decline from 30 m/sec to 29 m/sec is very rapid.

Change of Shape

Once a sediment has been laid down, it changes in texture by the processes of diagenesis; these are discussed elsewhere. Here we are mainly concerned with changes, such as rounding, that occur before particles are deposited, but we need to remember that the disintegration process also affects the final particle shape. At the outset, two observations are in order.

1. Man's natural habits make it easy for him to overestimate the importance of rounding in sedimentation, for two reasons. Firstly, on account of cushioning, small particles suffer less rounding than large ones, but the small particles are hard to see. With his naked eye, man is constrained to study only a part of the whole population of particles, and the part he can study is more subject to rounding than the remainder. Secondly, angularity in the detritus of hot and cold deserts is as conspicuous as rounding is on beaches and in river beds, but man prefers to avoid deserts and live close to rivers, and so becomes biassed in his range of experience. (This observation, while possibly trivial in the context, is an example of an important generalization: *nature rarely offers us a random sample.* For another example, recall that outcrops do not fairly represent the rocks of a region; they selectively represent the *more resistant* rocks. Similarly, fossil populations selectively represent the more durable species from the initial accumulation of remains.)

2. Particles of a given shape may become dominant in a group by *sorting* as well as by shape change. A sand that contains well rounded grains is not necessarily a sand that has suffered much abrasion—as has been mentioned, wind can select the round particles from a population and concentrate them to leeward. By such a selection, the "average shape" of a group can be changed without change in the shape of any individual particle. The generalization here is: *however sure we are that process X leads to effect Y, we still may have no confidence that effect Y implies the action of process X.*

To proceed with discussion of shape change, it is convenient to group particles by origin and size:

1. Chemical and biologic;
2. Detrital, silt size and smaller;
3. Detrital, sand size and larger.

The shape of particles of chemical and biological origin is determined more by origin than by transport: particles such as oolites, sponge spicules, fecal pellets and small shells have shapes determined at their time of formation; larger shells are modified mainly by simple fracture; and much reef detritus is shaped by the borers that break down the reef. The particles are rarely much modified by physical processes before deposition and chemical change before deposition is usually trivial in comparison with later diagenesis.

The shape of detrital particles below sand size is also established before transport. At the weathering stage, some minerals in a rock break down chemically, their constituents usually converting to flaky grains of sheet silicates. The more stable grains are thus loosened and separated from each other, and large grains may be parted into fragments by alteration localized in cracks. If these separate grains are small, their momentum during transport is never sufficient to cause destructive impacts nor does their weight lead to grinding—except for matters such as flocculation, they are deposited in much the same condition as when they were taken into suspension.

Detrital particles of sand size or larger are more liable to change by physical processes during transport, but even here it must be recognized that the disintegration process has great influence. Especially in the pebble size fraction, this influence is very clear: isotropic rocks give equidimensional pebbles, while flaggy rocks and shales give flat pebbles. In fact, to a first approximation, it may be said that sphericity is controlled by disintegration, and only roundness by abrasion (*sphericity* and *roundness* are distinguished in Fig. 16-7). To a second approximation, it is recognized

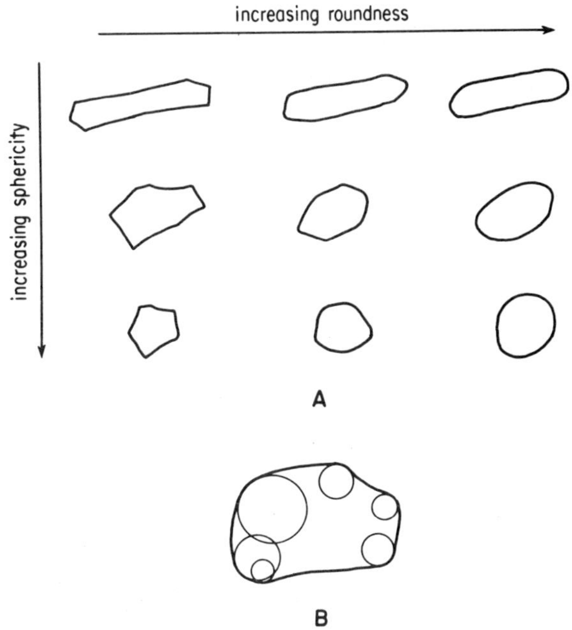

Fig. 16-7 Sphericity and roundness. Sphericity is related to the difference between the various diameters of the particle; roundness is related to the sharpness of the corners, i.e., to the radii of circles which touch the surface on the inside, as in diagram B.

that extreme rounding affects sphericity, but it is only on surf beaches and a few river beds that the sphericity originally conferred on the particles in disintegration is much increased by rounding.

DIAGENESIS

Diagenetic changes can be described largely in chemical terms, as in the latter part of Chap. 13. Even the loss of water during compaction is properly regarded as a change in the bulk chemistry of the sediment, but it is with regard to loss of water that principles from physics have their most direct application to diagenesis. Loss of water leads to loss of porosity, that is, to *compaction* of the sediment; some generalized information on the subject is presented in Fig. 16-8.

There is, as shown, a general tendency for porosity to diminish as depth increases, but the smoothness of the change is subject to interference. Firstly, the textures of a set of beds may be such that the lower ones are naturally more porous than the upper ones. Secondly, in simple circumstances, one bed has not only to give up its own pore fluid but also to transmit all the pore fluid that is given up from the beds below; this requirement may delay the compaction of the bed in question (if the beds below are very productive) or of the beds below (if the bed in question

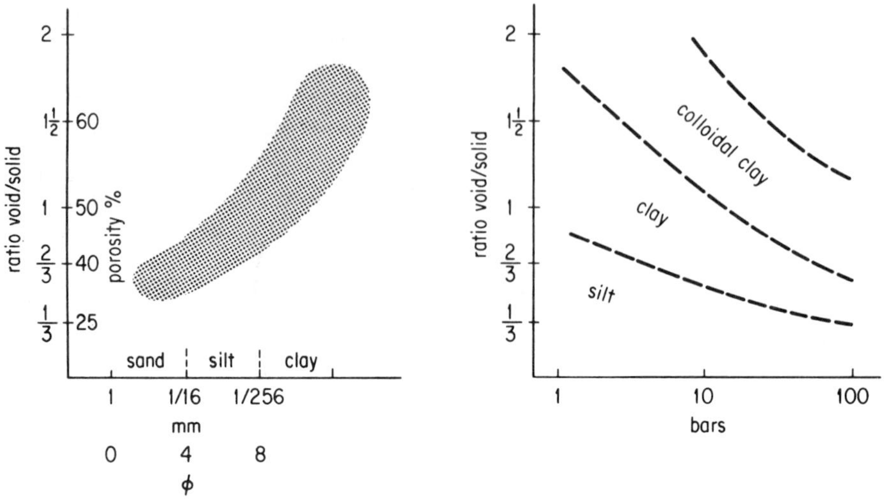

Fig. 16-8 Porosity of sediments. Diagram A shows the porosity of sediments as deposited, according to grain size. Diagram B shows the trend of decreasing porosity with increasing pressure of overburden (Chilingar and Knight, 1960; Shumway, 1960; Weller, 1959).

has low permeability). Such delays are magnified wherever kaolinite, montmorillonite and vermiculite are converted to less hydrous silicates with liberation of tightly adsorbed water as well as interstitial water; nor are the effects merely delays—a permanent imprint on the rock's character can be made. This is because a rock which reaches, say, 150°C while still rather hydrous is apt to undergo different mineral changes from a rock which is squeezed dry before reaching 150°C. The question of whether a mud will ever become a slate, for example, may hang upon such a difference as this.

It should perhaps be admitted that simple circumstances are not common in geology, and that the picture envisioned above, where all escaping water travels vertically, is apt to be much modified by tilting of beds and presence of aquifers. The entire petroleum industry derives from the fact that buoyant liquids beneath impermeable covers do not remain passively in place but migrate laterally toward the high points of any permeable layers below the barrier. Water behaves in the same way, so that, for example, shale below an inclined sandstone bed may be noticeably drier than similar shale just above it (see Fig. 16-9).

The manner in which physics and chemistry interact in diagenesis is well exemplified at this point: an observable fact is that when shale units are pierced by drill holes, the water that percolates into the hole from the shale is often very rich in solutes. A possible explanation derives from the

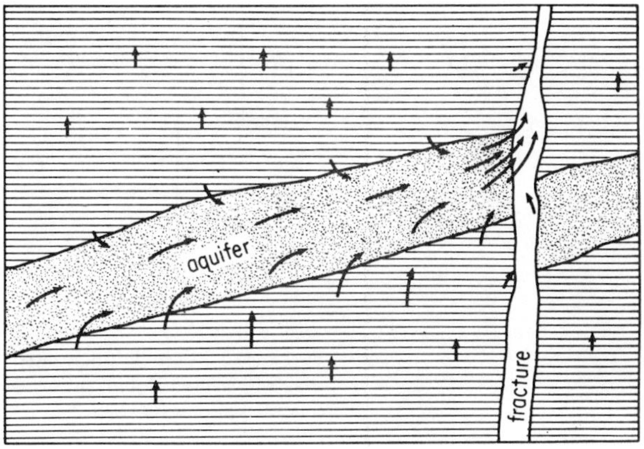

Fig. 16-9 Percolation pattern near an aquifer.

fact, emphasized in Chap. 15, that water close to a solid surface is apt to differ in solute content from nearby water that is farther from the solid surface (we shall call the latter "bulk water"). When a wet sediment is squeezed, it is probable that the bulk water is expelled more readily than the surface-adsorbed water; in fact, the adsorbed water may not be squeezed out at all by mere overburden but may remain, with an abnormal solute content, until drilling disturbs it. One might think that water which could not be squeezed out by overburden would not run freely into a drill hole, but here the time lapse between burial and drilling must be remembered: much water which was tightly adsorbed at the time of burial could be loosened by slow subsequent changes, and in particular by the increase in grain size that often results from prolonged burial. Of course, not even drilling liberates *all* the water from a shale; between any two crystals in the earth's crust there is probably some water to be found, although it may be best describable as merely a contaminant in the disordered surface phases that lie between one well-crystallized solid particle and the next. We should not ask, "Is water present?" but "How much water is present?"—a question of recurrent importance throughout all discussions of metamorphism as well as diagenesis.

In addition to reduction of porosity by simple rearrangement of grains, there can also be reduction by change in grain shape—see Fig. 16-10. The situation illustrated is hypothetical in that in practice one never knows the starting condition, and hence can never distinguish between redistribution of material and addition from a more distant source; but for some quartzites and marbles, the suggestion that they have lithified purely by internal rearrangement seems the most reasonable.

In addition to changes in texture, the structure of beds can be altered by overburdening and subsequent compaction. In the first place, uplifts and depressions can be generated by the "salt dome mechanism" (Fig.

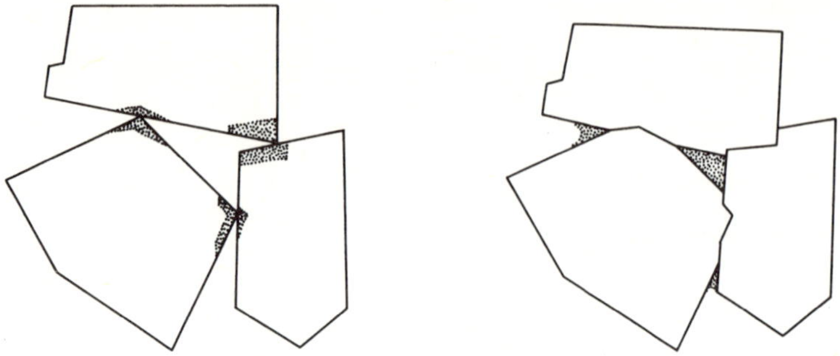

Fig. 16-10 Consolidation without addition of cement. Material from high-pressure sites within the original grains migrates to low-pressure sites between the grains; compaction and consolidation (right) ensue.

11-6); in unconsolidated sediments, these usually measure less than a meter in any direction and are known as *load structures* (load *casts* if only the overlying bed is preserved and observed). Secondly, compaction of different lithologies to different extents can generate quite strong dips—see Fig. 13-3. Beside the small structures illustrated, large structures such as reef aprons and deltas may be affected. In either of these, the coarser sediments are susceptible to less compaction than the finer sediments, because they have lower initial porosity (Fig. 16-8). Consequently after compaction the outward dip is greater than it was originally. In a reef apron, the change may be several degrees; in a large delta, both the initial dip and the change of dip are smaller, but compaction may nevertheless lead to strong marine transgression and the resubmergence of land that had emerged during active delta building.

Because the processes of diagenesis are largely continuous with those of metamorphism, further relevant discussion will be found in Part Four, especially in Chap. 22. The range of possible mechanisms by which material can move remains the same, regardless of the somewhat arbitrary distinction drawn between diagenesis and metamorphism. In fact, there is reason to look at metamorphic changes next: by the time a rock is moderately well metamorphosed, the individual grains are clearly distinguishable and often identifiable, so that changes can be discussed in quite precise terms; and knowing what probably happens during metamorphism, one can better speculate about what possibly happens during diagenesis. For example, the bulkiest and least well known of diagenetic products are the shales; they are the most difficult of all common rocks to study, and it is a help to know about schists as well as about clays, before trying to go further in this intricate field.

CHAPTER SEVENTEEN

METHODS OF STUDY OF SEDIMENTARY ROCKS

Following the format of Chap. 12, operations are divided into:

I. Operations on rocks
  (A) in the field
  (B) in the laboratory

II. Operations on rocklike materials (laboratory only)

Field Studies 1. **Observable Processes**

There is far greater scope for study of sedimentary processes than magmatic or metamorphic processes; particularly at sea, this is a vigorously expanding activity.

1. Disintegration. The transition from solid rock to detritus is sometimes observable; for example, a quarry wall may be kept under observation over a period of years as a guide to the suitability of the rock for engineering purposes. In a restricted sense, the process is also observable where a soil grades downward into the solid rock from which it formed. To reveal the disintegration process, not only the solids but the pore fluid also must be analyzed. Because interaction between pore fluid and air can alter crucial properties, quite elaborate techniques are needed to measure pH and Eh in situ.

2. Transport. Soil creep and erosion are studied by conservationists and related to slope, rainfall, vegetation etc. Hydrologists measure streams and rivers for gradient, volume, velocity and sediment load and type; the same kind of data for winds is collected to a lesser extent. Coastal transport can be studied experimentally, by building groins etc., and by introducing tracer materials at crucial localities. Submarine transport: tides and currents are measured by sailors; the resulting shifts of sandbanks etc. are charted, at least where shipping is active. Submarine slumping has been observed near the head of canyons on the continental slope. The progress of a slump or density current can occasionally be timed and mapped by the breaks it causes in submarine cables. Low-density turbidity currents (density < 1.1 g/cc) have not been observed on the ocean floor, but in lakes muddy inflow can be seen to sink at one point and eventually emerge at another.

3. Deposition. Deposits from ice, landslides, floods etc. (transient phenomena) are measured and analyzed after the event, but silting is measured in progress (e.g., feet per year). Deltas: growth of topset beds is directly observable over the years; growth of foresets is now measurable by repeated charting and sounding, although only a few deltas have yet been thoroughly studied in this respect. Deep sea deposition: recent deep-sea deposits are sampled by coring devices; the rate of accumulation is estimated by measuring the sediments' contents of ionium and thorium (Goldberg and Koide, 1962) or by dating a sample from a known distance below the surface. Whichever approach is used, the methods are based on one or another kind of radioactive decay.

4. Diagenesis. This is observable only in the sense that an alteration product and some unaltered material may be juxtaposed; as in disintegration, complete study includes the pore fluid as well as solid samples.

Field Studies 2. **Products**

1. Size and shape of sedimentary bodies. Methods are:
- **a.** ground survey
- **b.** air-photo survey
- **c.** geophysical surveys from the surface: (i) seismic refraction, using explosives; (ii) acoustic reflection and refraction ("echo-sounding"); (gravity, electric, magnetic, radiometric surveys occasionally useful)
- **d.** drilling and logging: (i) by inspection of core; (ii) by geophysical methods (seismic, radiometric, resistivity logs etc.)

Because large structures in sedimentary rocks are important in oil production, more money has been spent in perfecting this group of techniques than on any other aspect of geology.

2. Structures and textures. Features are useful according to different purposes:

a. to distinguish top from bottom: truncated cross-bedding, ripple marks, grading, mud cracks, load structures etc.

b. to determine direction of transport: cross-bedding, imbrication, orientation of pebble axes etc.

c. to determine nature of environment: thickness and regularity of bedding; truncations and discontinuities; color, coarseness and composition of the rocks; variability within and between outcrops; specialized indicators such as mud cracks, glacial grooves, salt pseudomorphs etc.

d. to determine history: angular relations of beds, unconformities etc.

3. Fossils. Much studied in laboratories; in the field, one may observe prevalent attitude and condition (erect, collapsed, fragmented etc.), relative abundance of species, horizontal and vertical distribution.

Laboratory Studies 1. **Geometrical**

1. Whole rock. The following and many other aspects can be studied:

a. grain-size frequencies

b. rounding and sphericity

c. porosity and the presence of cavities and cements

Grain-size frequency data helps to distinguish residues from transported deposits, and sometimes points to a particular transport agent. Rounding and sphericity are of limited usefulness directly, but may, with other properties, be measured at many localities and scanned for systematic trends (see 4 below).

2. Individual minerals. The question of whether a grain was detrital or grew in its present environment may have to be resolved, if at all, by observation of its geometry; euhedral overgrowths on rounded cores indicate that both processes have probably occurred. (Techniques for making textural observations easier include (1) separating the component grains—this can sometimes be done by alternate drying and soaking in appropriate solutions or by vibration; (2) using selective etches and stains on a flat surface, a microscope slide or a "peel"; peels are made by coating the rock with an adhesive that forms a skin and peeling it off when dry. The electron microscope is well adapted to revealing small textural features and has been used to distinguish, for example, windlain from waterlain sands.)

Laboratory Studies 2. **Compositional**

1. Whole rock. Methods are as for igneous rocks (Chap. 12). A greater proportion of sediments are fine-grained and difficult to analyze in terms of minerals even with a microscope, so that methods for estimating quantities of minerals by X-ray methods as well as their identities are currently expanding.

2. Part of rock. Whereas quartz from one locality is hard to distinguish from quartz from another, the heavy minerals a sediment contains (zircon, garnet, hornblende etc.) may clearly indicate its source area. The heavy minerals are usually concentrated by centrifuging a granular sample in a liquid of density 2.7 or more, and studied both for species and relative abundance.

3. Individual minerals. Examples of individual mineral studies are:

- **a.** oxygen isotopes in carbonates (a partial indicator of temperature of formation)
- **b.** boron content of illite (an indicator of paleo-salinity)
- **c.** potassium and argon content of glauconite (an indicator of age)
- **d.** magnesium content of calcite (sometimes an indicator of diagenetic history)

Laboratory Studies 3. **Fossils**

It is appropriate here only to remark that fossils can give indications of salinity, depth, temperature, cleanliness and turbulence of the water they lived in, and sometimes even more information (isolation of basins, existence of land bridges etc.); but they do *not* always reflect the conditions of formation of the rock in which they were entombed. Information from both fossils and host rocks is needed for a complete regional synthesis.

Laboratory Studies 4. **Regional Synthesis**

Compared with an igneous body, a sedimentary basin is more complicated both in its scope for local variations and in the variety of interactions between topography, source of deposits and their texture, composition and biologic aspects. The making of a regional synthesis can be laborious indeed. Computers are being increasingly used to isolate dominant trends and factors from data collected at scattered sample sites.

Laboratory Studies 5. **Physical**

Among physical properties of special interest, the permeability of sediments is important for oil and water supply, and their reaction to loading is important for all kinds of large-scale engineering.

Laboratory Study of Materials other than Rocks

1. Chemical. As in igneous rock studies, it is convenient to explore chemical effects in more simple mixtures than occur naturally. For example, many theories such as those in Chap. 15 can be tested in the laboratory as well as being compared with observation of nature.

2. Physical. In contrast to igneous processes, sedimentary processes can be imitated in wind tunnels, wave tanks, flumes etc., with natural materials or artificial ones with better-known properties (such as uniform spheres). Many processes can be reproduced on their natural scale; some, such as turbidity current flow, have to be scaled down, with due regard to dynamic factors as well as geometrical ones.

3. Mathematical. Beginning with the falling of a sphere through water or flow through a cylindrical pipe, there are many situations for which satisfactory theory can be devised. However, because experiments and observations of nature are comparatively easy, and the difference between natural detrital particles and any mathematical formulation is rather great, many formulas in sedimentation are generalized from observations rather than derived from postulates.

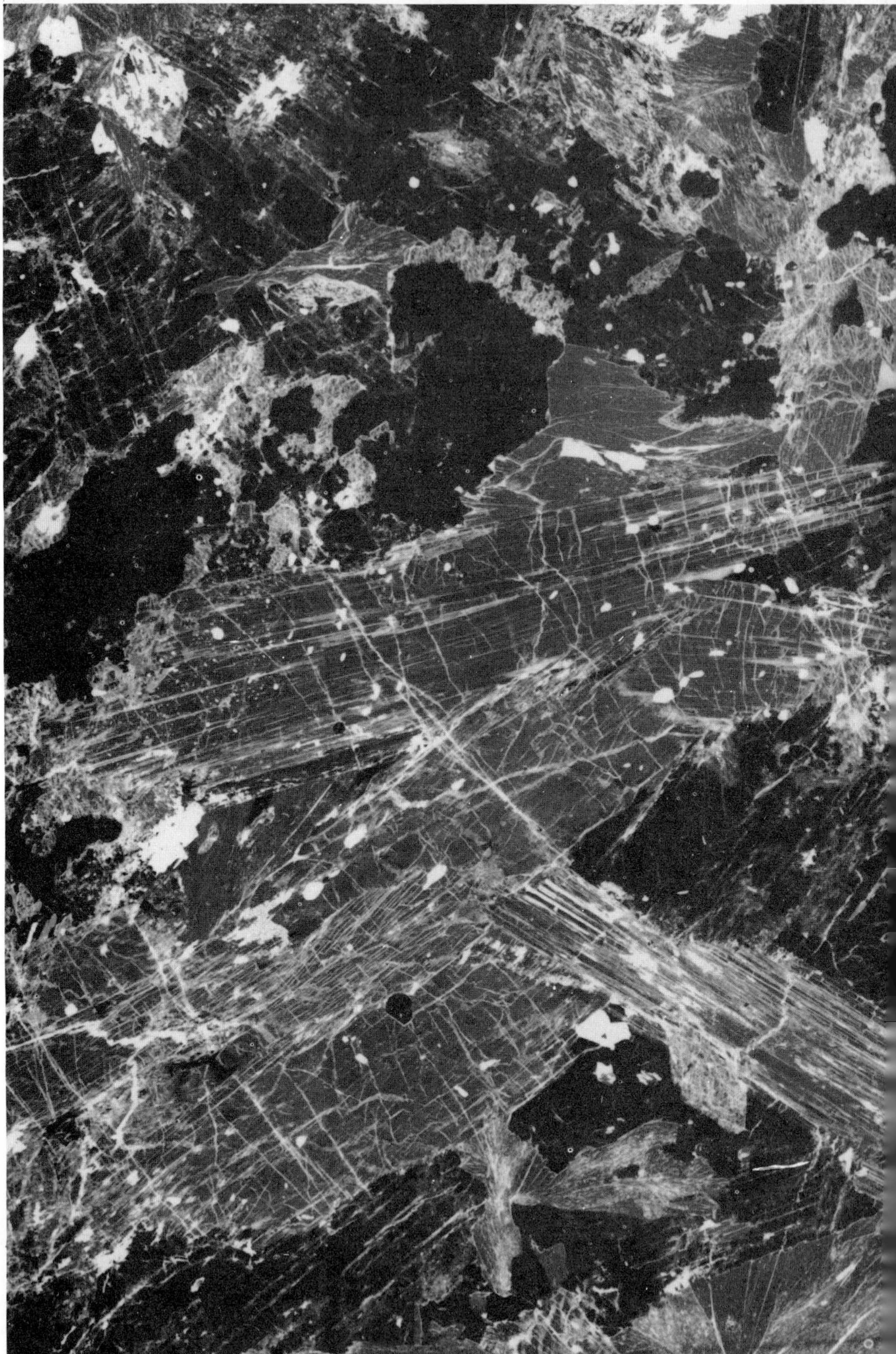

PART FOUR

METAMORPHIC ROCKS

And Subjects Between Metamorphism And Magmatism

The intention is pursued of segregating fact from hypothesis, but the alternation is on a smaller scale than in Parts Two and Three; after two short chapters designed to equip reader and author with a common terminology, Chaps. 20 through 23 each have their own sections of observable fact followed by tentative theory.

CHAPTER EIGHTEEN

VARIETIES OF METAMORPHISM

The variety of changes that can be induced in a solid rock without wholly melting it is enormous. The first need in approaching this wide field is to make some subdivisions.

Metamorphism, Diagenesis and Weathering

The processes of diagenesis and weathering are continuous with those of metamorphism, the differences being only of degree. If we wish for an arbitrary cut-off, we may class as diagenesis nondestructive processes occurring below 100°C and 1 kb by agency of surface waters. Destructive processes in the same conditions are called *weathering*. Processes occurring above 100°C or above 1 kb may fairly be called metamorphism. Processes occurring below these limits by agency of waters rising from within the crust are rather rare; it would probably be misleading to call them diagenesis or weathering, but whether we call them metamorphism or something else is not very important.

Isochemical Metamorphism and Metasomatism

Metamorphism is any change in the *appearance* of a rock by processes that leave it largely solid throughout, except for diagenesis and weathering. Metasomatism is any change in the *overall chemical composition* of a rock by processes that leave it largely solid throughout, except for diagenesis and weathering. A rock cannot be metasomatized without being metamorphosed, and, because of the lively activity of rock components, it is difficult to metamorphose a rock without some small effect of metasomatism. Thus isochemical metamorphism is a theoretical concept only; the field of possible metamorphic processes in the real world has to be divided into:

1. Metamorphism with only slight metasomatism;
2. Metamorphism with extensive metasomatism.

However, in the same way that behavior of a two-component melt is a useful guide to behavior of magmas, so behavior under isochemical conditions is a useful thing to study as an approximation to natural changes.

A given change may appear to be in class (1) or in class (2) above, according to the size of the rock body observed. In the change illustrated in Fig. 18-1, the volume *A* suffers metasomatism whereas the volume *B* suffers only an isochemical rearrangement of its contents.

Regional and Local Metamorphism

This is a simple division according to the extent of the area affected. Regional metamorphism is usually attributed to burial of the rocks involved, with accompanying increase in temperature and pressure. Local metamorphism may be due to deformation processes—the heat and stress generated by movement of a fault can metamorphose the immediate environment—or it may be due to intrusion of a hot magma body; the latter effect is called *contact metamorphism.*

Prograde and Retrograde Metamorphism

Changes in response to a higher pressure or temperature than the rock last adjusted itself to are prograde changes; changes in response to lower values are retrograde. Other possibilities exist—isograde metasomatism, adjustment to higher pressure but lower temperature, etc., but they are of less practical importance. At their extremes, prograde metamorphism borders on melting and retrograde metamorphism borders on weathering.

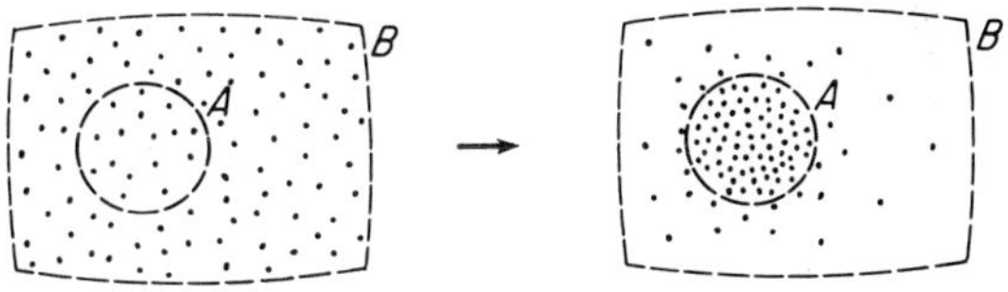

Fig. 18-1 Metasomatism and isochemical rearrangement.

Dynamic and Thermal Metamorphism

According to whether the effects of pressure or of temperature dominate, metamorphism varies between these two extremes. Burial leads to simultaneous increase in pressure and temperature, or *dynamothermal metamorphism*—most regional metamorphism is of this kind. The rise of hydrostatic pressure that can occur without accompanying heat effects is quite limited; in Fig. 18-2, therefore, the field of pure dynamic metamorphism is

Fig. 18-2 Ranges of metamorphism. Conditions represented by points in the blank field on left are unlikely to be met in nature; curves *C*, *D*, and *E* as in Fig. 1-5.

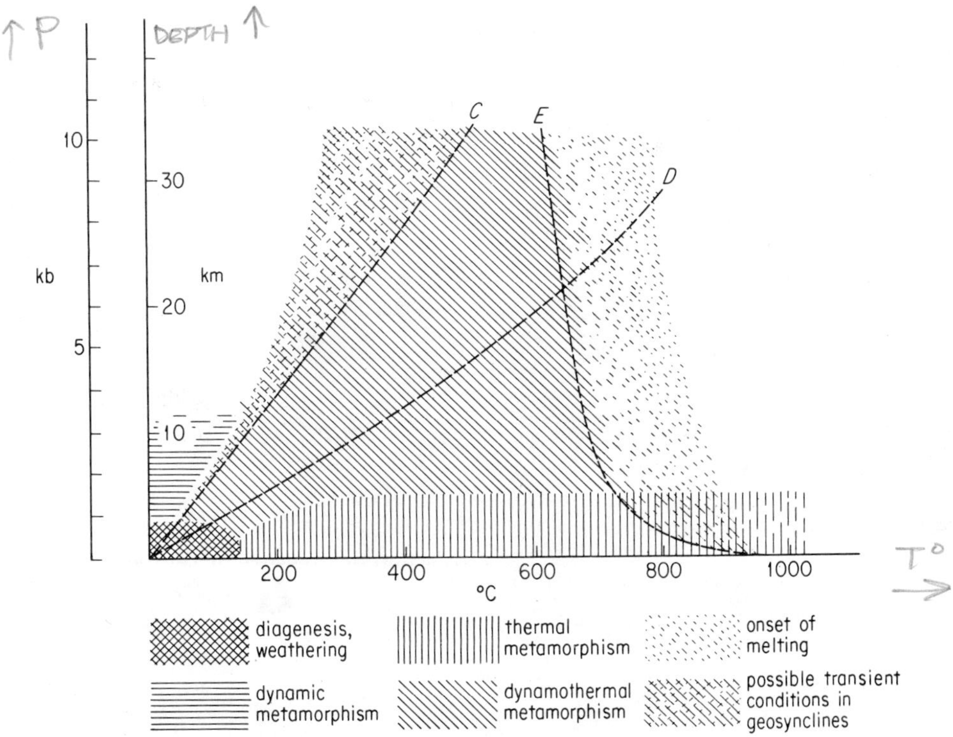

shown as rather small. However, rocks can sustain shear stresses or, in other words, can sustain departures from the hydrostatic condition; in fact, the compressive stress normal to some planes in a rock can exceed the hydrostatic pressure by 2 kb or more, and probably has done so in many examples of dynamic metamorphism. Wherever sound, solid rock has been fractured, departure to some such extent is indicated.

Autometamorphism

This is a special term for the changes that occur as a magmatic body cools in presence of its self-generated aqueous fluid: the changes include mineral reactions to more hydrous forms, rearrangement of crystal boundaries and adjustment of feldspar compositions.

CHAPTER NINETEEN

CLASSIFICATION AND NOMENCLATURE OF METAMORPHIC ROCKS

We wish to be able to discuss metamorphic rocks, as we do other rocks, in an unambiguous way. Whether we wish to summarize past discoveries or to propose a new hypothesis, a few names and classes of metamorphic rock will be a convenience. Before presenting a short list of useful classes, some precepts are repeated from earlier chapters:

1. In no field is it more important to distinguish between textural, compositional and genetic meanings of terms than in metamorphism.

2. If a classification of rocks is to be devised, every rock should have a place and no rock should have two places.

3. In describing the properties of a group of rocks, definition of the group must precede description.

4. In describing single rocks (defined by number or locality) it is not obligatory to give the rock a name; any term that expresses an observation or hypothesis exactly is acceptable; any term that confuses observations with hypotheses is not.

5. The choice between acceptable terms must be made on grounds of suitability to some purpose: the same outcrop could be described

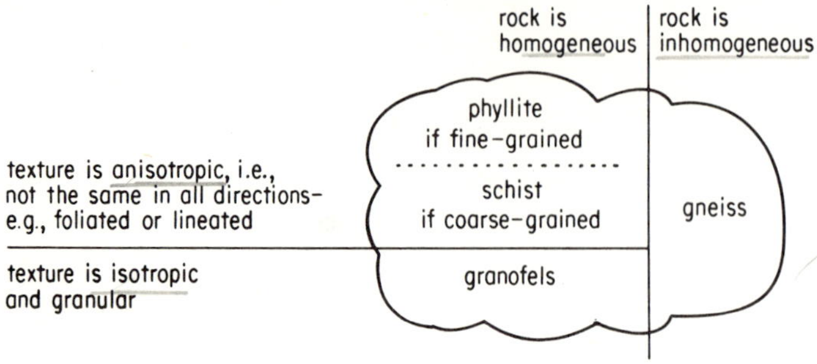

Fig. 19-1 A possible classification of metamorphic rocks by texture.

as *gangue, overburden, aquifer, cap-rock, B-tectonite, arenite* or *limestone*—these terms might all be equally correct, but according to one's different interests, only one is likely to be found suitable.

TEXTURAL TERMS

Four terms which constitute a classification almost compatible with (2) above are *schist, phyllite, gneiss* and *granofels*, as in Fig. 19-1. Different operators may set the boundaries at slightly different degrees, but in principle every metamorphic rock can be placed in this classification except an isotropic aggregate of flakes or rods. Such rocks exist, but are rare. Concerning Fig. 19-1 we note:

1. The terms are genetic to the extent that they imply metamorphism.

2. Whether a rock is homogeneous or not depends on how much of it is observed—it may be quite correct to say, "This specimen of schist comes from that outcrop of gneiss."

3. *Gneiss* and *granofels* are not restricted to any particular grain size but can be qualified by *fine, coarse* etc.

4. In this classification, *slate* is not a grain-size term in series with *phyllite* and *schist*. It is preferred to use *phyllite* for a metamorphic rock, however fine, rather than call a rock without slaty cleavage a slate.

5. There is no special term for rocks composed of flakes or rods matted in random orientation, such as some talc formations. The temptation to call these schists should be avoided—if the flakes have no preferred orientation, "talc rock" is preferable to the more technical, but wrong, term "talc schist."

For more exact description of a texture, a wide range of terms is available, such as:

foliated } unless *lineated* is stated, a schist is normally taken to be
lineated } foliated

layered
lenticular } a gneiss is normally one of these; for augen structure, see
clotted } Figs. 20-1B and 20-4E
augen

porphyroblastic: with isolated large crystals in a matrix of finer crystals, produced by metamorphism (cf. *porphyritic*, Chap. 5)

poikiloblastic: with crystals that have many inclusions, produced by metamorphism (cf. *poikilitic*, Chap. 5)

No term should be rejected as "ungeologic"—*spotted* is more direct than *maculose,* and *lumpy* serves many purposes better than *blastopsephitic.*

COMPOSITIONAL TERMS

The most concise composition terms applicable to metamorphic rocks are the igneous rock names in Fig. 6-1. *Granite gneiss* has been long used, but less common terms such as *diorite granofels* are equally useful: the latter implies that the rock has been metamorphosed, has granofels texture and has diorite mineralogy. It does not imply that the rock was a magmatic diorite before metamorphism. (At one time, these semi-igneous hybrid names did imply igneous parent rocks, but this sense is falling into disuse and is better abandoned.)

Beside the terms in Fig. 6-1, some terms from Chaps. 7 and 9 are available—charnockite, eclogite, anorthosite, pyroxenite and, less usefully, other ultrabasic terms. Since these do not imply magmatic origin in the first place, they may be used alone to refer to metamorphic rocks—they are composition terms of high purity, with no genetic implications whatever.

Unfortunately, metamorphism of sediments leads to many rocks whose composition cannot be described by any of the terms above. Figure 13-6 shows that these may be siliceous, calcareous, aluminous, ferruginous or magnesian, with respect to the main bulk of magmatic rocks. Thus *magnesian schist* is a useful unambiguous term. However, outside the igneous field, naming by minerals is an alternative: *talc schist* is more specific than *magnesian schist* and so for some purposes more suitable.

Finally, there are some special terms. In the following list, CT marks a term with compositional and textural meaning and CG marks a term with compositional and genetic meaning:

quartzite (CT): rock mainly quartz, not necessarily metamorphosed, that is sufficiently coherent to break across component grains rather than around them as in sandstone

marble (CG): metamorphosed rock, now mainly calcite or dolomite

calc-silicate rock = *skarn* (CG): metamorphosed rock, now mainly calcite and calcium-bearing silicates (diopside, wollastonite, Ca-garnet etc.)

hornblendite (mainly hornblende) and *epidotite* (mainly epidote): no textural or genetic meaning

amphibolite (CG): metamorphosed rock, now mainly amphibole *and plagioclase* (unmetamorphosed igneous rock mainly amphibole and plagioclase can usually be called hornblende gabbro or hornblende diorite)

pelite (*pelitic*): as commonly used, a pelite means an aluminous sediment metamorphosed, but if used systematically, as in Table 14-1, it means a fine-grained sediment metamorphosed. It is true that most aluminous sediments are fine-grained and most fine-grained sediments are aluminous, but there is no advantage in confusing the two classes; in fact, science comes to a standstill when *most* is not distinguished from *all*

psammite (*psammitic*): to a lesser extent, this term is also used in two ways, as is the corresponding sedimentary term *sandstone*, and so must be used with care.

GENETIC TERMS

To start with terms of general application, *epi-* and *meta-* are prefixes meaning metamorphosed; thus epidiorite, metagabbro, metavolcanic. (Since *epi-* strictly means *outside* or *after*, it is also used without metamorphic meaning as in *epizone*—the outer part of the crust, or *epiclastic*—a rock formed by aggregation of broken pieces, cf. *detrital.*) After such a prefix, the igneous terms retain their genetic sense; thus *diorite gneiss* tells only present mineralogy and texture, but *epidiorite gneiss* tells only origin and texture—its present mineralogy may have nothing in common with a diorite.

Ortho- and *para-* have extraordinarily wide application (*orthodox, parabola* etc.). Applied to metamorphic rocks, they indicate igneous and sedimentary parentage respectively; thus *orthogneiss, para-amphibolite.* (Exception: *orthoquartzite*—this term has two separate meanings, both purely sedimentary.) Now that the use of *granite* in *granite gneiss* and such terms has been freed of genetic meaning, *ortho-* and *para-* are exceedingly useful: thus *paragranite gneiss* unmistakably indicates genesis, composition and texture, *paragneiss* indicates genesis and texture but not composition, and *granite gneiss* indicates composition and texture but not origin.

First Digression: It would seem advantageous to use *para-granite* without a texture term to indicate origin (metasedimentary) and composi-

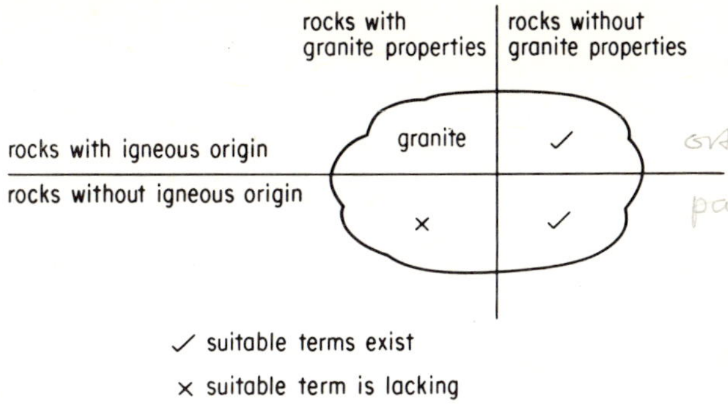

Fig. 19-2 An incomplete system of terms.

tion (granitic). For more than one hundred years, there has been dispute over the possibility of forming a rock with the properties of granite from sediments without extensive fusion, and controversy is still keen; but whether para-granites are abundant or nonexistent, it would be a convenience to be able to refer to them succinctly in such questions as "How could para-granites form?" and "Could the rock at Toe Head be a para-granite?"

Second Digression: Whether it is desirable to change current usage or not is an open question; geologists have managed with *granite* and corresponding terms having both compositional and genetic meanings for some time, and may continue to do so. Whatever the future, the question, "Why did difficulty arise?" is worthy of study, in order that we may avoid creating similar difficulties for ourselves in the future. The answer is that the adaptation of the old quarrying term *granite* to a specific geological meaning embodied an assumption. When *granite* was defined as a rock of igneous origin with such-and-such properties, no term was established for rocks with the same properties but not of igneous origin, because the proponents of the definition assumed there were none. The result was the incomplete classification shown in Fig. 19-2. The lesson is that every definition defines *two* classes, objects that satisfy the definition and objects that do not; it is as important to ensure that suitable terms exist for the latter class as for the former.

Beyond *epi-*, *meta-*, *ortho-* and *para-*, there are no genetic terms that can be said to form any kind of a classification. If one of the following terms happens to be appropriate it can be used; otherwise, one has to spell out genetic hypotheses thus, "thermally metamorphosed tuff" or, "formed by retrograde regional metamorphism with metasomatism from a peridotite."

hornfels: fine granofels produced by contact metamorphism

mylonite: fine-grained rock produced by dynamic metamorphism

cataclasite = *cataclastic rock*: rock containing angular fragments produced by dynamic metamorphism

(*fault breccia*: cataclasite at fault zone)

All these terms combine a textural meaning with their genetic meaning.

CHAPTER TWENTY

PRODUCTS OF METAMORPHISM

We are now in a position to consider the result of imposing different kinds of metamorphism on different kinds of igneous and sedimentary rock. Since it is local in scale and simple in effect, we begin with dynamic metamorphism.

DYNAMIC METAMORPHISM

Figure 20-1 shows some textures produced by dynamic metamorphism. Since the effects are not chemical, the range of possibilities is much the same whatever the preexisting rock. Recognition of the products is not always simple because of the following possibilities.

Cataclasites can be confused with other breccias, such as fanglomerates or agglomerates, particularly if subsequently metamorphosed. The margin against solid rock must be sought and checked for (1) sharpness—cataclasites usually grade into solid rock, while other breccias are usually sharply bounded, and (2) similarity between breccia fragments and the solid environment—if this is negligible, rock is not cataclasite; if some fragments match, the test is nondiagnostic. The fragments should also be compared with their matrix: if they are very close in composition, cataclasite is probable; if they differ markedly, cataclasite is less probable but still possible.

Augen-clasts can be confused with augen-blasts (augen-forming porphyroblasts), especially when composed of quartz or feldspar in a matrix that has recrystallized. If the matrix seems to be wholly clastic, the augen can be diagnosed as clastic also; if the augen are euhedral, they are more probably porphyroblasts; if neither of these lines of study is decisive, both possibilities should be retained in mind.

Mylonites can be confused with both sediments and tuffs in hand specimen; the field situation must be the criterion. If structural dislocation cannot be proved, a concordant mylonite is hard to confirm. Although mylonites with thickness measurable in centimeters constitute a very small fraction indeed of the earth's crust, the phenomenon of *slickensiding* is a common example of mylonite on a very small scale.

METAMORPHIC MINERALS AND THE PHASE RULE

Dynamothermal metamorphism is a subject of far greater complexity than dynamic metamorphism. Various combinations of high temperature with high pressure can cause many different mineral reactions to occur. Clearly, we cannot begin to trace these possibilities without first considering

Fig. 20-1 Textures of dynamic metamorphism. (A) Cataclasite; (B) augen formed by destructive processes (cf. Fig. 20-4E); (C) mylonite; (D) granulation on micro scale.

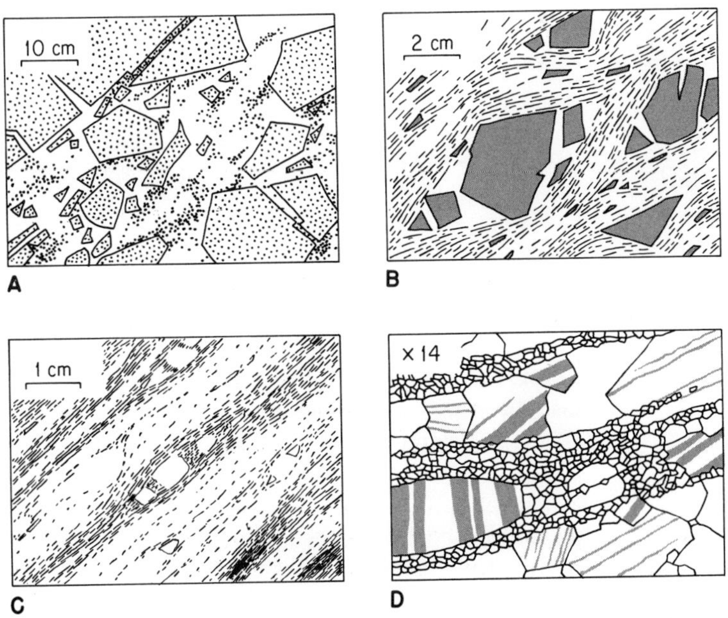

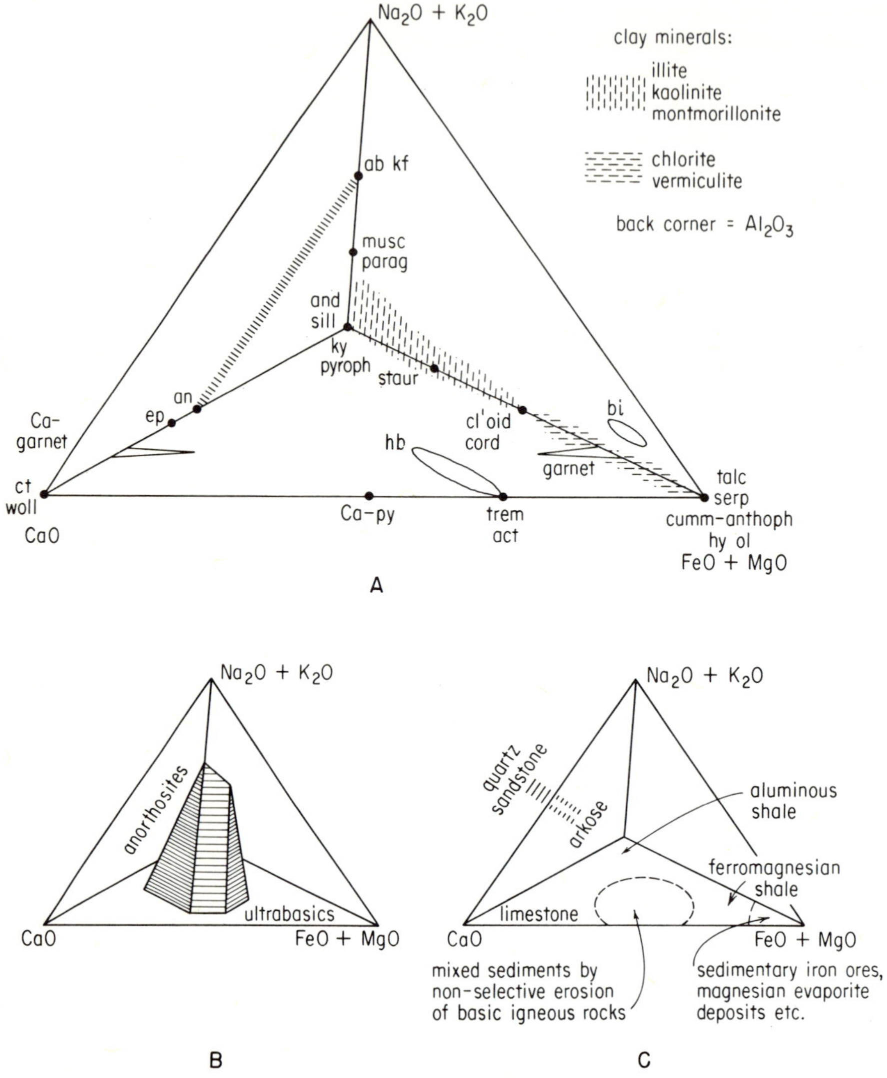

Fig. 20-2 Metamorphic minerals. The tetrahedral framework is as in Fig. 13-6, of which two parts are reproduced for comparison. For abbreviations see p. 307.

the minerals that may be expected to develop and then the number that may be expected to coexist.

The principal metamorphic minerals are shown in Fig. 20-2. The use of these diagrams is as follows: we assume a shale to undergo metamorphism that is close to isochemical, and ask what minerals are likely to be formed. Although there are over twenty minerals in the diagram, we cannot yet say of any that it will not form; we can say, however, using a version of the

Fig. 20-3 An extract from Fig. 20-2.

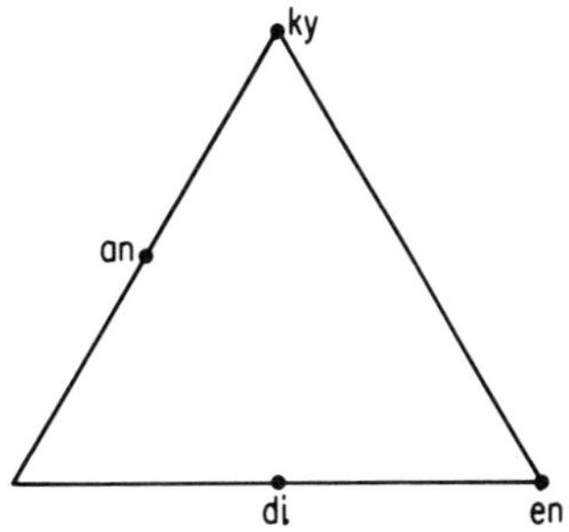

balance principle from Chap. 8, that, of anorthite or wollastonite for example, not much will form because these minerals are far removed from the bulk of the rock in composition. If anything is going to form in large amounts, it will be chloritoid, cordierite, staurolite or muscovite if the shale is aluminous, and chloritoid, cordierite, chlorite, biotite or garnet if the shale is ferromagnesian.

Such predictions are borne out by field observation, but observation also shows that not all twenty-odd minerals do develop. We next ask ourselves why this is so. As with other geological matters, the basic factor is energy. In the remainder of this section, we seek a link between the number of minerals coexisting and their energy relations, on theoretical grounds. We return to observable facts in the next section, on dynamothermal effects.

To trace the link, consider a rather simple rock containing no sodium or potassium; such a rock may be represented by a point on the floor of the tetrahedron in Fig. 20-2. Let it also contain no iron but only magnesium as the component in the right-hand corner, and to begin with, let it be so hot and dry that all hydrous minerals change over to anhydrous forms. Then among the minerals that may occur are anorthite, kyanite, enstatite and diopside (Fig. 20-3); but we notice that among these four, a reaction is possible:

$$\underset{\text{ANORTHITE}}{CaAl_2Si_2O_8} + \underset{\text{ENSTATITE}}{MgSiO_3} \rightleftharpoons \underset{\text{DIOPSIDE}}{CaMgSi_2O_6} + \underset{\substack{\text{KYANITE}\\ \text{OR ANDALUSITE}\\ \text{OR SILLIMANITE}}}{Al_2SiO_5} \qquad (1)$$

As a matter of fact, this reaction is unlikely to occur because the reaction

$$\underset{\text{ANORTHITE}}{CaAl_2Si_2O_8} + 2\ \underset{\text{ENSTATITE}}{MgSiO_3} \rightleftharpoons \underset{\text{GARNET}}{CaMg_2 \cdot Al_2Si_3O_{12}} + \underset{\text{QUARTZ}}{SiO_2} \qquad (2)$$

runs more readily, but reaction (1) is nevertheless a convenient example to discuss. In what follows, the assumption is made that reaction (2) does not interfere; then reaction (1) will occur at a fixed temperature for any given pressure—it is similar in this respect to the change from ice to water, or calcite to aragonite, or any other change of state. We may say that the mixture $Ca–Mg–Al_2–Si_3–O_{11}$ exists in the anorthite-enstatite state under some conditions and the diopside-kyanite state under other conditions, but it is only under very restricted circumstances that both states coexist.

Of course, a rock will not normally contain exactly equal numbers of molecules of anorthite and enstatite. Consequently, if conditions are changed so as to drive the reaction to the right, either anorthite or enstatite will be used up first and the product will be either

diopside, kyanite and enstatite (excess)

or

diopside, kyanite and anorthite (excess)

Similarly, if a diopside-kyanite rock is affected, it may react to give either

anorthite, enstatite and diopside (excess)

or

anorthite, enstatite and kyanite (excess)

Of the four possible minerals, any three can coexist indefinitely, but all four coexist only while reaction is in progress.

Up to this point, the silica content of the rock has not been mentioned. We have merely assumed that however much Ca, Mg and Al are present, the necessary Si is also available. To be realistic, we admit that the silica present may be either greater or less than this exact amount. In the former case, quartz will appear; in the latter case, some mineral must appear that is less siliceous than those mentioned, such as olivine or garnet; but because both sides of a reaction are not stable together, not more than four minerals will coexist indefinitely.

If mixtures of CaO, MgO, Al_2O_3 and SiO_2 give four-mineral assemblages, adding K_2O must give a fifth mineral—some K-bearing silicate, such as potash feldspar. Adding H_2O must give a sixth mineral, if conditions are made such that hydrous minerals can form; if they are not, and H_2O is forced into the system, a sixth *phase* will exist, to wit, some aqueous fluid. Thus we arrive at the *mineralogical phase rule: the number of minerals coexisting freely in equilibrium equals the number of components.* The limitation *freely* is made to allow for the fact that if we make the pressure and temperature just right, we can keep more minerals in coexistence than we have components (*analogy:* if we maintain 0°C and 1 atm we can keep a block of ice floating in water indefinitely—two phases, one component); *coexisting freely* is intended to mean coexisting without strict environmental control of this kind. The rule applies both to systems confined at such pressures and temperatures that there is no vapor phase and to systems where a vapor exists, as long as the vapor is counted as an extra mineral.

The mineralogical phase rule is simple to apply as long as the number of components is clear; but consider what would happen if, for example, iron were not excluded. In place of diopside, enstatite and Mg-garnet, mixed minerals would occur in the solid solution series diopside–hedenbergite, enstatite–hypersthene and pyrope–almandine. Thus the components would increase by one without the production of any new phases. On this account, some authorities rewrite the rule and say the number of minerals is equal to *or less than* the number of components; other authorities say that as long as iron can substitute for magnesium without the appearance of a new phase, it need not be regarded as a new component. The second seems more convenient. Thus, with titanium, for example, as long as the amount present is less than can substitute comfortably for other metals in the pyroxenes etc., we need not consider it; but in titanium-rich rocks, it can behave as an independent component and require the appearance of ilmenite, sphene or some other titanium mineral. This approach saves us even counting up metals such as boron and nickel—we merely assume that they occur in quantities sufficiently small to dissolve in the major-mineral lattices.

The major metals in most rocks are Si, Al, Fe, Mg, Ca, Na and K. In most of the ensuing discussion, we shall make three assumptions:

1. An excess of silica is available so that free quartz occurs in the rocks;
2. Plagioclase occurs and carries most of the sodium;
3. Iron and magnesium substitute for each other freely and so operate as a single component.

In addition to the Fe–Mg component, this leaves only Ca, Al and K to be kept in view—a total of four important variables, giving four or fewer major minerals at a time in addition to the quartz and plagioclase mentioned. The mineralogy of a metamorphic rock is thus apt to be less complex than Fig. 20-2 perhaps suggests.

Up to this point, oxygen is the only nonmetal that has been considered, but there are other nonmetals whose presence should be remembered—carbon, sulfur, phosphorus, fluorine and chlorine are the main ones. Apart from phosphorus, these are prominent in the volatile fraction as CO_2, SO_3, HF, HCl and as ions. To see what difference their presence makes to the phases present, it is therefore necessary to look in greater detail at systems with volatiles. The binary system $MgO–H_2O$ is a suitable starting point.

In the system $MgO–H_2O$ there are four pairs of phases which coexist freely:

$MgO + Mg(OH)_2$ (PERICLASE, BRUCITE)	$Mg(OH)_2 + H_2O$ liquid
$MgO + H_2O$ vapor	$Mg(OH)_2 + H_2O$ vapor

The assemblage [$Mg(OH)_2 + H_2O$ liquid $+ H_2O$ vapor] is also quite easy to prepare, in principle: a container is evacuated and then a sludge of brucite and water admitted through a stopcock. If the quantity of sludge does not fill the container, water will evaporate into the remaining space *until the water vapor pressure reaches a certain value* fixed by the temperature. We therefore find, not three phases coexisting *freely* but three phases coexisting at a particular pressure P_T; for example, if $T = 100°C$, $P_T = 1$ bar except for some small corrections.

Suppose next that one wall of the container is replaced by a freely-moving piston. As long as pressure P_T is applied, there will be no change, but if a pressure P_1 less than P_T is applied, water will continue to evaporate and the piston will move outward until all the liquid water is used up; after this point, a small movement of the piston will reduce the pressure in the container from P_T to P_1 at which point a stable situation will exist with two phases, brucite and water vapor. Alternatively, if a pressure P_2 greater than P_T is applied to the piston, water vapor will condense until there is none left, resulting in a stable vaporless situation, brucite with liquid water.

Up to this point, the brucite has played no part and behavior has not differed from that of a container with pure water. However, the pressure could conceivably be reduced until the brucite began to dehydrate; that is, there is another critical pressure P'_T at which the three phases periclase, brucite and water vapor coexist. If a pressure even slightly below P'_T were applied to the piston, all the brucite would dehydrate and the stable pair, periclase and water vapor, would result.

We are now in a position to consider the addition of a second volatile component, such as carbon dioxide, to the system. Suppose the system is at a temperature and pressure where the stable assemblage is brucite and water vapor, i.e., the pressure on the piston is P_1, between P_T and P'_T above, and let a *small* puff of CO_2 be introduced. The partial pressures P_{CO_2} and P_{H_2O} will adjust themselves until

$$P_{CO_2} + P_{H_2O} = P_1$$

$$\frac{P_{CO_2}}{P_{H_2O}} = \frac{\text{number of molecules of } CO_2}{\text{number of molecules of } H_2O}$$

That is, there will be a slight drop in P_{H_2O} and a slight increase in the volume, V, of vapor, subject to $P_{H_2O}V = \text{constant}$. The introduction of CO_2 will not lead to the formation of magnesite, $MgCO_3$, as long as P_{CO_2} is less than that needed to drive the reaction

$$Mg(OH)_2 + CO_2 \rightarrow MgCO_3 + H_2O$$

It now appears that we have three components and only two phases; but the position is just as with iron and magnesium, and we may similarly say that *as long as* the CO_2 mixes homogeneously into a phase already present, it need not be regarded as an extra component.

If a very large puff of CO_2 were introduced, P_{CO_2} could well rise sufficiently to convert all the brucite to magnesite as above, leaving two phases, magnesite and mixed vapor. This situation resembles the pair magnesite and CO_2 in the binary system $MgO–CO_2$ and it may again be argued that the third component, water, is not *acting* as a component because it is dissolving passively in the CO_2.

Intermediately, there is the possibility that some magnesite will form and some brucite remain. This resembles the situation mentioned above where titanium is too abundant to dissolve in the silicate phases already present (pyroxene etc.) and so generates an independent phase as well (e.g. ilmenite). We may say here that CO_2 is too abundant to dissolve in the water present and generates an independent phase, magnesite.

The upshot of the foregoing discussion is that volatile substances, like nonvolatile ones, may or may not function as independent components, according as they can or cannot dissolve in phases already present without causing new phases to appear. The lack of rigor in this concept of a component is implicit in the mineralogical phase rule—it can be avoided only by abandoning the mineralogical phase rule and using the phase rule proper, as proposed by Gibbs.* The additional concept involved, of "degrees of freedom," permits greater precision in one's statements about solutions, but it is still customary to use the rule given above (that what does not generate extra phases need not be counted as a component) to justify omitting seventy or eighty minor elements from one's list of components. Thus, in spite of its lack of rigor, the mineralogical phase rule is a very useful guide, and it is important to realize that it is applicable to volatile components as well as nonvolatile ones. In fact, whether a component is volatile or nonvolatile makes very little difference to thermodynamic theory—whether changes are commonly called evaporation, reaction, freezing, dissolving, precipitation or unmixing, they are all subject to the same rules and are all simply changes of phase.

DYNAMOTHERMAL METAMORPHISM

The preceding discussion persuades us that in any set of circumstances, only a few minerals coexist stably. It should therefore be possible to state how these minerals react if the pressure or temperature is changed. In this section and the next, observations on the sequence of reactions are assembled; in Chap. 21, we try to discover reasons for the sequence observed.

Sediments are discussed first and reference made to Fig. 20-2. Since the minerals in an arkose are stable over a wide range of crustal pressures and temperatures, under dynamothermal metamorphism little mineral change occurs. If interstitial clay has crept in, it will form mica but the principal change is in texture (Fig. 20-4A). The other sedimentary varieties are more reactive. For descriptive purposes, five subdivisions are made (Fig. 20-5A);

*J. Willard Gibbs (1839–1903).

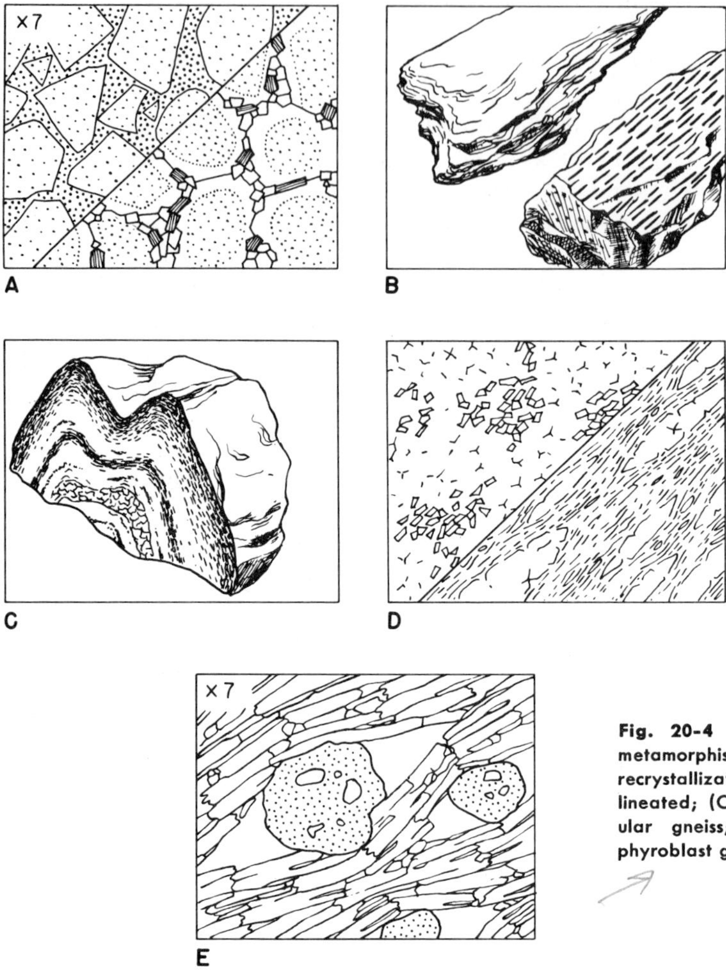

Fig. 20-4 Textures of dynamothermal metamorphism. (A) Arkose before and after recrystallization; (B) schists, foliated and lineated; (C) gneiss; (D) clotted and lenticular gneiss; (E) augen formed by porphyroblast growth (cf. Fig. 20-1B).

their transformations are traced in Fig. 20-5B–E and Table 20-1. The five numbered columns in Table 20-1 correspond to the five numbered composition fields in Fig. 20-5A. The levels B–E correspond to the levels of metamorphism illustrated in Fig. 20-5B–E and Fig. 20-6. For explanation of abbreviations, see list on p. 307.

As an example, consider the metamorphism of a common brown clay (one with considerable iron and magnesium), such as would fall in Field 4 of Fig. 20-5A. In its sedimentary state it consists of aluminous sheet silicates,

Fig. 20-5 (A–E) Stages of dynamothermal metamorphism. Dominant minerals are shown by filled spots, possible minerals by open spots. Quartz present throughout. (F, G) Stages of thermal metamorphism. Dominant minerals are shown by filled spots, possible minerals by open spots. Quartz or tridymite present throughout.

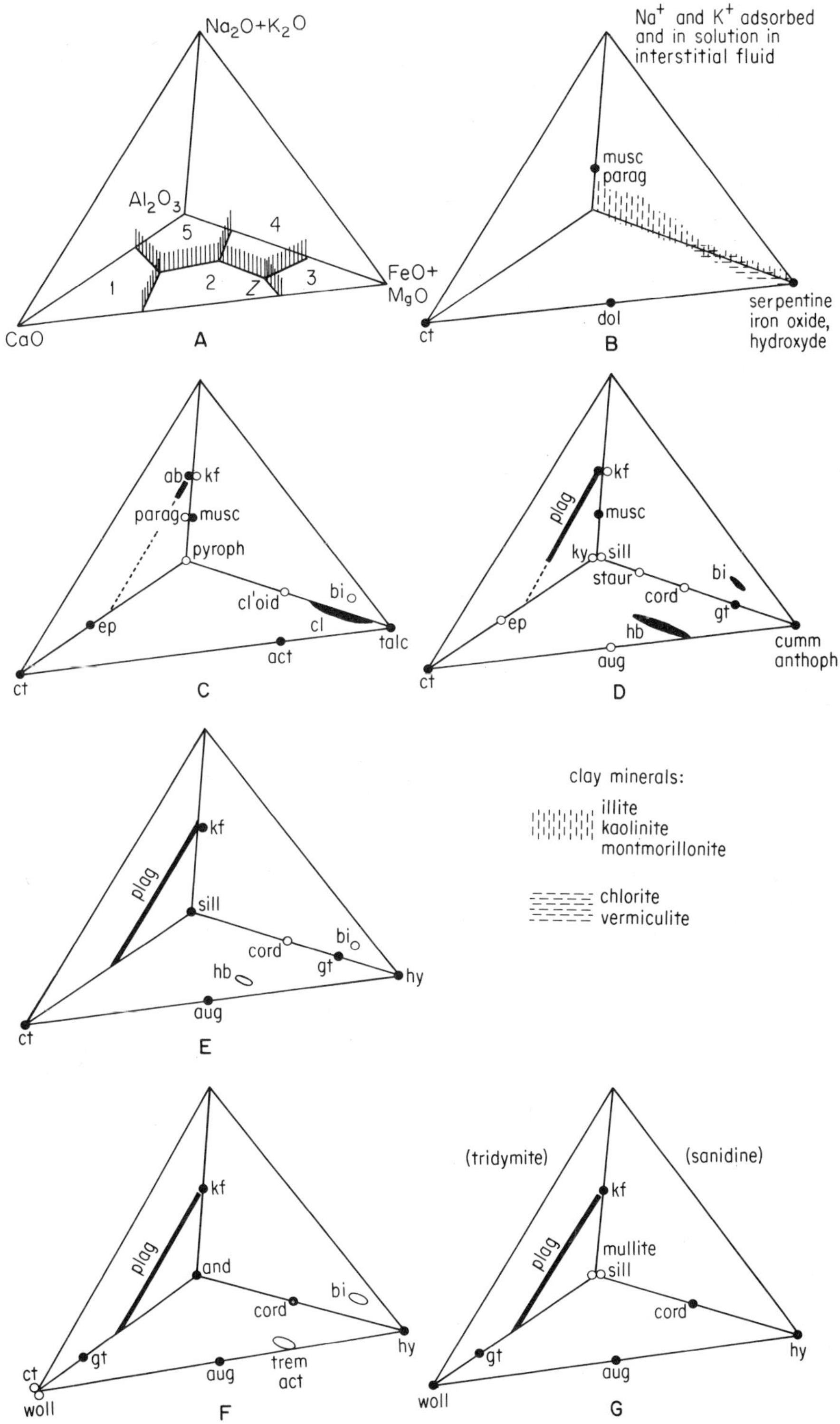

Na2O+K2O
Al2O3
5
4
1
2
Z
3
FeO+
MgO
CaO
A
Na+ and K+ adsorbed
and in solution in
interstitial fluid
musc
parag
ct
dol
serpentine
iron oxide,
hydroxyde
B
ab
kf
parag
musc
pyroph
cl'oid
bi
ep
cl
talc
act
ct
C
kf
plag
musc
ky
sill
staur
cord
bi
gt
ep
hb
cumm
anthoph
aug
ct
D
kf
plag
sill
cord
bi
gt
hb
hy
aug
ct
E
clay minerals:
illite
kaolinite
montmorillonite
chlorite
vermiculite
kf
plag
and
cord
bi
gt
hy
ct
woll
aug
trem
act
F
(tridymite)
(sanidine)
kf
plag
mullite
sill
cord
gt
hy
woll
aug
G

TABLE 20-1 DYNAMOTHERMAL METAMORPHISM

1	*2*	*3*	*4*	*5*
		B		
ct + clay → ep ct + musc → ep + kf	dol + qz → ct + talc dol + qz + K → ct + phlog	Mg–cl + clay → Al–cl		clay + K → illite illite + K → musc illite → musc + pyroph
	dol + cl → act + {ct / ep}			
		cl + K → bi	cl + musc → bi + {cl'oid / pyroph}	parag → ab
		C		
ep + qz + {musc / pyroph} → an (in plag)	act → hb	cl + qz → gt		
		talc → {cumm / anthoph}	cl'oid → staur	pyroph → ky
			staur → gt + {ky / sill}	
		D		
	hb + ct → aug	cumm / anthoph → hy		ky → sill
	hb → hy + plag		musc + gt → cord + kf	musc → sill + kf
	hb + plag + water → magma	bi + qz → hy + kf		
		E		
		plag + ol → clinopyroxene + gt		

TABLE 20-2 THERMAL METAMORPHISM

1	2	3	4	5
		C*		
ep + qz + {musc / pyroph} → an (in plag)	act → aug	talc + cl → ol + cord		pyroph → and
		talc + ol → en–hy	cl → cord	
		talc → en + qz		
plag + ct + qz → Ca–gt		bi + qz → hy + kf		musc → and + kf
		F		
ct + qz → woll				and → {mullite / sill}
				kf → sanidine
		G		
		qz → tridymite		

*Up to level C, changes under thermal metamorphism are identical with those under dynamothermal metamorphism.

magnesian sheet silicates, and iron oxides and hydroxides; its calcium content occurs as carbonate; its sodium is in the clay particles or adsorbed onto them or in solution in the pore water; even a smooth mud carries appreciable silica as quartz. During diagenesis, potassium enters the aluminum silicates giving illite and so muscovite; sodium begins to be fixed as albite or paragonite; the magnesium sheet silicates take in iron and aluminum and develop chlorite structure. Further reactions in columns (3)–(5) of Table 20-1 bring the rock to a mineralogy with chlorite, muscovite, quartz and albite; in addition, pyrophyllite or chloritoid may appear if Al/K is high, and potash feldspar or biotite if Al/K is low. The rock will by now have the texture of a slate or phyllite; it is about at stage C in Fig. 20-5.

The next conspicuous changes are the replacement of chlorite by garnet and, if present, chloritoid by staurolite, and pyrophyllite by kyanite. By this stage, whatever minor calcite or epidote was present will turn to anorthite and dissolve in the albite. Muscovite will certainly persist and biotite will do so if formed above. The rock thus becomes a garnet mica schist as at stage D. Beyond this stage, increase of intensity of metamorphism results in disappearance of the hydrous minerals (staurolite, muscovite and eventually biotite). If the water-vapor pressure is low, they turn to potash feldspar,

Fig. 20-6 Estimated stability fields for mineral assemblages in Fig. 20-5. (Broken lines as in Fig. 1-5.) The diagram is drawn assuming that the water pressure approximates the pressure of the rock overburden; if the water pressure is less than that of the rock overburden, all fields C, *D*, *E* and *F* shift to the left and the melting curve shifts to the right.

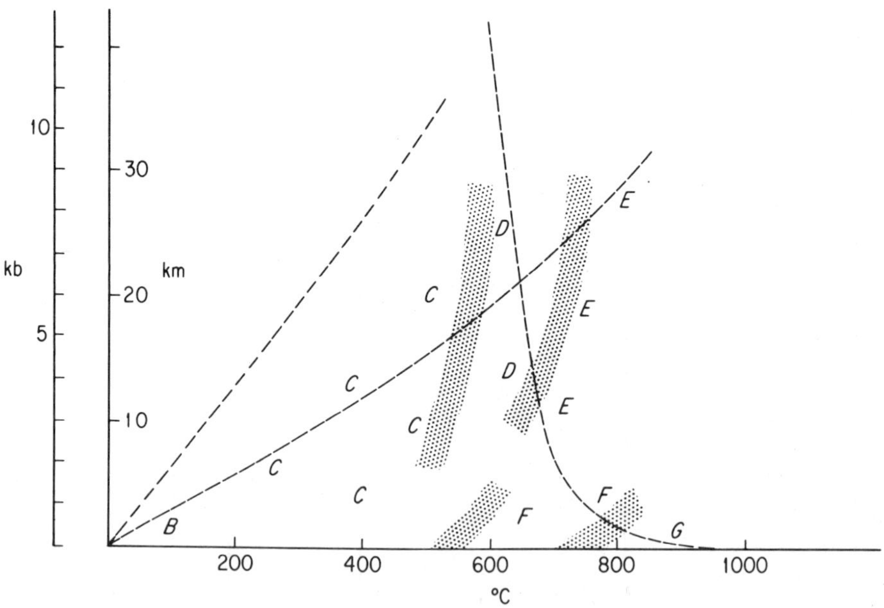

sillimanite and garnet; the rock loses much of its schistosity and becomes more of a granofels or gneiss. If the water-vapor pressure is high, melting may begin before biotite breaks down; after this point the history of the rock can no longer be discussed as a quasi-isochemical metamorphism—the physics and chemistry of assimilation take control (Chaps. 9 and 11).

Transformation of sediments from other composition fields can similarly be followed. Of course, the diagrams present only a skeleton of the process; there are numerous other reactions that can occur under special circumstances, and the order in which the reactions occur is not necessarily exactly that of the table. However, the dominant trends in dynamothermal metamorphism can be recognized; they are:

1. Progressive dehydration (it is instructive to examine Table 20-1 for reactions which do *not* involve escape of water);

2. Progressive increase in density (usually associated with loss of water, but apparent also in some of the anhydrous reactions, e.g., plagioclase + olivine $\longrightarrow$ clinopyroxene + garnet);

3. Progressive increase in the amount of solid solution—in the sedimentary minerals, substitution of one metal for another is slight, but with rise of temperature, the tolerance of the few high-density structures for metals of all sizes becomes great; rocks near *Z* in Fig. 20-5A, which melt with difficulty, eventually reduce to only two minerals, pyroxene and garnet, but the garnet contains Ca, Fe, Mg, Fe^{+3}, Al and Si, while the pyroxene contains all these and Na as well (see eclogite minerals, Chap. 9).

The manner in which the dehydration and solid-solution processes interact to give one sequence of reactions at one point, and a slightly different sequence at another, is common to all kinds of metamorphism; it is therefore reserved for discussion in Chap. 21.

Dynamothermal metamorphism of magmatic rocks follows a course determined by (1) the extent of weathering and autometamorphism, and (2) the intensity level maintained. Thus mild dynamothermal metamorphism of a fresh diorite causes mainly retrograde changes, but in more severe metamorphism of a much-altered diorite, prograde changes dominate. It is also possible that a fresh diorite will be subject to severe metamorphism, or an altered diorite to less severe metamorphism. In these situations little change occurs—at intensity level E in Fig. 20-5, the main diorite minerals simply remain stable, and at intensity level D, the minerals of an altered diorite remain without change. The diagrams, in fact, continue to show the minerals stable at a given intensity level, and to a great extent the retrograde reactions are simply those of Table 20-1 reversed.

As an example, the retrograde metamorphism of a diorite is briefly outlined. In the original mineralogy the six dominant metals, p. 237, are contained in plagioclase (Ca, Na, Al), potash feldspar (K, Al) and pyroxenes

(Ca, Fe, Mg). Mild retrograde effects are growth of biotite and hornblende from the pyroxenes and the beginning of alteration of plagioclase to mica and epidote; most coarse diorites show these effects by autometamorphism alone—it is mainly in more quickly cooled diabase sills that the original mineralogy can be seen fresh. More extensive retrograde effects are not produced by autometamorphism under normal circumstances (cool environment etc.); reaction rates are slower at low temperature and a diorite often cools through stage C (Fig. 20-5) without adjusting to it, and so reaches surface conditions, where, though manifestly out of equilibrium, the rock is too cold to change. However, if cooling is particularly slow, or the environment particularly hydrous, or if a cold diorite is mildly warmed (to c. 300°C), it may adjust to intensity level C, the condition of the *greenstones* or *greenschists*, so called because of the dominance of the three green minerals epidote, chlorite and actinolite. Of course, to accommodate the sodium and potassium originally present, muscovite and albite or paragonite must occur if the change is isochemical; but the change is not likely to be isochemical—if the permeating fluid is carbonated, the alkalis will tend to be dissolved, or if it has marine connections considerable additional sodium may be introduced.

It is convenient at this point to introduce the concept of *metamorphic convergence*. A sediment with composition *Z*, such as a dolomitic marl, can also be converted to an epidote-chlorite-actinolite rock by mild sustained warming. In other words, the same temperature episode may convert a diorite (retrogressively) and a marl (progressively) to indistinguishable rocks—from opposite directions, the magmatic assemblage of minerals and the sedimentary assemblage *converge* on the same metamorphosed product. This is one of the facts that makes metamorphic terrains so enigmatic—it is possible to know a rock unit in great detail, in respect of extent, field relations, present mineralogy, chemical constitution and economic potential, and yet to be ignorant as regards the fundamental genetic question of whether it was a sediment or a magma. Rocks of similar composition at level D, the amphibolites, are another group well known for the difficulty that exists in assigning them a genesis. The greatest promise of solution to such problems seems, at present, to lie in studies of isotope composition or trace elements; for example, plagioclases reconstituted from marine sediments might differ consistently from plagioclases that crystallize from melts in respect of, say, lead content. Studies of this kind are described by McIntire (1963).

Textures of Dynamothermal Metamorphism

The outstanding feature of dynamothermal textures is a tendency to preferred orientation of non-equant grains. In mica-rich rocks, this leads to schistosity or slaty cleavage, but even in rocks with more nearly equant grains, some degree of orientation can often be distinguished. On the other hand, in a calcite-pyroxene-plagioclase rock, for example, there is no mineral that strongly tends to give a directed texture, so that however anisotropic the stresses imposed, the texture that results is often granular and isotropic, i.e., granofelsic (the grains' invisible internal features—symmetry axes, optic axes etc.—are not discussed here as part of the texture, but see Chap. 25). Table 20-3 gives these generalizations in a little more detail. It must be understood that the table shows only *usual* correlations—a rock can have abundant mica and yet be neither a phyllite nor a schist.

A partial discussion of the physics underlying these correlations is given in Chap. 22. Illustrations are given in Fig. 20-4.

TABLE 20-3 TEXTURES OF COMMON MINERAL ASSEMBLAGES

Mineralogy	*Usual Texture*
abundant fine mica or pyrophyllite	phyllite or slate
abundant coarse mica or chloritoid	foliated schist
abundant amphibole (less often, kyanite or sillimanite)	lineated schist
abundant talc	foliated schist or isotropic talc rock
abundant quartz and feldspar	
1. minor ferromagnesian minerals clustered	gneiss
2. minor minerals not clustered	granofels
abundant calcite	usually granofels, sometimes schist or gneiss by arrangement of minor minerals
abundant pyroxene and feldspar	granofels
abundant garnet	
1. in schistose rocks	foliation disturbed by the garnet porphyroblasts, often leaving augen
2. (less often) in granular rocks	granofels
quartz and feldspar AND ferromagnesians both abundant	layered gneiss or augen gneiss (felsic augen)

THERMAL METAMORPHISM

The dominant trends in thermal metamorphism differ from those of dynamothermal metamorphism in predictable ways:

1. At the lower pressures involved, dehydration reactions occur at lower temperatures (as does the boiling of water, for similar reasons);

2. Loss of carbon dioxide accompanies loss of water;

3. There is less restriction to minerals of high density—both low-density and high-density minerals form freely as the chemistry requires them;

4. On account of (3), rocks under thermal metamorphism rarely achieve the bimineralic simplicity of some dynamothermally metamorphosed rocks.

Other differences are:

5. The scale is quite different: thermal metamorphism is almost always contact metamorphism—whereas the transition from unmetamorphosed rocks to the heart of a dynamothermally metamorphosed region is commonly over 10 km (cf. thickness of continental crust), even intense thermal metamorphism usually fades to imperceptible effects in 1 km and may be localized within 100 m or less of a contact.

6. The time involved is also much smaller: thermal metamorphism is distinguishable from dynamothermal only within c. 4 km of the surface (Fig. 18-2); at such shallow depths, contacts cool quickly in a geological sense—the contact temperature may be maintained for 10,000 to 100,000 years compared with a million or ten million for a dynamothermal episode.

The consequences of trends (1)–(4) are summarized in Fig. 20-5F, G and Table 20-2; these replace the data for dynamothermal metamorphism only beyond stage C in Fig. 20-5. Up to stage C, distinct thermal and dynamothermal trends cannot readily be separated. The relations between the physical states that lead to the different assemblages are shown approximately in Fig. 20-6.

The first difference from dynamothermal metamorphism—more rapid dehydration—leads to almost complete disappearance of amphiboles: tremolite or actinolite forms before stage C, but soon thereafter breaks down directly to pyroxene without highly aluminous amphibole ever forming. Similarly, cummingtonite and anthophyllite may fail to appear or have only brief lives. The micas also break down at lower temperatures, enabling andalusite to form from muscovite, instead of its polymorph sillimanite.

The second difference—loss of carbon dioxide—leads to the appearance of lime garnet and wollastonite, both minerals extremely rare under dynamothermal conditions. On the other hand, in the absence of high pressure, lime-free garnets are not common—the low-density mineral cordierite occurs instead. The extremes of thermal metamorphism—as suffered by xenoliths enclosed in basic magmas, and the first few centimeters of wall rock—are characterized more by new polymorphs than by new reactions: thus tridymite, sanidine and mullite or sillimanite may appear in place of their lower-temperature equivalents (Fig. 20-5G).

As with dynamothermal metamorphism, the upper limit to a tabulation of "solid-state" chemical reactions is the onset of melting. Thus, in quartzofeldspathic rocks, glass can form before or with tridymite, although on account of the short duration and extent of the metamorphism, its volume is not expected to be great. The higher-melting composition Z in Fig. 20-5A does not change mineralogy even at the higher temperatures it is capable of withstanding—in thermal metamorphism, it is calcareous rocks in Field 1 that are most reactive at high temperatures. If silica is abundant, the formation of wollastonite occurs as shown, but if, for once, we turn to silica-deficient rocks, such as limestones at basaltic contacts, we find a profusion of new reactions and minerals—monticellite, melilite, tilleyite, spurrite, rankinite, merwinite, larnite etc. Since these all form from the very restricted list of available components CaO, MgO, SiO_2 and CO_2, the situation is excellently suited for reproduction in the laboratory (Chap. 25). In fact, the establishment of pressures and temperatures for a dozen or more of the reactions apparent at such contacts is one of experimental petrology's great achievements to date. Comparison with behavior predicted from the thermodynamic properties of the minerals involved is also well advanced, so that the marriage of theory, laboratory experiment and petrography is a model for studies on other compositions. (An additional facet, of striking if marginal interest, is the occurrence of the same complex assemblage of minerals in some Israeli limestone horizons apparently heated, not by magmatic activity at all, but by the oxidation of sulfides and graphite in adjacent shales—a kind of progressive autometamorphism of astonishing vigor.)

Textures of Thermal Metamorphism

The outstanding feature of thermal textures is their lack of preferred orientation, compared with dynamothermal textures (Fig. 20-7). Two effects combine to minimize any such tendency: (1) the much-reduced importance of the main flaky or elongated minerals, the micas and amphiboles, and (2) the lower overall pressures, leading to smaller stress differences between one direction and another. It was noted that metamorphism to stage C in Fig. 20-5 is common to both dynamothermal and thermal trends, and that this brings the rock to the condition of a slate or phyllite; but in thermal metamorphism beyond this stage, no schist develops—only granofels. (To the finer-grained varieties, the term *hornfels* is applicable, Chap. 19.)

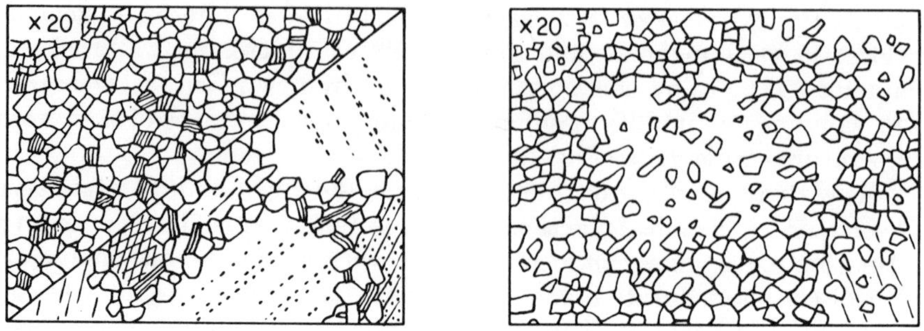

A B

Fig. 20-7 Textures of thermal metamorphism. (A) Granular texture from a fine-grained and a coarse-grained original; (B) granular texture with ill-defined porphyroblasts such as cordierite.

A characteristic feature in hornfels and thermal phyllites is the development of *spots*. These may be more conspicuous in hand specimen than in thin section, where it is often difficult to determine the exact mineral change involved. Some such spots are certainly incipient cordierite crystals, poikiloblastic to the extent of being 60 or 80 per cent inclusions (Fig. 20-7B); but growth of chloritoid or andalusite is probably responsible for the spottiness in a minority of instances.

RECOGNITION OF METAMORPHIC ROCKS

There are three subdivisions of this activity:

1. Recognition that the rock in question has been metamorphosed;
2. Recognition of its premetamorphic nature—for example, mud, lava flow, granite, arkose or tuff;
3. Recognition of the kind and degree of metamorphism suffered.

Only the first two parts are touched on here; further remarks on part (2) and a discussion of part (3) are given in Chap. 21.

As with igneous rocks and sediments, decisions about genesis are based more on associations than on the characteristics of individual hand specimens. If the rocks of a region measured in tens or hundreds of miles are crystalline but thoroughly inhomogeneous, the *region* has been metamorphosed and, unless there is evidence to the contrary, all rocks in the region are also considered metamorphosed. [Evidence to the contrary would be, for example, a boundary cutting across the regional structure and enclosing homogeneous crystalline rocks (possibly igneous) or stratified detrital rocks (probably sedimentary).] Although it is necessary in a sedimentary sequence to watch for isolated lava flows or tuffs, it is not necessary to watch for isolated metamorphic units—the nature of the metamorphic process precludes this possibility except for local dynamic metamorphism, p. 232.

Of course, there is difficulty in deciding which rocks to call *metamorphosed* and which simply *indurated* etc., and there is also difficulty, when one has found and recognized, for instance, a charnockite or a granitic gneiss, in deciding whether it is a product of solid-state metamorphism, partial fusion or total fusion; but these are difficulties in recognizing degrees of metamorphism. Genuine uncertainty as to whether a rock has been well metamorphosed or not at all is rare. It may exist, for example, where:

1. Basalt intrudes mudstone: because the resulting hornfels can resemble basalt, it may be difficult to say of an isolated outcrop whether it lies within or just outside the igneous perimeter; this problem can usually be solved by making a thin section;
2. Unmetamorphosed granite intrudes well metamorphosed granite: because granite is so insensitive to metamorphism, a small amount of weathering may mask the crucial differences;
3. A concordant mylonite is suspected.

Recognition of Premetamorphic Nature

A powerful regional indicator has already been intimated: *where units are large and homogeneous they are probably meta-igneous; where they are small and inhomogeneous they are probably metasedimentary.* The number of *ifs* and *buts* to which this remark is subject should not be allowed to mask its fundamental value; the controls on sedimentation are complex and locally variable to an extent that controls on igneous processes are not (see Chap. 13). In fact, when one looks for the most locally-variable igneous rocks, one turns to the layered sills and lopoliths where gravity-settling, i.e., sedimentation, has been at work. Of course, both *large* and *homogeneous* are matters of degree; but where an assemblage of small, highly-contrasted units has a discordant boundary against an assemblage of larger, less contrasted units, it is proper to suspect an igneous contact.

Beside the obvious objections that some sediments are bulky and invariant and some igneous assemblages complex in detail, interesting possibilities exist that metamorphism may increase or decrease a region's homogeneity. Some authorities believe that the homogenizing effect of intense shearing (Fig. 20-8A) is sufficient to turn layered sediments into bodies as homogeneous as metamorphosed stocks. Others regard shearing as a power-

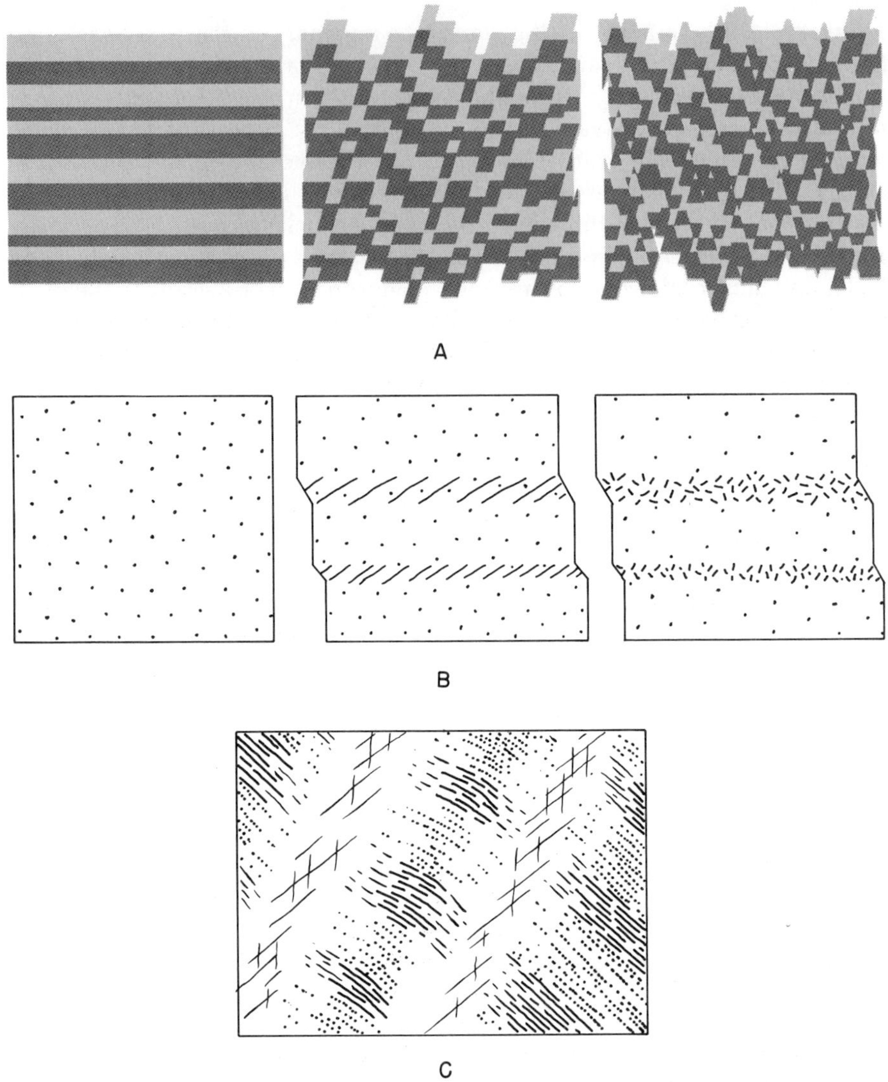

Fig. 20-8 Effects of shearing on homogeneity. Inhomogeneity may be reduced by shearing (A) or enhanced by shearing (B). In (C), the effect of shearing is to reduce the layer contrasts and to promote a new layering parallel to the shear zones.

ful agent for producing contrasted layers from homogeneous rocks (Fig. 20-8B). Of course the two views are not incompatible and differ mainly in the size relation of the rock unit observed to the rock unit within which shearing is intense. Figure 20-8C shows shearing in process of eliminating one set of layer contrasts and generating another. The actual power of these processes is in dispute; Chap. 22 attempts to introduce the factors on which a quantitative estimate must be based.

Textural Indicators. The following small-scale features, if sufficiently preserved, can reveal the igneous or sedimentary origin of a metamorphosed rock:

lava "pillows"	cross bedding
fossils	graded bedding
mud cracks	load structures

However, to use, for example, cross bedding as a criterion, it is insufficient merely to be familiar with cross bedding in sediments; one must also understand how closely structures in metamorphosed igneous rocks can *resemble* cross bedding. Observers who lack this double experience should be aware that, when they reach a conclusion, they have no means of estimating the probability of its being wrong, and that conclusions subject to this uncertainty are hardly better than no conclusion at all.

More common than the presence, or suspected presence, of any of the above features are rocks that are divisible into *lumps* and *matrix*. The lumps may be as large as boulders or as small as sand grains. The first thing to discover is whether they are single crystals or aggregates. Single crystals may be porphyroblasts, phenocrysts, crystals in tuff, detrital crystals or authigenic crystals (i.e., crystals which grew in their present situation but in a sedimentary environment rather than during metamorphism). Features that may lead to a decision between these possiiblities are:

Mineral identity;

Shape and size (including augen form, Figs. 20-1B and 20-4E);

Fresh or altered condition;

Cohesion with matrix;

Nature of matrix.

However, metamorphosed rocks containing isolated crystals where no sign is present to indicate whether the rock was a greywacke or a tuff are quite common.

Where the lumps are polycrystalline, their degree of rounding should be noted. Round lumps can be pebbles, concretions, amygdules, volcanic lapilli or bombs (much less probably, they might be orbicular structures of plutonic origin or fragments that have become rounded by plastic deformation); the lumps' mineralogy, and resemblance or difference from lump to lump, will usually discriminate between these possibilities. Angular lumps are likely to be volcanic debris, xenoliths or sediments fractured by nonmagmatic processes; sometimes the mineralogy is decisive, but where the composition is in the range common to igneous rocks and sediments, metamorphism can make a decision by "internal evidence" impossible. In any case, the shape and marginal relations of the lumpy rock unit must be studied if possible; they will often provide an indication where the petrological evidence is inconclusive.

Even in rocks that are not conspicuously lumpy, there may be textural features inherited from the premetamorphic state. Ophitic texture, for example (Figs. 5-8A and 21-6E) may still be recognizable even where the pyroxene is altered to matted amphibole shreds. One might expect that fine textures would be more readily obliterated by metamorphism than coarse textures, but this is not necessarily so: the extent of rearrangement is probably much influenced by the water present, and a compact anhydrous rock such as trachyte or basalt may retain its fine texture while more porous rocks nearby are extensively rearranged.

Compositional Indicators. The basis for recognizing premetamorphic nature from composition appears in Fig. 13-6—most igneous compositions can be imitated by sediments but some sediment compositions cannot be imitated by igneous rocks. In particular, rocks excessively rich in silica, alumina or lime can be diagnosed as not igneous (excepting, as usual, the carbonatites). In mineral terms, it is convenient to distinguish:

1. Minerals which indicate nonigneous nature if present at all;
2. Minerals which indicate nonigneous nature if present in sufficient quantity.

It is prudent to qualify *nonigneous:* the metasomatism that goes on at the margin of some igneous bodies generates nonigneous compositions such as china clay (mainly kaolinite). Consequently, if kaolinite is included in list (2) its abundance in a rock shows only that the rock is not a *straightforward* igneous rock—it still may be a metasomatized igneous rock (or a metasediment or a nonmagmatic vein). A second restriction arises from the possibility of contaminated magmas: if aluminous xenoliths are thoroughly digested, a magma may well be emplaced while sillimanite crystals are floating in it (presumably undergoing reaction but not wholly absorbed). In such a case, again, the rock is not a *straightforward* igneous rock, but it has nevertheless more igneous character than any other. The presence of one of the minerals listed below therefore indicates only that the rock in question is not a normal igneous rock; it is incumbent on the observer to decide, using field relations as well as mineralogy, whether metasomatism or contamination or a wholly nonigneous genesis is indicated. These provisos apply to both of the columns in Table 20-4.

A slightly different kind of indicator is the Fe/Mg ratio in a rock: if a magnesian rock is dead white, it is probably not igneous. Even in peridotites, which are the most magnesian of possible igneous rocks, there is enough iron to give the rock a color, whereas dolomite can accept magnesium and reject iron to such an extent that, on metamorphism, the amphibole or pyroxene that forms is milky white.

TABLE 20-4 INDICATORS OF NONIGNEOUS ORIGIN

Minerals That Indicate Nonigneous Nature If Present At All	*Minerals That Indicate Nonigneous Nature If Present In The Quantity Stated*		
staurolite	quartz	if over 50%	
chloritoid	muscovite	25%	
cordierite	lime garnet	25%	except in eclogite
andalusite kyanite	calcite	25%	except in carbonatite
sillimanite	kaolinite montmorillonite pyrophyllite	25%	
talc			
wollastonite			

CHAPTER TWENTY-ONE

REACTION CONDITIONS IN METAMORPHISM

The question examined in this chapter is: why do not the reactions in Tables 20-1 and 20-2 always occur in the same order? To answer this question we must consider the quantities that determine whether a reaction shall proceed or not. These are principally the change in volume on reaction, the heat absorbed or evolved and the extent to which the reagents are involved in solutions, liquid or solid.

DRY REACTIONS

It was stated in Chap. 3 that one solid arrangement is apt to change to another when the *free energy* of the second arrangement is the smaller. The example discussed there was the quartz–coesite transition but the same is true of polymineralic transitions such as the change from anorthite + enstatite to garnet + quartz mentioned in Chap. 20—see Fig. 21-1. Diagram A shows the change in free energy with temperature at a fairly high pressure P_1—the free energy of both forms falls as temperature rises but that of (an + en) falls faster, so that above T_1 it is the stable form (lower free energy). Diagram B shows the change with pressure—the free energy of both forms rises, but again that of (an + en) changes faster so that above

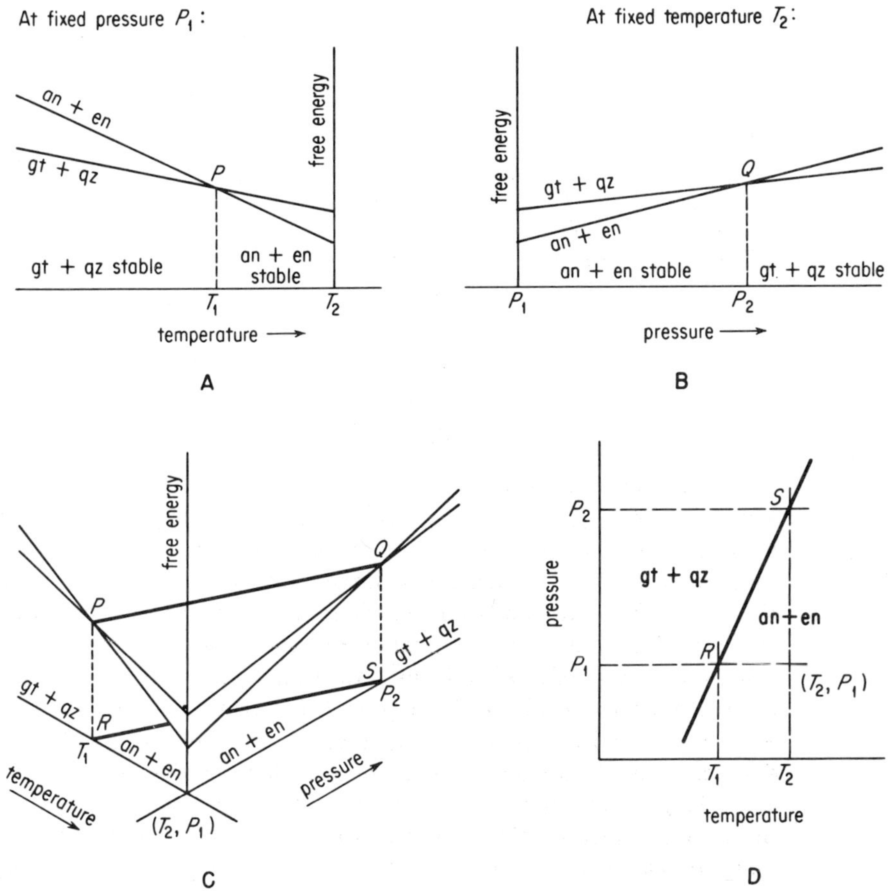

Fig. 21-1 Change of equilibrium with pressure and temperature.

P_2, (gt + qz) is the stable form. The change of free energy with both pressure and temperature is shown in diagram C, in which it appears that the field where (gt + qz) is stable joins the field where (an + en) is stable along a sloping line PQ. The projection of PQ in the pressure-temperature plane is shown at RS in diagram D.

The slope of RS is clearly determined by the sharpness with which the lines converge in the two "walls" of diagram C. If the two forms differ greatly in volume, the two lines in the pressure-energy plane will have very different slopes and P_1P_2 will be very short. Similarly if the two forms differ greatly in entropy, the two lines in the temperature-energy plane will have very different slopes and T_1T_2 will be very short. If we write ΔV for the volume difference per mole and ΔS for the entropy difference per mole between the two forms, in fact, $P_1P_2 \propto 1/\Delta V$ and $T_1T_2 \propto 1/\Delta S$. These serve

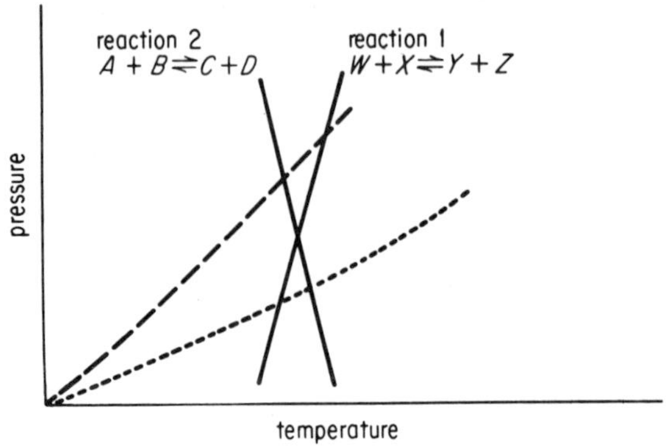

Fig. 21-2 A change in the order of occurrence of two reactions.

as a reminder of Clapeyron's Equation in the form

$$\text{slope of } RS = \frac{P_1P_2}{T_1T_2} = \frac{\Delta S}{\Delta V} = \frac{S_{\text{products}} - S_{\text{reagents}}}{V_{\text{products}} - V_{\text{reagents}}}$$

The slope may be positive or negative; heating encourages high-entropy forms, so that $S_{\text{products}} - S_{\text{reagents}}$ is always positive for changes provoked by heating, but $V_{\text{products}} - V_{\text{reagents}}$ may have either sign. When the product formed by heating is of greater volume than the reagents, the slope is positive; in the less common situation where the product formed by heating is of smaller volume, the slope is negative (the transition ice ⟶ water is the most common example of this kind; the transition andalusite ⟶ sillimanite is another). In both situations, the slope of RS is a rather large—over 20 bars/°C—for most solid reactions among silicates, or 2 kb/100°C. This means, for example, that to keep a low-temperature form stable, we should have to raise the pressure on it by 2 kb for every 100°C rise in temperature. Now at most points in the earth's crust the pressure seems to rise less than this—at many points only 1 kb/100°C. Thus at greater depths, higher-temperature assemblages tend to occur even though they may be a little less dense than a lower-temperature equivalent.

Figure 21-2 shows that this behavior is a possible explanation for some variability in order of reaction. At a site where the geothermal gradient is steep in degrees/km (dotted line in figure), reaction (1) occurs "before" reaction (2), whereas at a site where the geothermal gradient is less steep (dashed line), reaction (2) precedes reaction (1). Bearing in mind that measured geothermal gradients have varied by a factor of 25, we realize that it would not be surprising to find some reversals of order on this account.

REACTIONS INVOLVING FLUIDS

There are two reasons why these are more variable than "dry" reactions: the volume of most fluids is more variable with pressure than the volume of most silicates, and, in geological situations, doubt exists as to how much of any fluid evolved will be retained at the reaction site. Two extreme situations can be envisaged: (1) all fluid evolved escapes from the reaction site, and (2) all fluid evolved remains at the reaction site. In situation (1), change of pressure at the reaction site will not affect the energy of the fluid evolved at all; hence, in Clapeyron's Equation the volume of fluid must not be included in the calculation of ΔV. On the other hand in situation (2), where the fluid is retained, the volume of fluid must be included in the calculation of ΔV and will often be the dominating term. Consider a reaction in which water is evolved, such as:

$$\begin{array}{llcll} \text{chlorite} & + \text{ quartz} & \longrightarrow & \text{garnet} & + \text{ water} \\ 7.8\text{ g} & 1.7\text{ g} & & 8.5\text{ g} & 1.0\text{ g} \\ \underbrace{2.6\text{ cc} \quad 0.7\text{ cc}}_{3.3\text{ cc}} & & & 2.5\text{ cc} & \ldots \end{array}$$

If the water escapes entirely from the reaction site, the proper value for ΔV is -0.8 cc, but if the water is wholly retained at the reaction site, its volume must be taken into account. The volume of 1 g of water under geophysical conditions is shown in Fig. 10-1; it is more or less always greater than 1 cc, so that the products have greater volume than the reagents, but the difference ΔV changes from 10 cc or so at low pressure and high temperature to only 0.4 cc under dynamothermal conditions. Points at which the reaction is in equilibrium therefore lie on curve (i) or curve (ii) in Fig. 21-3, or between the two curves if water escapes to an incomplete extent from the reaction site.

By contrast, the reaction

$$2\text{ actinolite} + \text{epidote} + 2\text{ chloritoid} \longrightarrow 3\text{ hornblende}$$

proceeds, if it proceeds at all, with negligible change in volume or water content; it is a change driven principally by the tendency toward increasing solid solution as temperature rises. It will therefore be insensitive to changes in pressure or water behavior (Fig. 21-3, curve iii). Thus we see that if water escapes freely, chlorite will go to garnet before actinolite goes to hornblende; if the water is confined, the order will be reversed.

In thermal metamorphism, where carbon dioxide as well as water escapes, there is additional variability. The temperature of the wollastonite reaction (calcite + quartz $\longrightarrow$ wollastonite + carbon dioxide) ranges from 400°C to 900°C according to the ease with which carbon dioxide escapes, while the temperature of dehydration reactions is controlled more by the behavior of the water. The ease of escape of one fluid is to some extent independent of the ease of escape of the other, so that in some situations decarbonation is easy and dehydration not, while in others dehydration is easy and decarbonation not.

A third fluid that influences metamorphic reactions is *oxygen*. The idea that free oxygen gas should be present at the site of a metamorphic reaction

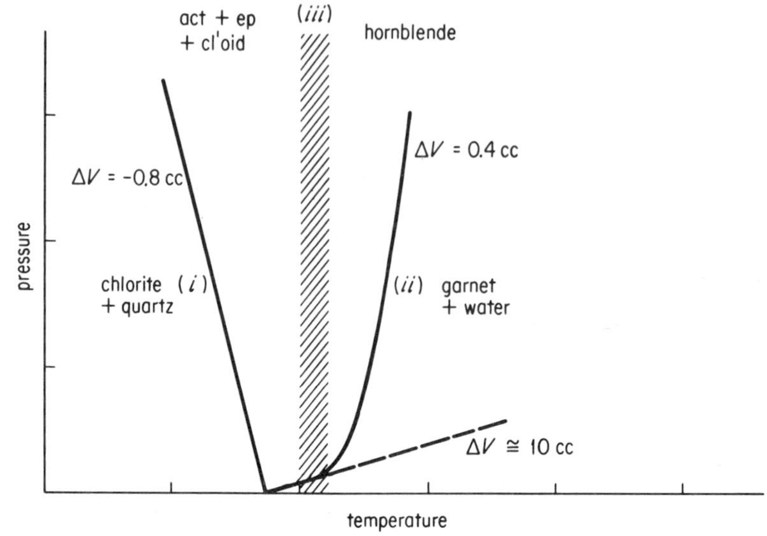

Fig. 21-3 Equilibria involving water. (i) The reaction boundary, chlorite + quartz ⇌ garnet + water, if the water generated is wholly lost from the reaction site; (ii) the same reaction boundary if the water generated is wholly retained at the reaction site; (iii) the reaction boundary for the conjectural reaction, actinolite + epidote + chloritoid ⇌ hornblende, in which the volume change is very small.

may seem, at first sight, as improbable as that water should be present, but note the following points:

1. The air contains both oxygen and water (with nitrogen etc.);
2. The sea also contains both oxygen and water, only the proportions being different;
3. If conditions are such that the liquid–gas distinction does not exist, the fluid phase should still be expected to contain both oxygen and water, in some proportion.

In fact, in extension of the discussion of mixtures of water and carbon dioxide in Chap. 20, it is always permissible to ask of a metamorphic reaction site, "What is P_{H_2O} at the site? What is P_{CO_2}? *And* what is P_{O_2}?"

Being satisfied that this is a permissible question, we may ask whether it is one of interest to petrologists. The water pressure affects dehydrations and the carbon dioxide pressure affects decarbonations, but what kind of reaction is affected by the oxygen pressure, to make this quantity an important one? Briefly, any reaction involving iron, or any other metal of variable valency, can be affected by oxygen pressure. In fact, the reader should recall the section on iron minerals and oxidation potentials in Chap.

15, and notice that the oxidation potential and the oxygen pressure in a fluid, while not the same, are closely related quantities. The relation between them derives from the equation:

$$2\ H_2O \rightleftharpoons O_2 + 4\ H^+ + 4\ e$$

It follows that

$$\frac{[O_2]\cdot[H^+]^4\cdot[e]^4}{[H_2O]^2} = \text{constant at fixed } P,\ T$$

For a fluid of given acidity and water pressure, which fix $[H^+]$ and $[H_2O]$, this equation in activities means that, as $[e]$ goes down, $[O_2]$ goes up; and although the activity of oxygen is not exactly proportional to its pressure, the two measures increase or decrease together.

Because of the abundance of iron-bearing silicates, and the wide stability range of the iron oxides, reactions of the form

$$\text{iron-rich silicate} + \text{oxygen} \rightleftharpoons \text{iron-poor silicate} + \text{iron oxide}$$

are numerous. It was mentioned in connection with laterites that on a tropical land surface, oxidation generates hematite, Fe_2O_3, with such vigor that all iron silicates are decomposed. In metamorphism, magnetite is the more common oxide and a typical reaction is

$$\text{hypersthene} + \text{water} + \text{oxygen} \rightleftharpoons \text{talc} + \text{magnetite}$$

Here the juxtaposed products, one snow-white and the other jet-black, powerfully demonstrate the way in which Mg and Fe, almost inseparable when both are divalent, can be completely separated by partly oxidizing the iron.

REACTIONS INVOLVING SOLID SOLUTIONS

The behavior of solid solutions of silicates obeys the same laws as that of liquid solutions or liquid mixtures, whether silicates or not. Figure 21-4 shows the effects that various solutes have on equilibria. The diagrams, all of the same type, differ only in the scales. Diagrams B and C were mentioned in Chap. 8; diagrams D and E are closely similar but illustrate two important metamorphic transitions.

The behavior in diagram D rests on the fact that manganese substitutes freely for magnesium in the garnet structure but not in the chlorite structure, i.e., manganese mixes with garnet-borne magnesium but not with chlorite-borne magnesium, as albite mixes with liquid quartz but not with solid quartz. The effect is to lower the transition point for manganiferous garnet below that for manganese-free garnet. In other words, chlorite dis-

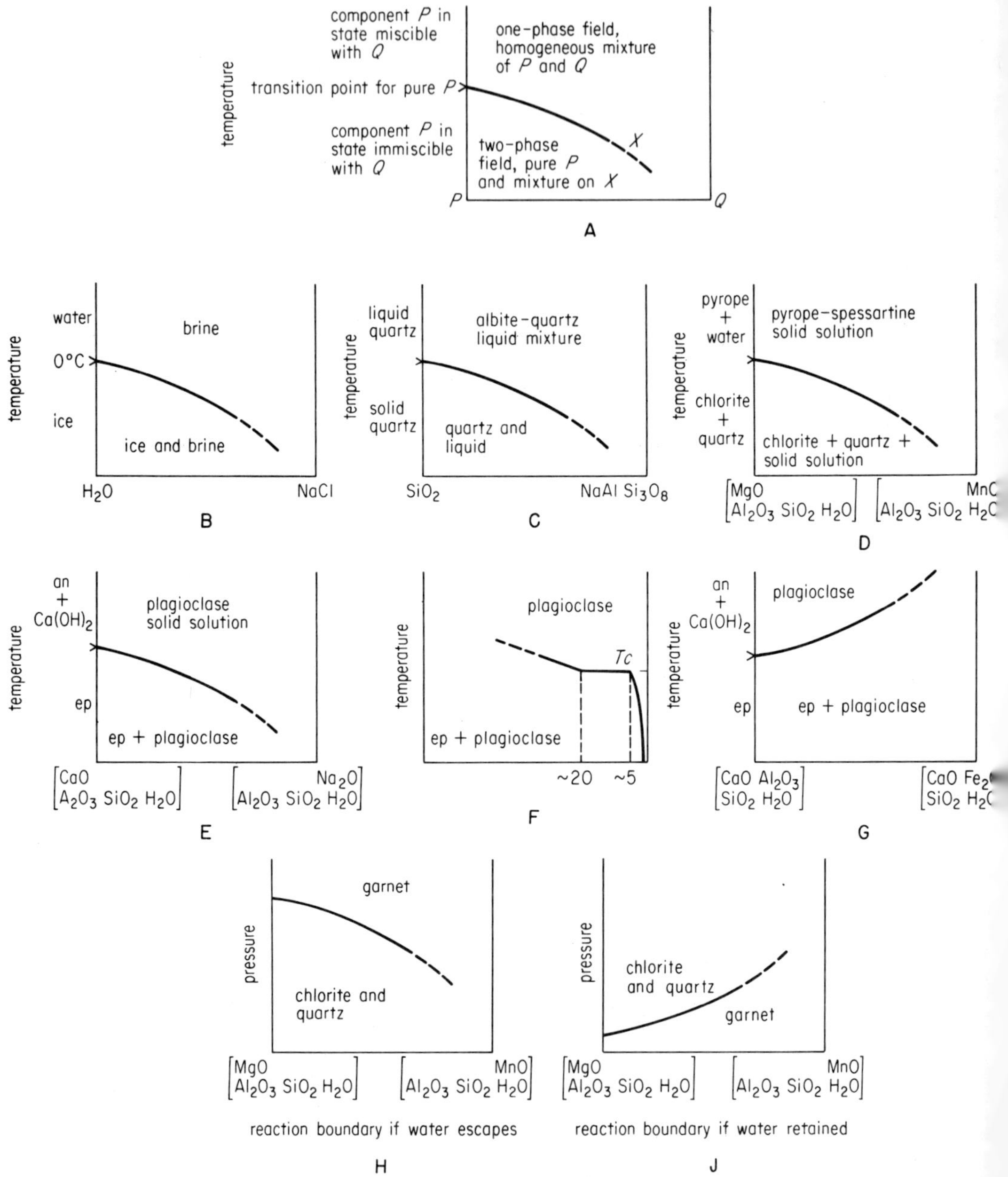

Fig. 21-4 Some effects of solutions on equilibria. Diagrams A through G are drawn at constant pressure and diagrams H and J at constant temperature. (Diagram D from Ramberg, 1952; diagrams E and F from Christie, 1962.)

appears from manganese-rich terrains at a lower temperature than that at which it disappears from manganese-poor terrains, insofar as the disappearance is indeed due to conversion to garnet. (Field work suggests that garnet's appearance in manganese-rich terrains also occurs at a lower temperature than in manganese-poor terrains, but the reason is different. Using heating in the diopside–albite–anorthite system, Fig. 9-2, as an analogy, we note that the temperature at which the last solid disappears is affected by the ratio of feldspar to diopside in the initial mixture, but the temperature at which the first liquid appears is affected only by the ratio of albite to anorthite. Similarly, the temperature at which garnet first appears must depend on the composition of the unknown manganese-bearing mineral it forms from, but not on the manganese mineral's abundance.)

The behavior in diagram E rests similarly on the fact that NaSi substitutes for CaAl in plagioclase, but not in epidote. Actually the petrological interest of the system lies more at the albite end, diagram F, where, because of a peculiarity in the internal structure of plagioclase, the curve has a flat portion. The effect is that, during diagenesis or very early in metamorphism, epidote forms in presence of almost pure albite; on heating, a very little epidote dissolves in the albite (with separation of $Ca(OH)_2$ which usually converts to $CaCO_3$); then rather suddenly at T_C, the reaction surges until the plagioclase has been brought up to about 20 per cent anorthite; thereafter the anorthite percentage rises more steadily with temperature until all the epidote is used up. Since epidote and plagioclase are so common, and the composition of the plagioclase is recognizable optically without any chemical operations, this behavior is a convenient guide to the history of a rock: if epidote and albite coexist, the rock has adjusted to a temperature below T_C, while if epidote and some plagioclase more calcic than An_{20} coexist, the rock has adjusted to some temperature above T_C. Two conditions must be added:

> **1.** It is only when in presence of epidote that the plagioclase composition is indicative;
>
> **2.** The temperature T_C and the width of the interval are somewhat variable with total pressure, water vapor pressure and the amount of Fe^{+3} in the epidote.

The last-mentioned effect is like that of Na in the plagioclase, except that the phase involved in the solid solution is the low-temperature phase. The corresponding diagram is shown at G. The reader should satisfy himself that the temperature of the chlorite–garnet transition can similarly be raised by altering the Mg:Al ratio, which is assumed to be exactly 3:2 in diagram D.

Up to this point, effects have been studied with respect to temperature at constant pressure. The alternative is equally important: diagrams H and J show the effect of composition on reaction pressure, on the alternative as-

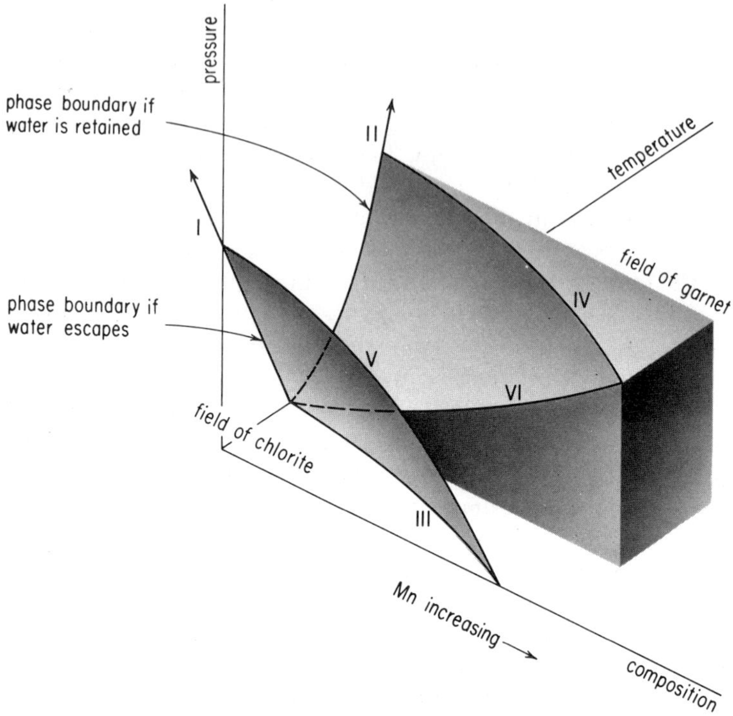

Fig. 21-5 Change of equilibrium with four variables. The side wall of the diagram (on pressure-temperature axes) repeats Fig. 21-3. Curves (iii) and (iv) are as in Fig. 21-4D, at two different constant pressures. Curve (v) is as in Fig. 21-4H where water escapes and garnet is favored by high pressure. Curve (vi) is as in Fig. 21-4J—the less common situation where water is wholly retained and chlorite is favored by high pressure.

sumptions that the water evolved escapes or is wholly retained. Figure 21-5 combines all the effects discussed in this section.

There is almost no reaction in petrology whose occurrence is not affected by some solution effect, as well as by temperature and pressure; it is therefore fortunate that a single rule covers all situations: *any component present at a reaction site enlarges the stability field of the phase it dissolves in most freely.* As far as the magnitudes of the effects go, we notice that an extensive solid solution, such as that in the plagioclases, may affect the balance of a reaction by as much as a change of 100°C or 2 kb. Some textures that may result from incomplete transition are shown in Fig. 21-6; when such textures are seen, it must be remembered that they may have developed because of change of temperature *or* change of pressure *or* change in the availability of some component.

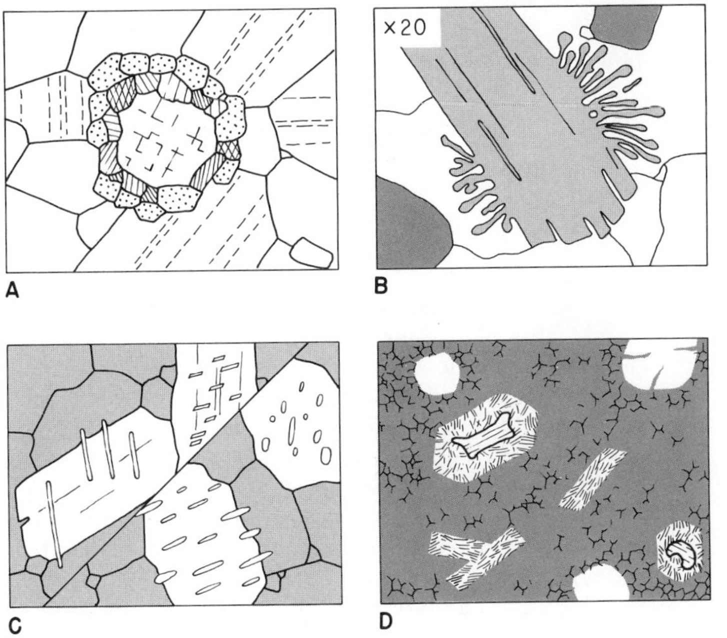

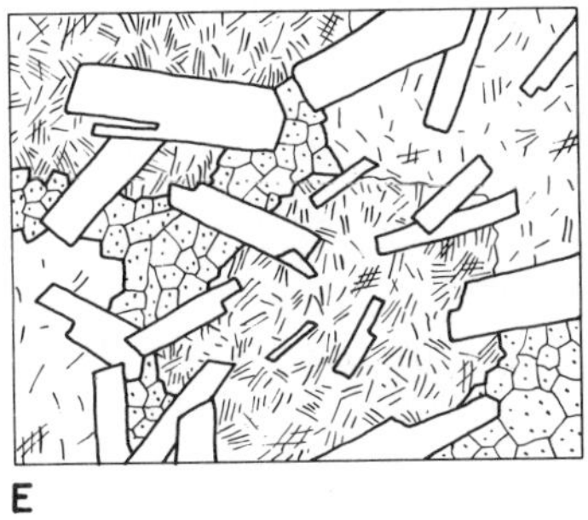

Fig. 21-6 Reaction textures. (A) Reaction rim; (B) vermicular intergrowth (called *myrmekite* when components are quartz and feldspar); (C) projecting inclusions in pyroxene and in perthite; (D) pseudomorphs after phenocrysts (cf. Fig. 5-4H)—two pseudomorphs carry relics; (E) recrystallized ophitic texture (cf. Fig. 5-8A), with fine matted amphibole in place of pyroxene.

RECOGNITION OF METAMORPHIC INTENSITY

Is it possible, by observing the minerals now present, to say with what severity a rock has been metamorphosed? To some extent it is, but the procedure is not as simple as Tables 20-1 and 20-2 suggest. Complications arise for the following reasons.

1. A rock's past is not simply a single former condition. It is misleading to say "The rock now at the surface was formerly at 500°C and 3 kb." A rock now at the surface has been in an endless series of former conditions (see Fig. 2-1), several of which may be recorded simultaneously by features now preserved. Consider, for example, a sediment with thin alternating sandy and muddy layers that is metamorphosed until garnet begins to form and is then exposed by erosion. An observer may find:

a. Detrital kyanite in the sandy layers, recording the metamorphic condition of the sediment's source rock;
b. Detrital chlorite which did not react sufficiently quickly to form garnet during the metamorphism—this records the most recent sedimentary episode in the rock's history;
c. Garnet, recording the most recent metamorphic climax;
d. Retrograde chlorite, recording the rock's reversion to milder conditions since the climax.

Clearly, it is insufficient merely to note a mineral's presence, find at what level in Table 20-1 or Table 20-2 the mineral appears and then read the corresponding physical condition from Fig. 20-6. This procedure has to be supplemented by the enquiry, "About what stage of the rock's history does this mineral inform me?" Three ways of answering this enquiry are:

a. Observe the textures and structures in the rock for signs of abrasion and reaction (see Fig. 21-6);
b. Compare what one mineral seems to indicate with what others do;
c. Compare the history as indicated by one hand specimen or rock type in a region with what is indicated by other rock types immediately adjacent.

To emphasize point (c), recall that igneous and sedimentary environments are best reconstructed by study of a complete assemblage of rock types and field associations; metamorphic environments require no less. In fact, if a region is monotonous in lithology, its metamorphic history probably cannot be worked out with precision; our most detailed knowledge of the progress of metamorphism comes from regions where a rock suite is persistent and uniform along strike but has chemically-contrasted types closely spaced across strike.

2. A second reason that a simple record of minerals present does not permit us to proceed directly to a good estimate of a level of metamorphism in Table 20-1 or Table 20-2 is that a mineral's range of stability is variable. It depends firstly on the solid solutions involved and secondly on what is available in the environment for reaction. In fact, it is as important to notice mineral pairs as to notice individual minerals. As an example, suppose we find talc in a rock and can satisfy ourselves that it is a prograde, not a retrograde, product: then its presence shows simply that metamorphism did not reach the level at which talc dehydrates to enstatite and quartz. However, if we find talc *touching olivine*, this conjunction shows that metamorphism did not reach the level at which talc reacts with olivine to give pyroxene, and this is a more severe temperature limit. Thus, in general, the noting of minerals present is only a means to an end. What has to be established by observation is a list of reactions that have occurred and another list of reactions that are chemically possible but have not occurred. Two such lists can be used in conjunction with Table 20-1 or Table 20-2 to delimit a level of metamorphism.

3. Even if it is established that, for example, chloritoid has given way to staurolite while in a neighboring rock hornblende has not reacted with calcite, still the range of possible conditions is rather wide—see Zone *D* in Fig. 20-6. Even this long zone fails to show the full range of possibilities remaining, on account of the water vapor condition to which the diagram is subject. Both reactions

$$\text{chloritoid} + \text{quartz} \rightleftharpoons \text{staurolite} + \text{garnet} + \text{water}$$

and

$$\text{hornblende} + \text{calcite} + \text{quartz} \rightleftharpoons \text{pyroxene} + \text{plagioclase} + \text{water} + CO_2$$

generate volatile compounds and in this respect resemble the chlorite–garnet transition already discussed (Fig. 21-3). The limits drawn for mineral assemblage *D* in Fig. 20-6 are of the same type as curve (ii) in Fig. 21-3; but if the volatiles escape freely from the reaction site, the limits for mineral assemblage *D* would be more like curve (i) in Fig. 21-3, that is, much further to the left. The freedom with which volatiles *were* able to escape is not now observable. The result is that even when favorable observations have been made, still the temperature and pressure may not be indicated within 200°C and 5 kb.

In view of these rather extensive complications, it is well to note two favorable aspects. The first has been mentioned already: although one or two reactions fail to indicate the metamorphic history of a region with precision, where different rock types are closely associated it may be possible to observe many reactions (including dehydrations, isovolumetric changes and so on) and thus to reach a more precise conclusion. Secondly, although absolute measures are elusive, comparisons within a region are subject to far fewer uncertainties; and for many purposes, such as understanding the genesis of a given ore, knowledge of the gradients of temperature, pressure and other variables can be as important as knowledge of their absolute values.

Facies and Isograds

The subdivision of the pressure-temperature field into zones or pigeonholes as in Fig. 20-6 has been done in more formal ways by several authorities (see Fig. 21-7), but there is some doubt among petrologists as to exactly how the boundaries should be defined. Possibilities are:

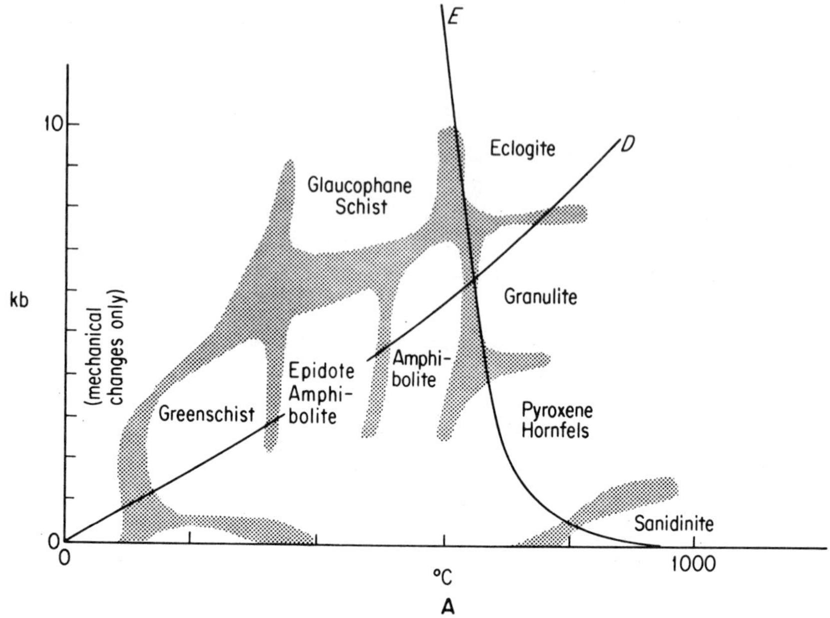

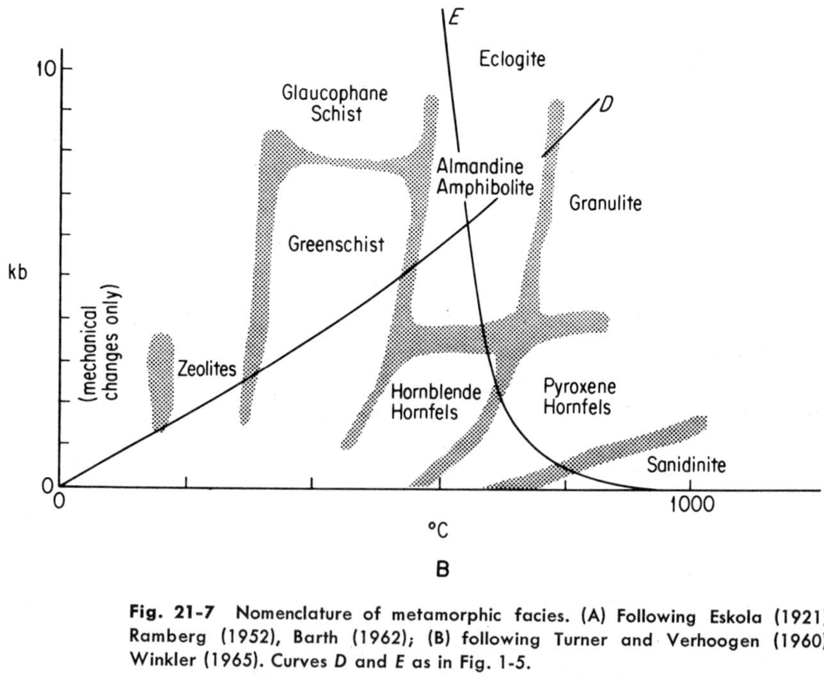

Fig. 21-7 Nomenclature of metamorphic facies. (A) Following Eskola (1921), Ramberg (1952), Barth (1962); (B) following Turner and Verhoogen (1960), Winkler (1965). Curves *D* and *E* as in Fig. 1-5.

1. The boxes could be defined by specified pressures and temperatures thus: "Box 1: 500–700°C, over 5 kb." It would then be impossible ever to say, "I see that this rock belongs to box 1"—the assignment of a rock to a box would always be a hypothesis.

2. The boxes could be defined by a few specific reactions thus: "Box 1: rocks in which staurolite has converted to sillimanite but muscovite has not yet converted to sillimanite, and any other rocks metamorphosed to the same degree." Under this system, some rocks could be classified unarguably by observation but many could not.

3. The boxes could be defined by many specific reactions thus: "Box 1: rocks in which staurolite has converted to sillimanite but muscovite has not, plus rocks in which calcite has converted to pyroxene but biotite has not, plus . . . ." Under this system we should lose assurance that all rocks in one box had suffered the same degree of metamorphism.

Geologists are undecided between system (2) and system (3) as well as between the two systems in Fig. 21-7. Consequently, no more will be said about the definitions and names involved—the names need detailed study for exact use and, for inexact use, simpler terms are already available. In many regions, phrases like "severe dynamothermal metamorphism" or "higher pressure than further west" express all that can be deduced with confidence.

The use of the term *facies* for the subdivisions in Fig. 21-7 is different from its use in sedimentology (e.g., "littoral facies of the Bovey Beds"). A *facies* is a part of a larger whole that differs from other parts in appearance. In metamorphism, the larger whole, of which the Granulite Facies (Fig. 21-7), for example, is a part, seems to be the set of all metamorphic rocks. Consequently, the Granulite Facies is not a certain set of conditions; it is a set of rocks that look as if they have suffered those conditions—a facies is not a pigeonhole but the rocks in the pigeonhole. This point has little practical significance but keeps sedimentary and metamorphic usage in conformity and so helps one geologist to understand another.

A comparable system for recording observations involves isograds. An *isograd* is a line on a map that divides locations where a particular mineral does occur from locations where it does not; examples are the garnet isograd, staurolite isograd and sillimanite isograd. Since these minerals require successively greater intensities of metamorphism, the isograds tend to form closed loops, one inside the other, like contours (Fig. 21-8). A record in terms of isograds is purely observational and unambiguous, since no element of conjecture enters except what is involved in drawing a smooth contour between the scattered points at which observations can be made. The drawing of isograds is thus a very useful step in the description of a metamorphosed region. It is not so useful, however, if one's purpose is to compare the

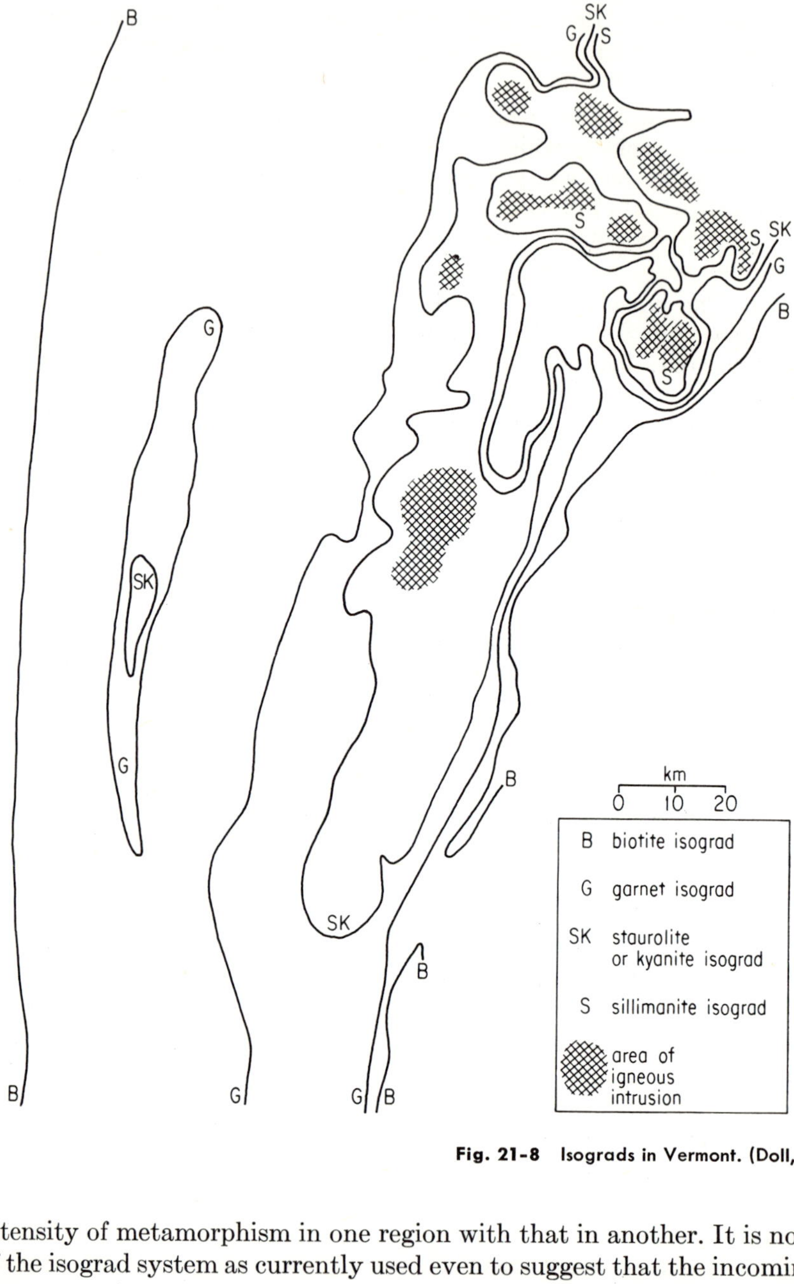

Fig. 21-8 Isograds in Vermont. (Doll, 1961.)

intensity of metamorphism in one region with that in another. It is no part of the isograd system as currently used even to suggest that the incoming of, say, garnet in one region is due to the same physical conditions as the incoming of garnet in another region; whereas some such intention does underly the facies classification. Whether one maps isograds or attempts to recognize facies, therefore, depends on one's purpose; an exhaustive study may involve both.

The isograds most commonly used are those of chlorite, biotite, garnet, staurolite, kyanite and sillimanite. The last three can be recognized only in aluminous rocks, and the first three are usually mapped using aluminous rocks also. Since biotite or phlogopite can form from dolomite in ferruginous limestones, we should distinguish the biotite isograd based on aluminous rocks from a possible biotite isograd based on carbonate rocks, and similarly for chlorite and garnet. This adds precision without marring the pure observational character of the terms.

The use of isograds has hitherto been mainly in mapping regional metamorphism. Concentric mineral zones are just as common in local metamorphism but, for historical reasons, the term *isograd* is not often applied in this circumstance.

CHAPTER TWENTY-TWO

MOVEMENT OF MATERIAL IN METAMORPHISM

Metamorphic reactions were discussed in Chap. 21 in isochemical terms. We assumed that both reagents and products were at a reaction site and neglected the movement of material which reaction necessarily involves. These movements are the next topic of interest; we shall inquire in turn what distances, what mechanisms and what driving forces are involved

MAGNITUDE OF THE MOVEMENTS

Let us begin with an example where the distance material moves is easily established. Figure 22-1 (cf. Fig. 21-6A) illustrates a reaction in which highly aluminous, iron-free plagioclase is reacting with aluminum-free ferriferous hypersthene to give two minerals, both intermediate in aluminum content and in iron. Apparently, iron and magnesium move outward and calcium and aluminum inward; thus plagioclase is turned to garnet and hypersthene to hornblende. The further the reaction proceeds, the further the ions have to travel; for the reaction to continue beyond the stage illustrated, the effective journey of each ion is *PQ*, which in the example figured

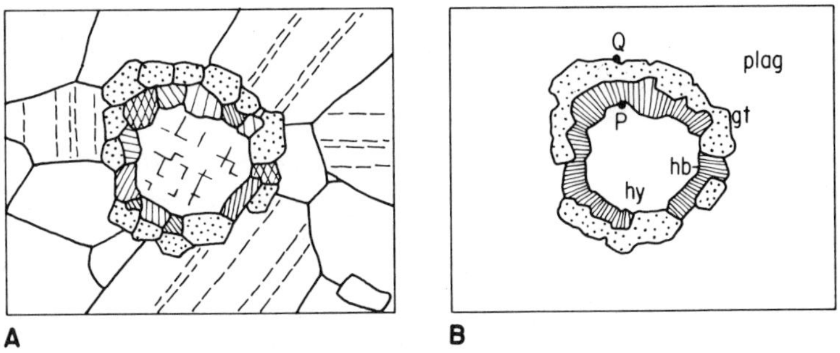

Fig. 22-1 Movement of components during reaction. (A) Reaction rim redrawn from Fig. 21-6A; (B) [stippled] zone where plagioclase has turned to garnet by loss of Ca, Al and gain of Fe, Mg; [hatched] zone where hypersthene has turned to hornblende by gain of Ca, Al and loss of Fe, Mg.

Thus

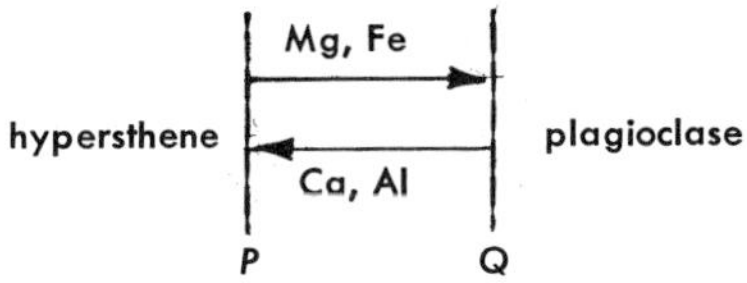

is a millimeter or two. Movements of the same order of magnitude are involved in the growth of porphyroblasts; if a chloritic phyllite, in which the chlorite is uniformly dispersed, changes to a garnet schist in which the garnets are 1 cm apart and of radius r mm, the last of the iron and magnesium must travel $(5 - r)$ mm or more before incorporation in a garnet lattice. Grain growth without change of mineralogy also involves transfer of material over distances of the same order of magnitude as the grain's final radius. Thus, journeys of a few millimeters are standard for most components of a metamorphic rock of medium grade.

Some components in some situations travel much farther, however. Barth (1962, p. 300) shows an example of garnet development around a xenolith in granite where iron ions must have moved 6 m, and in the contact silicification of limestone, where the limit of silicification is a minimum distance for the migration of silica, journeys of over 15 m can be proved (e.g., Burnham, 1959). Compounds more soluble in aqueous fluids, such as boron and fluorine, travel further—distances of the order of a kilometer—while the aqueous fluids themselves may travel 10 km in one episode (the chlorite-garnet transformation referred to above generates water at a depth of 10 km or more; although some may be absorbed by hydration reactions at shallower depth, if the overlying rocks are already fully hydrated some of this water may reach the surface).

It will be noticed that confidence about the migration range diminishes as the component's concentration gets less or the component involved more common. If we find five per cent of fluorine in a rock 500 m from a granite we are entitled to say migration over 500 m is "proved"; but if we find one per cent at 1000 m, the evidence is less convincing, because concentrations up to one per cent can occur without local magmatic activity. With a mineral such as quartz, transfer can rarely be proved, because what is added cannot be distinguished from what might have been present initially—there is a problem of distinguishing "signal" from "noise" or "background." Where a localized source is evident, as in the contact silicification referred to, the problem is sometimes solvable, but when regional migration is postulated, an alternative explanation is usually possible involving an original patchy distribution of the component in question.

A particularly common uncertainty in highly metamorphosed terrains is the behavior of potassium. There are many localities where the following transition can be observed:

At A, normal metasediments (potash-poor gneisses);
At B, potash-rich gneisses;
At C, gneiss that is like granite in composition and in homogeneity.

The distance A–C may be $\frac{1}{2}$ km to 5 km. Such a situation can be interpreted in two ways:

1. Sediments as at A were initially uniform over the region; magma was emplaced from elsewhere at C; escape of potassium from the magma led to potassium enrichment at B.

2. Sediments were initially nonuniform over the region, with a high potassium content around C, perhaps by admixture of rhyolite tuff; on metamorphism, partial remelting occurred where the potassium concentration was favorable.

There are other possible interpretations also (see p. 251) but one of two basic postulates must be made: that the transition is due to initially nonuniform potassium distribution *or* that the transition is due to nonuniform introduction of potassium into sediments from some external source.

To sum up, the verifiable facts leave a petrologist great latitude in choosing how much or how little movement of material he will postulate by nonmagmatic processes. The next section attempts to set some limits on the possibilities by considering the physics of the situation.

MECHANISMS OF THE MOVEMENTS

There are four ways in which a component can move through a rock that retains a solid fabric throughout:

1. Diffusion through the crystal lattices;
2. Diffusion over crystal surfaces;

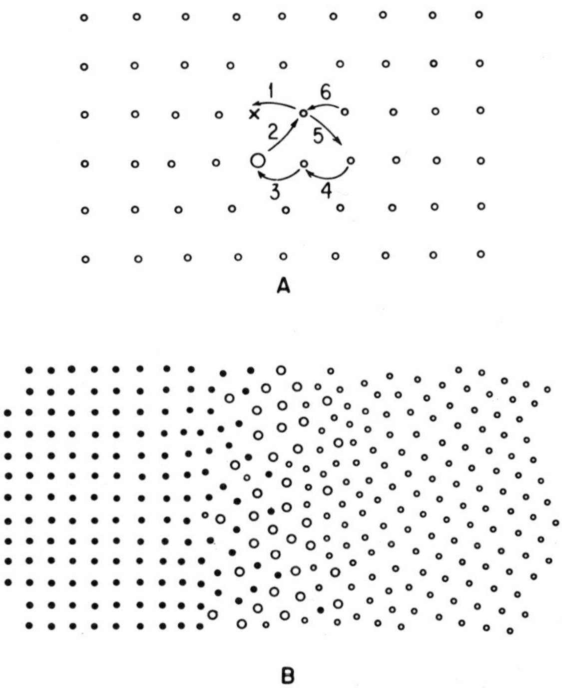

Fig. 22-2 Diffusion situations. (A) Migration of ions through a solid lattice. Migration depends on the existence of lattice defects. In the diagram, before move 1, a large foreign ion (open circle) is causing distortion of the lattice that leaves a cavity at x. In move 1, an ion enters this cavity; after move 6, the situation is as at the outset except that the foreign ion and associated cavity are further to the right. Such moves are random in direction and do not result in mass migration unless the foreign ions are nonuniformly distributed; but if there is a concentration gradient, more ions will move down the gradient than up it, and overall transfer results.

(B) Disorganized ions at a crystal interface. The surface layers of a crystal are believed to be less well organized than the interior, and foreign ions are more easily accomodated. If the open circles were water molecules, the terms "surface phase," "adsorbed water" and "interstitial film" would all be applicable, and interstitial adsorbed water is transitional into interstitial bulk water (see discussion of surface effects, Chap. 15).

3. Diffusion through a stationary intergranular fluid;
4. Migration of an intergranular fluid carrying the component as a passive solute.

Diffusion through a lattice is illustrated in Fig. 22-2A in a highly oversimplified way. Diffusion over a surface is similar but occurs in the disorganized surface layers of a crystal where imperfections are abundant (Fig. 22-2B), rather than in the interior of the crystal where they are rare by comparison (though still very numerous: 10^{12} to 10^{17} per cc). Diffusion through

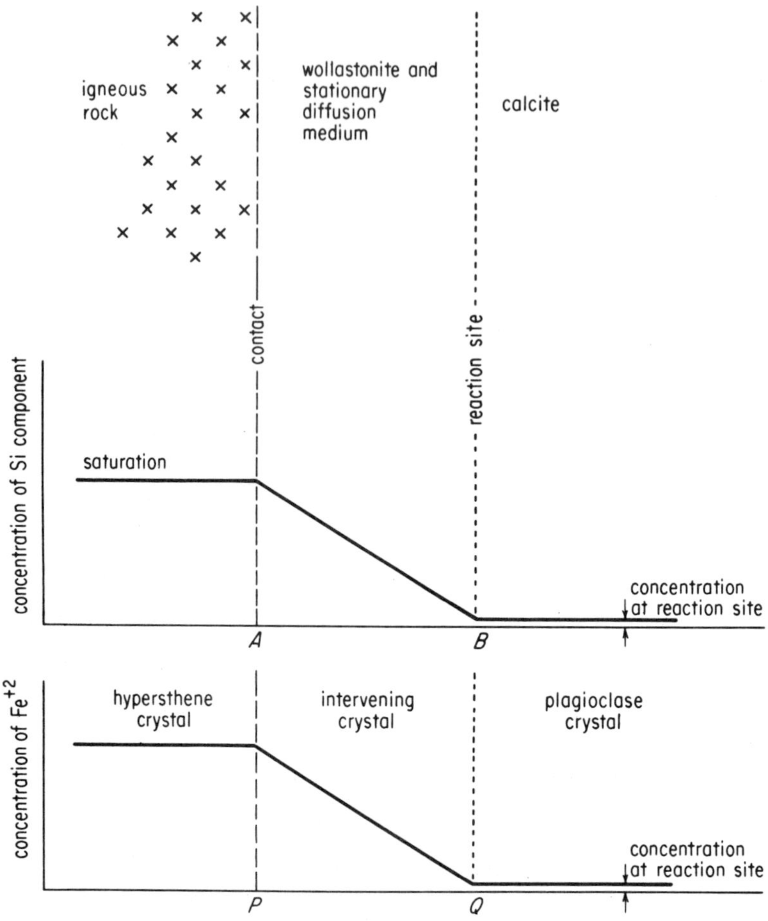

Fig. 22-3 Concentration profiles during reaction. The upper profile relates to diffusion of silica through a stationary intergranular pore-fluid; the lower profile relates to diffusion of iron or magnesium through crystal lattices. The two profiles differ in little but scale.

the fluid and diffusion of the fluid are familiar in principle, although it should be noticed that the diffusion rate through an intergranular fluid may not be the same as it would be if the same fluid were contained in broad channels—the internal structure of a fluid phase close to a crystal boundary is likely to be somewhat different from that of the fluid in bulk.

All four mechanisms are driven by the tendency of material to move away from high-energy sites. In mechanism (4), diffusion of the fluid, these are simply high-pressure sites (a high-pressure site is a high-energy site by virtue of the PV term in $F = U + PV - TS$). In the other three mechanisms, high concentration, as well as high pressure, leads to high energy as in aqueous solutions (see Chap. 15). Where the concentration and pressure gradients both fall in the same direction, there is no doubt which way material will go, but where the two gradients are opposed, either may prevail: a component can be made to run up a slight concentration gradient by pressure or up a slight pressure gradient by concentration (osmotic effects at a semipermeable membrane are analogous, if water containing a solute is regarded as water at a lower concentration than pure water).

The rate at which material is transferred across a reference plane in the absence of pressure gradients is given by an equation similar to that for heat flow (Chap. 11):

$$\text{diffusion rate} = D \times \text{concentration gradient}$$

where D = diffusion coefficient (cf. also Chap. 16). However, there are two diffusion situations which need to be distinguished:

1. The diffusing component, after travelling some distance, takes part in a reaction and is so removed from the fluid phase;

2. The diffusing component takes part in no reactions.

Situation (2) is closely similar to heat flow, and a curve showing concentration at any distance from an infinite plane source after a finite diffusion time has the same form as those in Fig. 11-3. In situation (1), however, the concentration at the reaction site never rises above whatever fixed minimum is needed for the reaction to proceed (see Fig. 22-3) and the concentration curve at any moment is effectively straight; in the example illustrated, silica diffuses through the inert wollastonite already formed, reaches the calcite interface at B, converts calcite to wollastonite and, in so doing, shifts the reaction interface to the right. The rate of advance of the reaction interface can be estimated as follows.

Let the concentration difference between A and B be C g/cc; when $AB = x$ cm, diffusion rate $= DC/x$ g/cm^2-sec. If every cc of wollastonite formed requires Z g of silica, wollastonite will form at a rate of DC/xZ cc/cm^2-sec. The reaction interface B will therefore advance at a rate dx/dt equal to DC/xZ cm/sec. Integration gives the distance AB after t seconds from the start of the diffusion episode as $\sqrt{2DCt/Z}$. This result is applicable equally well to all three ion-diffusion mechanisms on pp. 274–75, except for the following point: the cross section of the diffusion path is different in each of (1), (2) and (3). Consider diffusion from one side to the other of a 1 cm cube of rock: in lattice diffusion, more or less a full square centimeter of area is available for diffusion; in surface or intergranular diffusion, only a small fraction, f, is available. We should therefore write $AB = \sqrt{2DCft/Z}$ and remember that $f = 1$ only for lattice diffusion. For diffusion of a solvent, we replace *concentration* by *pressure* and the *diffusion coefficient* by a *permeability*.

Although both the theory outlined and knowledge of the quantities involved are grossly inadequate, it is interesting to insert some trial values:

1. Try mechanism (3), intergranular diffusion, in the wollastonite-calcite situation of Fig. 22-3, where the diffusing component is silica, probably travelling as $Si(OH)_4$. At 2 kb and 400°C, $D = 2 \times 10^{-4}$ cm²/sec (estimate by Walton, 1960); C = saturation value at A, $= 2 \times 10^{-3}$ g/cc (Figs. 10-1 and 10-2); $Z = 1.5$ g/cc. From Larsen's data, Fig. 11-3, a tabular basic body 1 km thick could maintain a contact temperature around 400°C for 10^4 years, $= 3 \times 10^{11}$ sec. If we try f = one per cent = 0.01, we find AB = 40 cm.

2. Try mechanism (4), diffusion of a fluid, in the same situation. Here the amount of wollastonite formed is limited by the amount of silica soluble in whatever water the cooling igneous body evolves (cf. Fig. 10-3). Assume the magma delivers two per cent by weight of aqueous fluid on crystallizing, and that the fluid carries 0.2 per cent silica by weight. If all the fluid escapes via one wall, e.g., the hanging wall, then

$$\begin{array}{c}\text{thickness of}\\ \text{wollastonite}\end{array} = \text{thickness of body} \times \frac{2}{100} \times \frac{0.2}{100} \times \frac{\text{density of magma}}{\text{density of } SiO_2 \text{ in wall}}$$

For a 1 km body, thickness of wollastonite = 8 cm.

If we continue to ask how 15 m of silicates form we find it as easy to modify the second calculation as the first. Possible changes are: (1) make a less siliceous silicate—monticellite or larnite instead of wollastonite; then Z becomes 1 g/cc; (2) assume a higher temperature—an acid fluid at 600°C might carry 0.01 g/cc of silica and give $D = 5 \times 10^{-4}$ cm²/sec; (3) assume a thicker body, say 5 km. Then mechanism (3) gives AB = 10 m. Mechanism (4) still gives only 3 m of deposit, if uniformly distributed, but of course this is an unlikely condition; if the volatile emissions from a dome-shaped body were concentrated mainly over one-third or one-fifth of its roof, a 10 or 15 m capping could result (Fig. 22-4).

It is not easy to envisage circumstances in which diffusion over a crystal surface could be more effective than diffusion through an intergranular fluid, and diffusion through a crystal lattice is severely limited by the low values of D involved. Tests with metals show that even the most highly diffusive systems, such as lead in gold, hardly reach $D = 10^{-6}$ cm²/sec. Such work as has been done on geological materials points to lower values—more like 10^{-10} cm²/sec even for a favorable situation such as sodium in feldspar at 900°C (Sippel, 1963). Other ions diffuse more slowly, with values probably in the range 10^{-13} to 10^{-19} cm²/sec at geological temperatures (Naughton and Fujikawa, 1959). With $D = 10^{-14}$ cm²/sec, a 2 mm reaction rim as in Fig.

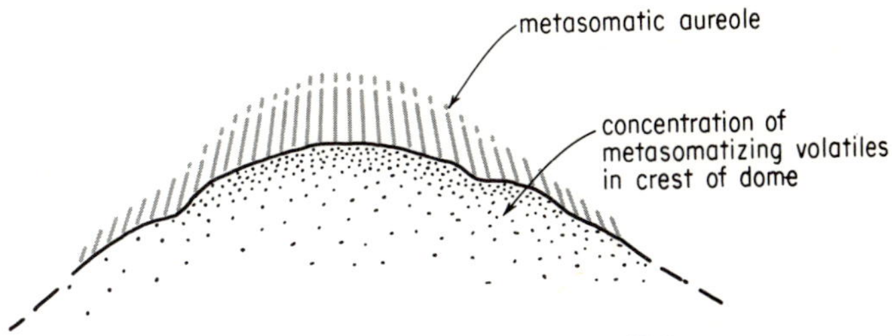

Fig. 22-4 Concentration of metasomatism at the crest of a dome.

22-1 would take one million years or so to develop. There are some changes, such as migration of impurities from the heart of a crystal to the rim, that cannot involve any intergranular fluid; but most changes would seem to occur faster by intergranular than by lattice diffusion—if $D_{\text{lattice}} < D_{\text{intergranular}}$ by a factor of 10^{10}, C and f can both be very small indeed for the intergranular fluid, and yet its effect will predominate.

If we are so bold as to attempt calculations for the regional movement of potassium (or sodium) described on p. 274, we find that all mechanisms verge on the inadequate. To increase the potassium content of a rock by three per cent through a zone up to 160 m from a potassium source such as a cooling intrusion, we have to assume either diffusion of a fluid with solute, where

solubility of K = 0.1 per cent by weight;
solubility of water in magma = 10 per cent by weight;
thickness of magma body = 5 km;
all fluid escapes through $\frac{1}{10}$ of the top surface;

or diffusion through an intergranular fluid, where

solubility of K = 0.1 per cent by weight;
$D = 0.1\ \text{cm}^2/\text{sec}$;
$f = 10^{-2}$;
$t = 1$ million years $= 3 \times 10^{13}$ sec;

or similar quantities. If we wish to affect a zone 500 m thick, it is very hard to extend diffusion of a fluid plausibly to fit; whether diffusion through an intergranular fluid can be extended depends very largely on what D actually is—the value of D for potassium ions in water is not known in these conditions; the value 0.1 above is merely the value it is necessary to assume to achieve the desired result.

DEVELOPMENT OF PREFERRED ORIENTATION

A kind of movement in metamorphic rocks that cannot easily be overlooked is that which leads to the preferred orientation of grains. We note at once two extremes (Fig. 22-5A):

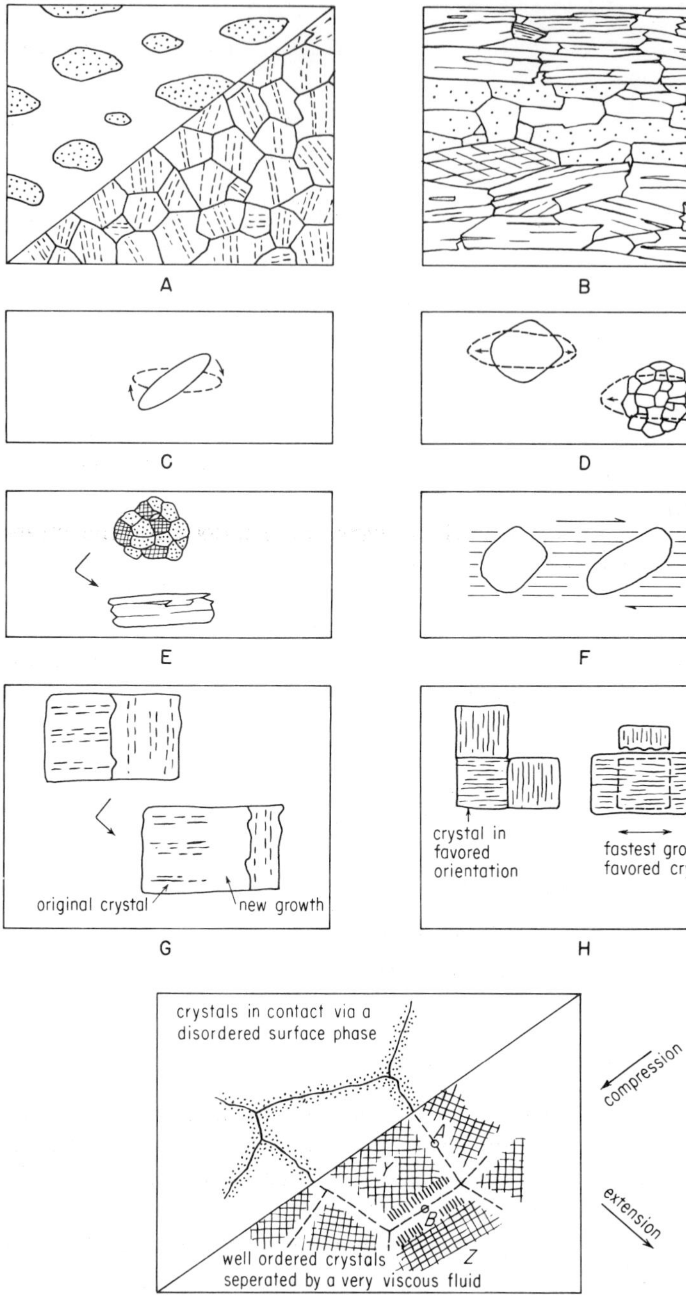

Fig. 22-5 Development of preferred orientation. (A) Orientation of shape but not lattice, and orientation of lattice but not shape; (B) orientation of shape and lattice (layering is also pronounced); (C–F) mechanisms for developing preferred orientation of shape; (G) mechanism for developing preferred orientation of lattice; (H) mechanism for developing preferred orientation of lattice and shape; (J) details of Mechanism D: either description of a crystal aggregate is permissible; on account of its viscosity, the "fluid" will be at a higher pressure at *A* than at *B* and will therefore migrate and lead to growth of crystals *Y* and *Z* on the faces marked.

1. Shape orientation;
2. Orientation of lattice directions such as symmetry axes.

In theory these need to be carefully distinguished, although in rocks one almost always accompanies the other to some extent (diagram B). There are four main ways in which orientation by shape arises (diagrams C–F):

1. Rotation of particles without change of shape;
2. Change of shape under pressure without growth of a new mineral;
3. Growth of a new mineral in elongated form;
4. Change of shape by asymmetrical shearing movement.

The fourth has been recognized mainly in conglomerates and is not easily distinguished from the second in minerals. The first is simple in principle and is limited only by the freedom of one grain to slip past its neighbors; this freedom will be high in rocks where the disordered intergranular layers are thick, and low where the grains are dry and angular; it may go on concurrently with mechanism (2). In mechanism (2) it is important to distinguish changes of shape with and without change of size; change of shape without change of size can never be proved in minerals, and most well-oriented grains are clearly larger than those from which they developed (compare the grain size in schists and gneisses with that in sediments). The main problem, then, seems to be to explain the transformation of many small unoriented grains into a few larger, well oriented grains.

To solve this problem we have to ask why grains grow. Answers are:

1. For the same reason that fine froth spontaneously changes to a coarser froth—a saving in energy can be made by eliminating some of the surface tensions;
2. Because if the stress in a rock is not isotropic, then in some lattice orientations the crystals contain less energy than in other orientations; thus a saving in energy can be made by transfer of ions to a favorably oriented crystal from an unfavorably oriented one as in diagram G.

The two factors together cause the total disappearance of grains with unfavorably oriented lattices, and growth of fewer larger grains with favorably oriented lattices. Finally, we note that on anisotropic crystals, growth rates are different in different directions—even a spherical mica nucleus is apt to grow to a plate-shaped grain—so that shape orientation is a customary concomitant of lattice orientation (diagram H).

The above analysis requires only that the stress be different in different *directions*. It is also possible that it will be different at different *points*. Since the disordered surface layer of a crystal behaves as a very viscous fluid, there may be a higher hydrostatic pressure at some points in a rock than at others. In diagram J, for instance, the pressure at *A* would be greater than that at *B*. Such a pressure difference would cause movement of material from *A* toward *B*, with extension of crystals *Y* and *Z* in the indicated direction, by growth. Here the favorably oriented crystal would grow faster, and so gradually become dominant in bulk.

In summary:

1. Stress difference in different *directions* gives lattice orientation and derivatively a favored shape orientation (Fig. 22-5G, H);
2. Pressure difference at different *points* gives shape orientation and derivatively a favored lattice orientation (Fig. 22-5J).

The same controls operate when new minerals grow; the only difference is that then chemical energy as well as mechanical energy is available to drive the process.

METAMORPHIC AND METASOMATIC DIFFERENTIATION

The mechanisms outlined in the previous section explain schistosity but nothing has yet been said that explains *gneissosity* or uneven distribution of minerals in the rock. Without question, some composition contrasts in gneisses are directly inherited from bedding in sediments or xenoliths in plutons. Thus in Fig. 22-6C, although it would be difficult to say which is the trace of the bedding surface, it could still be argued that the composition difference between *Y* and *Z* is a consequence of an initial composition contrast produced by sedimentation. We may in fact ask whether any composition difference can arise in a metamorphic rock that is not a consequence of a premetamorphic difference. In other words, can a homogeneous rock be made inhomogeneous by metamorphism alone? Because of their abundance in gneissic terrains, inhomogeneities involving quartz and feldspar are of particular interest.

We remember that homogeneity depends on scale. We call a hand specimen homogeneous if it is a uniform mixture of 1 mm grains, whereas we call it inhomogeneous if it contains a 3 cm porphyroblast, even if such porphyroblasts are uniformly distributed through the outcrop from which it came. Thus the growth of a porphyroblast is an instance of metamorphism making a rock inhomogeneous. The growth of augen is a better one, because three different compositions are often generated: the central crystal, the groundmass and a pocket of quartz or feldspar in the corner of the eye.

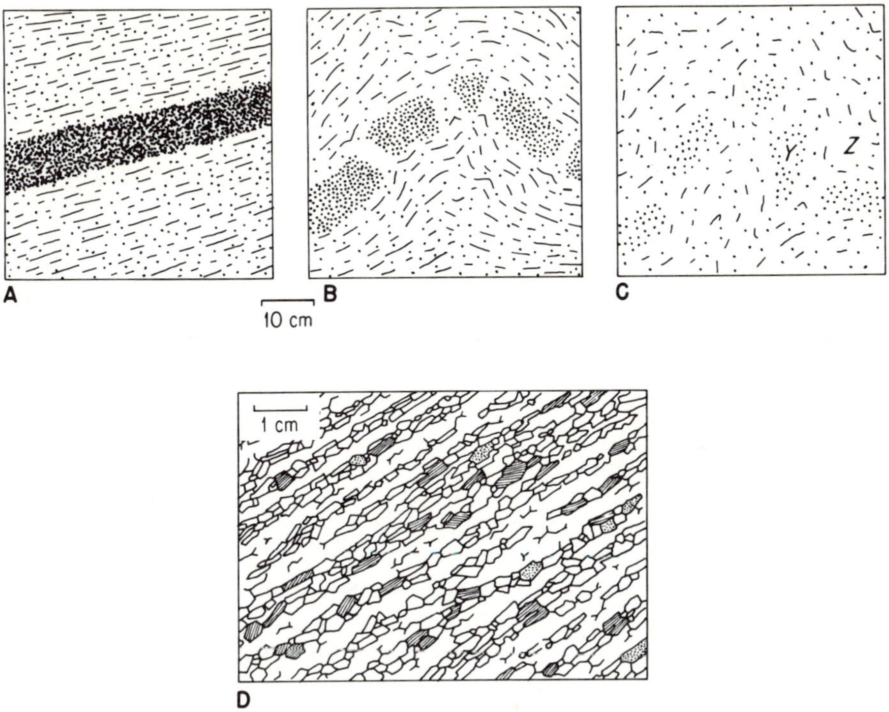

Fig. 22-6 Composition contrasts and bedding. (A–C) The three stages show that a chemical contrast inherited from layered sediments can persist even when the bedding is no longer a recognizable structural feature; (D) in a laminated gneiss it may be hard to find evidence to establish whether the composition contrasts are inherited or caused by metamorphism.

Here we have an example of an inescapable truth: not only are Na, K and Si concentrated in an aqueous pore-fluid at the end of a magmatic episode; they are also the components that move most readily in it during metamorphism. If a disturbance produces a low-pressure region in a metamorphic rock, it is most often quartz or feldspar (or calcite) that creeps preferentially into it. This can be plausibly attributed to their high solubility, and means that a chemical inhomogeneity can be expected to arise at any marked mechanical inhomogeneity. Saddle reefs of ore are a comparable phenomenon.

It was also remarked that partial remelting usually begins with separation of a quartzo-feldspathic melt. Is it true, then, that every quartz-feldspar concentrate in metamorphism is either a premetamorphic inheritance or a pressure-controlled segregation or a pocket of partial remelting? Probably not, because there is abundant gneiss in which layers of quartz and feldspar a few millimeters thick alternate with layers of ferromagnesian minerals (Fig. 22-6D); the layering is too uniform to be by partial melting, quite the wrong shape for pressure control, and rather unusual in terms of composition for a sediment or tuff. The alternative is to suppose that even without a pressure hollow or a silicate melt, the quartz and feldspar have a positive tendency to congregate with each other. No adequate theory exists to explain why they should do so, but three points may be made:

1. Experiments with kaolin/sulfide mixtures by Roberts (1965) appear to establish a tendency of this kind in sulfides;

2. The surface energy of quartz-feldspar interfaces seems to be somewhat less than that of interfaces with ferromagnesian minerals —otherwise granophyre and myrmekite would be less abundant (Figs. 5-8B and 21-6B);

3. If quartz and feldspar differ mechanically from the ferromagnesian minerals, a laminated rock might be more stable under shearing than a homogeneous one.

Whatever energy saving the lamination is attributed to, it would presumably be achieved by movement of quartz and feldspar via the pore-fluid.

Although it is true that Na, K and Si seem to be the most mobile of the common metals in metamorphism, differentiation in terms of other components can be striking. An example is the enhancement of a lithologic contrast by ore formation. Some metal ores are precipitated to high concentration at particular horizons and negligibly in neighboring rocks of very similar characteristics; the marked economic difference finally observed may be the magnified effect of a quite trivial difference in porosity or some such property. Similarly, joint planes can become planes of mineral concentration; in an extreme case, presumably a sediment with no inhomogeneity

Fig. 22-7 An example of metasomatic differentiation—the concentration of iron at an igneous contact.

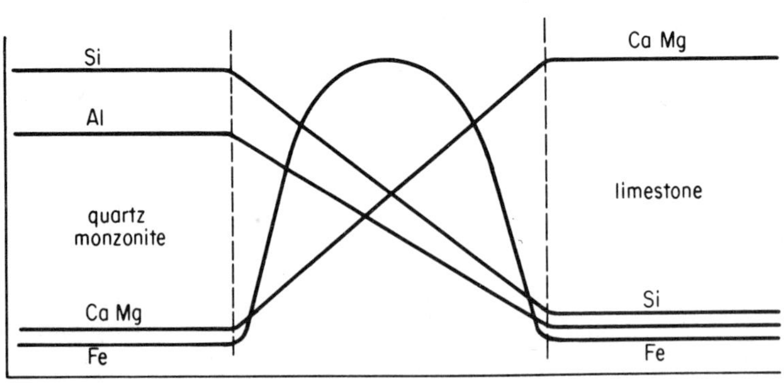

except bedding-plane partings could become a rock with alternating layers of contrasted composition.

The ore just discussed would be an example of selective metasomatism, but concentration of an economic mineral can occur in metamorphism that is isochemical when viewed on a moderately large scale. The iron ores at limestone contacts are such concentrates. In terms of most components, the contact zone is of composition intermediate between the reacting rocks (Fig. 22-7), being composed of calcium and magnesium aluminosilicates, such as pyroxene and lime garnet. The capacity of these minerals for iron, however, leads to its occurrence in far higher concentration than in the igneous rock or the limestone. If concurrently or later subjected to oxidation, this iron can form rich magnetite pockets. It is thus a characteristic of contact effects that they may include enrichment in a component that is not at all abundant on either side.

CHAPTER TWENTY-THREE

VEINS AND VEINING

Some veins are purely igneous in origin; the country rock in which they were emplaced was chemically inert and merely parted to admit the vein; such veins were described in Chap. 10. Other veins appear from their field relations to be purely metamorphic, and it is reasonable to ascribe their formation to one of the mechanisms outlined in the preceding chapter. Other veins again, however, may result from magmatic activity combined with metamorphism of the environment. We are now in a position to consider such veins, combining ideas from Chap. 10 with those from Chap. 22. As elsewhere, a section concerned with field observation is followed by a more speculative section, on the nature of the vein-forming fluid.

CLASSIFICATION AND NOMENCLATURE OF VEINS

Four varieties of veining have already been distinguished:

1. Veins filled with magmatic material emplaced as a melt of similar composition (veins that are small dikes);

2. Veins filled with magmatic material emplaced by an aqueous fluid of magmatic origin (the calculation in Chap. 10 concerns this type);

3. Veins filled with material derived from the wall rocks by agency of an aqueous fluid of magmatic origin;

4. Veins filled with material derived from the wall rocks by agency of an aqueous fluid also from the wall rocks.

In (3) and (4), the term *leaching* is convenient for the selective removal of some components from a solid by solution in a percolating fluid. The material leached can be moved along the vein before being deposited, so that the transport distance may be as great in (3) as it is in (2). It is sometimes possible to show that the vein filling is not of local origin, but if it is a common mineral such as quartz it will be difficult to go on and show either that it is magmatically derived or that it is derived from a distant sediment. It may be more practical, therefore, to divide veins as follows:

1. As before;

2. Veins filled with solids of distant source by agency of a fluid of distant source;

3. Veins filled with solids of local source by agency of a fluid of distant source;

4. Veins filled with solids of local source by agency of a fluid of local source.

(A fifth category—solids of distant source emplaced by a fluid of local source—seems to be of negligible importance.)

Figure 23-1 shows that both these systems give comprehensive sets of mutually exclusive classes, in accordance with the precepts in Chap. 4.

It is important to notice also that material may be removed in the process of vein formation. The vein filling may thus *replace* rather than *displace* the preexisting rock. *Dilation* is the term commonly used for any moving apart of the walls that occurs during vein formation, so that a vein can be intermediate between a pure dilation vein (no replacement) and a pure replacement vein (no dilation). In Fig. 23-2A, if grains P and Q were originally in contact, the vein is a *dilation vein;* if they were originally x cm apart, the vein is (barring freak circumstances) a *replacement vein*, and if they were originally $\frac{1}{2}x$ apart, then some dilation and some replacement have occurred. Clearly, a vein of type (2), filled with nonlocal solids, can be anywhere between the two extremes, and Fig. 23-2B, C, D shows that veins filled with local solids can also be of either kind, depending on whether the material is removed from the vein site by diffusion (diagram C) or by rearrangement of largely undissolved grains (diagram D).

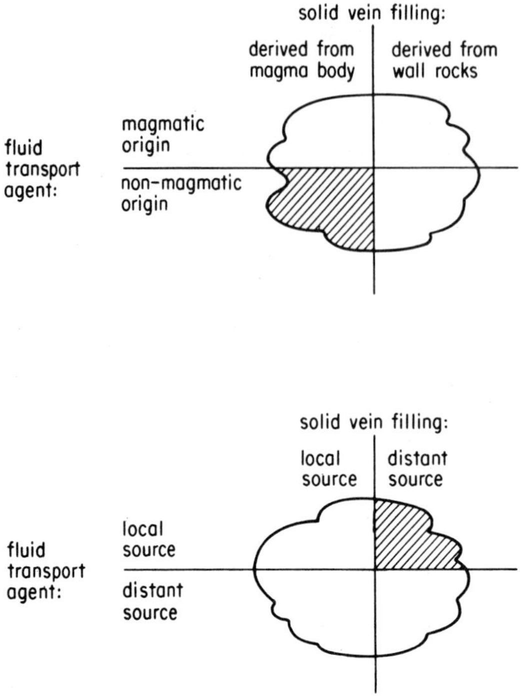

Fig. 23-1 Two possible genetic classifications of veins. Veins belonging to the hatched quardants are likely to be very rare.

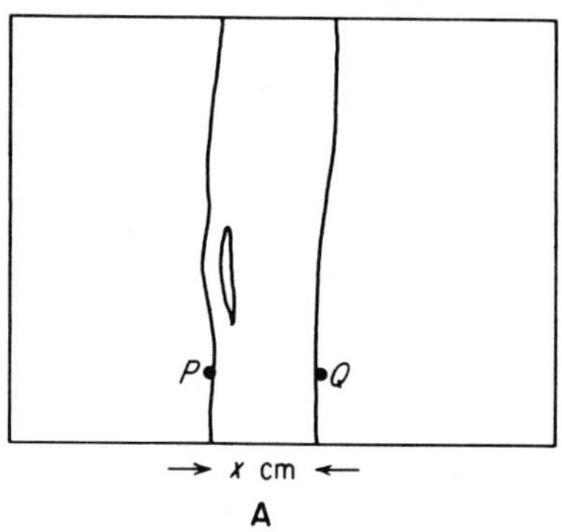

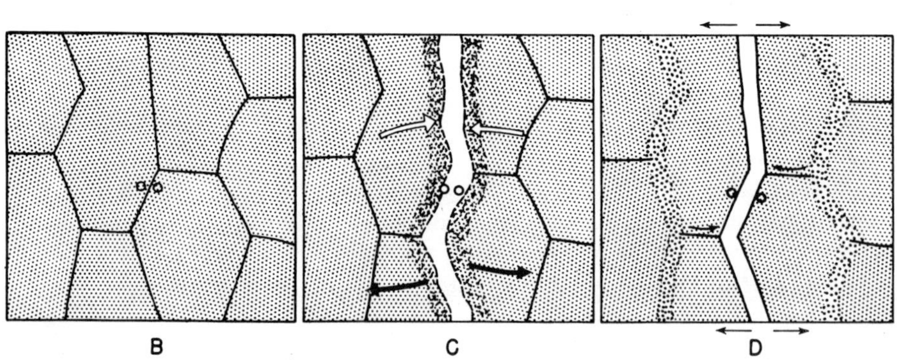

Fig. 23-2 Dilation and replacement veins. (A) The vein is a dilation or replacement vein according as *P* and *Q* were in contact before the vein formed or *x* cm apart; (B–D) Idealized schemes for development of a vein by lateral secretion, without dilation (C) or with dilation (D). In the change B ⟶ C, only black matter moves outward, by diffusion, an equal amount of white matter diffusing inward simultaneously; the two particles marked **o** remain contiguous throughout. In the change B ⟶ D, grains bordering the vein move bodily outward and the two particles marked **o** become separated; the vein is filled by inward diffusion of matter from intergranular boundaries within the wall rocks.

It is now time to ask to what extent the different mechanisms discussed can be recognized by their products, i.e., to what extent the genetic classes established correspond with classes based on observable features. Observable features are:

1. Geometrical: (a) fit between vein walls and (b) character of vein walls;
2. Chemical.

The main geometrical features to be looked for are shown in Fig. 23-3A–D. The main chemical features are (1) impoverishment of the walls in the vein-filling constituent (Fig. 23-3E) and (2) incompatibility between the vein minerals and the walls. A quartzo-feldspathic vein in a dolomite, for

Fig. 23-3 Dilation and replacement criteria. (A) Fit of such quality is unlikely to arise by replacement and therefore is probably by pure dilation; (B) moderately good fit could be by dilation or replacement; offset of bedding shows that some movement has occurred in plane of vein, and so *probably* some movement has also occurred perpendicular to plane of vein, i.e., there has been some dilation, with or without replacement; (C) this arrangement could arise by pure dilation, with *X*, *Y* and *Z* initially in contact; even if fit is not exact, movement could still be pure dilation, with some movement perpendicular to plane of section; in this figure, replacement is made unlikely by the angularity of the boundaries; (D) more typical outline for replacement—curved continuous boundaries rather than straight with sharp angles; right-hand vein is more or less proved to be by replacement because of undisplaced fragment of layer *W*, but an initial slight dilation episode is not ruled out; (E) composition profile revealing that the vein material was very probably derived, at least in part, from the wall rock.

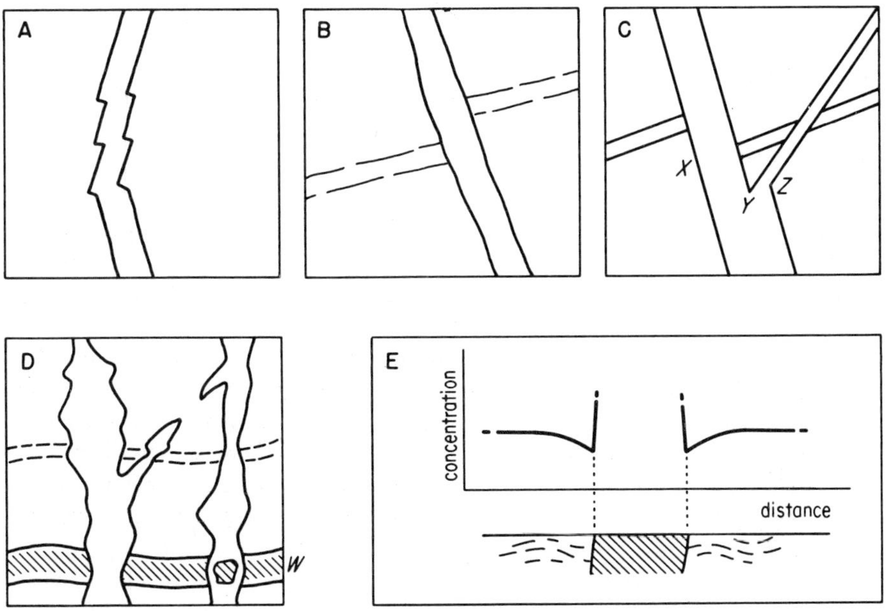

example, is probably not locally derived, especially if the incompatibility is underlined by the development of marginal reaction products such as talc or tremolite. Mere scarcity of the vein constituent in the environment does not prove a nonlocal source since scarcity in the environment may in fact be due to concentration in the vein, as in diagram E. With many metals, the concentration in shale, for example, is very small, but it requires a program of sampling and analysis of considerable magnitude to show that a given ore *could not* have been derived by concentration from local sediments.

We note that the geometrical criteria mainly inform us about the amounts of dilation and replacement, while the chemical criteria distinguish between local and nonlocal source, if anything. Of course, the absence of the chemical features mentioned is wholly uninformative, and usually both features are absent. Even if a concentration profile as at E is determined, it only shows the location of the vein material immediately before the vein formed; it does nothing to show whether the component was disseminated in the environment during the environment's formation or subsequently. Thus there is usually a great deal about a vein's origin that cannot be discovered by simple observation; even complex observation, by drilling, sampling, trace-element analysis, age determination and so on, often fails to reveal a coherent history. There is, in short, no clear correspondence between classes of vein based on observation and classes of vein specified by origin. On the other hand, every vein reveals one or another part of its history—there is no telling what kind of conclusion may be derivable about a particular vein encountered in the field, until one has discovered what features that particular vein shows.

THE NATURE OF THE VEIN-FORMING FLUID

The Occurrence of "Steam"

The properties of water in geological environments were summarized in Chap. 10. We now look at these in more detail, in order to get a clearer picture of the fluid so often referred to in discussing the different classes of vein.

Figure 10-1B shows that in rocks where the geothermal gradient is "normal" (20–40°C per km) any aqueous fluid in the pores has density far closer to water than to steam. In what circumstances, then, does something like steam occur, say with volume over 100 cc/g? The required circumstance is presumably the approach of magma close to the surface. Let us observe conditions at a point 2 km below the surface as magma rises toward it.

For convenience, we refer to the aqueous fluid as steam when its density is less than 0.01 (100 cc/g) and as water when its density is over 0.10 (10 cc/g). While the magma is still distant, the temperature at 2 km depth is 60–100°C and the water pressure at least 200 bars, because even if the rock overburden does not contribute to the pressure, we can presume that there is a 2 km head of water reaching to the surface. As the magma approaches, the temperature at 2 km will reach, say 300°C; by this time the water at 2 km will have density around 0.80 g/cc, but water at the surface will still have density close to 1.00 g/cc, so that the pressure is reduced to about 180 bars. When the 350°C isotherm reaches 2 km, the water's density there may drop to 0.50 g/cc, but if the surface water remains near 1.00 g/cc, the pressure at 2 km is still enough (c. 150 bars) to prevent further expansion. Before the 400°C isotherm reaches 2 km, however, the stable condition passes, expansion at the foot of the column of water pushes out the cool, dense water near the surface, and the pressure at the foot is thus reduced, permitting further expansion—in short, the water channels to the surface fill with steam from the bottom upward.

Naturally, the foregoing is an oversimplification and not quantitatively accurate for a variety of reasons, but the main point is valid: there will be no fluid like steam at all until the crisis, and thereafter there will be no fluid like water, until magmatic activity is over; any aqueous fluid still dissolved in the magma will separate and reach the surface in the low-density steam-like range. (To make water turn to steam above, while not turning to steam below, requires conditions something like the following:

over 200°C at $\frac{1}{2}$ km depth;
under 300°C at 1 km depth.

Such conditions seem unlikely to arise in nature.)

From the foregoing, it appears that aqueous fluids resembling steam will be rare in geology, and that fluids intermediate between steam and water will be rarer—shortlived as well as localized. Now the solubility of solids in steam of over 100 cc/g is negligible, even compared with the solubility of silicates in water. Two possibilities therefore remain: solids may be emplaced as veins during the sudden expansion—in fact, as a consequence of it—or they may be emplaced by the agency of the waterlike fluid (< 10 cc/g). The expanding fluid can do no more than drop its solute content because its life is far too short for any diffusion mechanisms to operate; on the other hand, fluids in the waterlike range may have long lives at high temperatures and permit diffusion as well as carrying solutes, as discussed in Chap. 22.

The Formation of Pegmatites

There seems little doubt that most metal ores and vein quartz are transported by what is effectively very hot water (enhanced in activity, no doubt, by presence of acids), because movement of a corresponding melt would require unrealistic temperatures. The emplacement of silicates, however, is possible by either melt or aqueous solution, and an outstanding problem is why veins of these two very different origins do not differ recognizably in appearance. On account of their abundance, the problem is most acute over quartz-feldspar veins. It is true that two classes of quartz-feldspar vein that differ greatly in texture have long been recognized, namely, aplites and pegmatites (aplite vein = quartz-feldspar vein with uniform fine grain, usually 1 mm or less; pegmatite vein = any vein, usually but not necessarily quartzo-feldspathic, whose grains are large compared with those of its environment). However, the view that a solidifying granitic melt generates only aplites directly, and that all pegmatites have been emplaced by agency of the aqueous phase alone is not widely accepted at present. In any case, it is clearly appropriate to look more closely at the conditions where silicate melt and aqueous fluid coexist.

Fig. 23-4 **The plagioclase–water system. A largely hypothetical diagram for the system as it might be at approximately 6 kb water pressure (left-hand side based in part on Tuttle and Bowen, 1958; each dotted line is horizontal).**

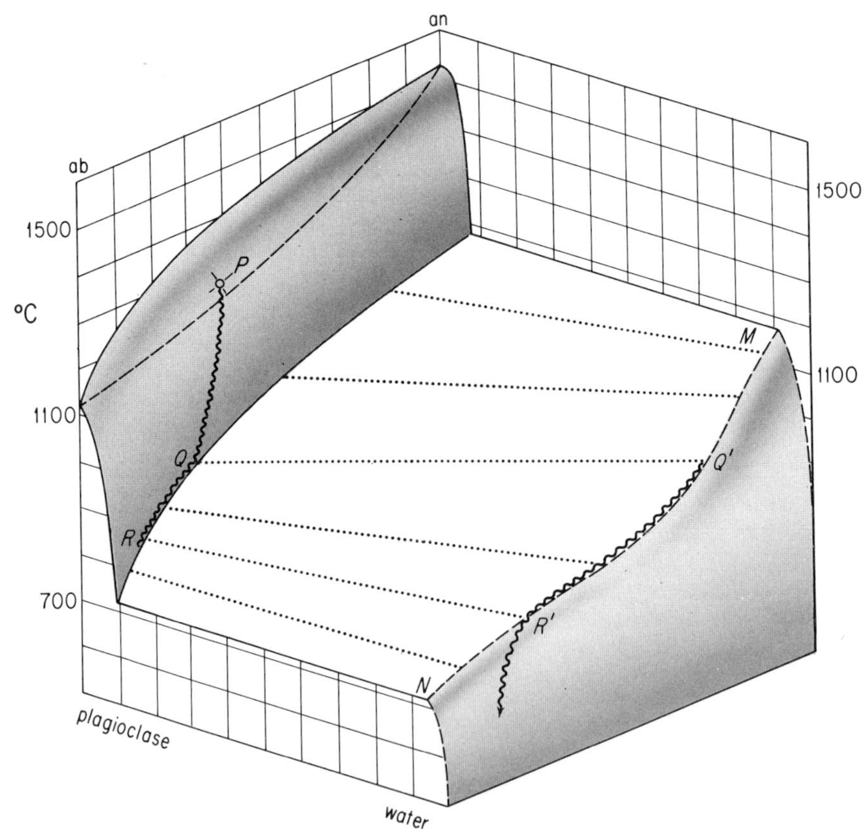

In Fig. 23-4, some relevant binary diagrams are combined by the same process as was used to construct Figs. 9-2 and 9-4. (The part drawn in broken lines is not accurate in dimensions or shape although MN is likely to have a flat portion similar to that in Fig. 21-4F.) Reference to the behavior of melts in the binary systems quartz–water and albite–anorthite reminds us that a slightly hydrous melt of composition P (say 30 per cent An, 65 per cent Ab, 5 per cent water) will, at onset of crystallization, become enriched in both albite and water. Its composition will change with falling temperature along some curved line such as PQ until the melt becomes saturated with water at Q. Thereafter, separation of more plagioclase crystals will be accompanied by separation of an aqueous phase saturated with plagioclase, represented by Q'. As continued crystallization of plagioclase permits the silicate fluid to travel down QR, the coexisting aqueous fluid will travel down $Q'R'$, until no more silicate fluid remains. Thus the stage PQ, where only a silicate fluid is present, and the stage after R, where only an aqueous fluid is present, are separated by the stage QR where two fluids coexist (this stage is sometimes referred to as the *pegmatitic stage*). The lines such as QQ' are horizontal, since Q and Q' coexist at the same temperature, but apart from this, their actual trend is very little known. In fact a purpose of this discussion is to emphasize that the Ab–An ratio in Q' can be markedly different from that in Q.

Although more complicated to draw, the diagram for the Ab–Kf–water system would be similar, because the Ab–Kf binary diagram is much like two plagioclase diagrams put together (Fig. 9-3). Thus we may expect that as a silicate melt approaches its minimum-melting composition (Fig. 9-4), it will be accompanied by a *variable* aqueous fluid. The proportions of Na, K, Si, Al and Ca in the aqueous fluid will change not only with temperature, pressure and the exact proportions of the coexisting melt, but also with abundance of CO_2, HCl and other components. It is therefore to some extent explicable that quartz-albite pegmatites, quartz-muscovite pegmatites, quartz-microcline pegmatites and so on should all occur.

We now revert to the question of texture: the coarseness of pegmatites makes it apparent that they are not formed merely by magma—an aqueous fluid is involved at some stage. Possibilities are:

1. The aqueous fluid and silicate fluid are both present at the same time, while the pegmatite solidifies directly to its present texture;

2. The silicate fluid comes first and crystallizes; then the aqueous fluid comes mainly afterwards and changes the initial texture to its present form;

3. No silicate fluid is involved at all, pegmatites being emplaced solely by agency of the aqueous fluid.

(Possibilities (1) and (2) grade into each other, but possibility (3) is wholly separate.) None of these hypotheses is wholly disproved by facts now known,

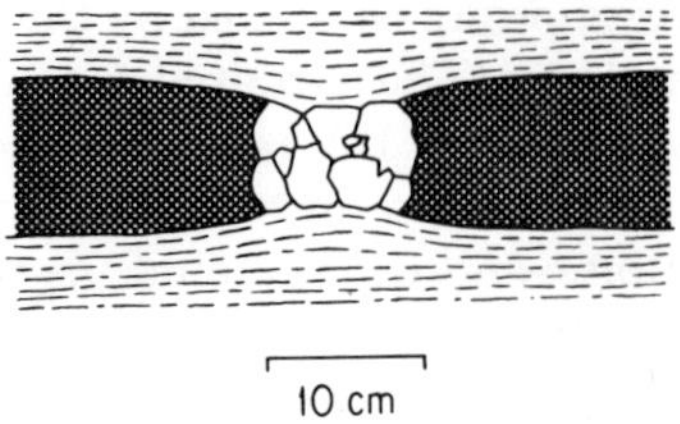

Fig. 23-5 Boudinage of a stiff layer. It is supposed that the stiff layer (stippled) was once continuous and has parted during deformation; the resulting low-pressure region is expected to fill with the most mobile materials available nearby.

and possibility (1) has been corroborated on a tiny scale by experiment. While retaining them all as possibilities, we may note the following facts:

1. It is not necessary that all pegmatites have the same origin; although they all have coarseness in common, within the class defined by this property there are many varieties and field situations (close to stocks, far from stocks, etc.); study may show that some form by each of the three methods listed.

2. The peculiar texture of pegmatites is found again in the tension cracks of boudinage structures (Fig. 23-5), where a silicate melt is highly improbable; thus *some* pegmatites seem to form by mechanism (3).

3. The bulk composition of a pegmatite is very hard to measure; the coarseness requires that a large sample be treated, and techniques for such a task are not at present satisfactory.

4. Although the silicate fluid and aqueous fluid undoubtedly coexist in time, it seems mechanically improbable that they are much associated in space: the viscosities are over 10^4 and under 10^{-2} poises respectively, so that in any veining situation, one is apt to travel to quite a different locality from the other.

5. Positive evidence in favor of the belief that a silicate melt is involved in pegmatite formation has not often been recorded. Proponents of the nonmagmatic view have tried to disprove magmatism by arguments such as (2) above, but proponents of the magmatic view have not, for historical reasons, been comparably active.

MINERALS IN VEINS

For purposes of description, we distinguish silicate veins without sulfides, silicate-sulfide veins and carbonate veins.

Silicate Veins Without Sulfides

There are three main kinds, quartzofeldspathic veins, quartz veins and epidote veins. The last-named are not as widespread as the other two kinds, being confined mainly to rocks rich in the necessary components, such as basic volcanic rocks, especially tuffs, and amphibole-rich metamorphic rocks. Quartz veins are abundant in host rocks of all kinds, in fields of low-grade metamorphism; quartzofeldspathic veins become the dominant variety where the metamorphic grade of the environment is moderate or high.

The quartzofeldspathic veins are the most variable in mineralogy. Beside the ratios of quartz, plagioclase and potash feldspar being widely variable, there are apt to be unusual concentrations of mica and compounds of the elements of low atomic number—lithium, boron, beryllium and fluorine, viz. lepidolite, spodumene, tourmaline, beryl and topaz. Less commonly, unusual concentrations of some heavy metals occur, such as Zr, Nb, Ta, Ce, Y, Th. The rare and complex mineralogy achieved by some pegmatites is illustrated by those in Langesundfjord, Norway, where seventy-seven minerals have been identified, of which several have no other known natural occurrence. Quartz veins are more restricted in their accessory minerals (except sulfides—see next section), native gold being the best known, if not the most abundant.

Richness in water is commonly taken to be the cause of pegmatites' coarseness, and accords with their richness in hydrous minerals. However, anhydrous pegmatites containing hypersthene and potash feldspar are known, from regions of high metamorphic grade. This is a convenient reminder that what determines the appearance of an anhydrous assemblage or a hydrous equivalent (biotite plus quartz, as in Chap. 7) is not the *amount* of water available, but the energy levels involved. If hypersthene + K-feldspar + water is the lower-energy form, increasing the amount of water available will not make biotite form; this would require an increase in the *pressure* of the water, or some such physical change.

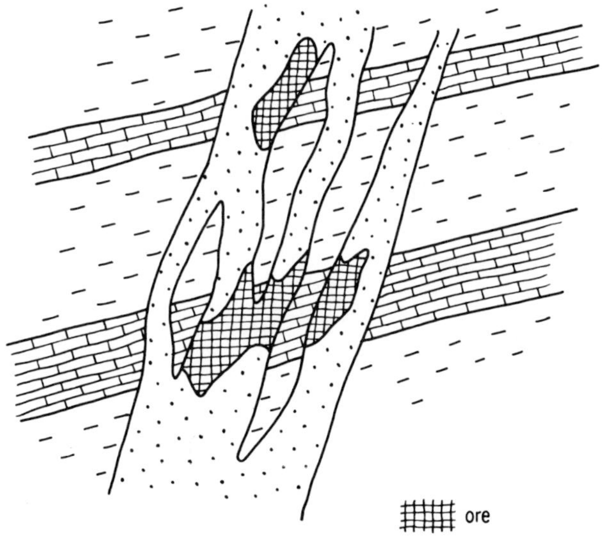

Fig. 23-6 **Stratigraphic control of ore. The ore is clearly localized at intersections of the vein with limestone horizons, but observation does not show whether the limestone or the vein is the source from which the ore has been concentrated.**

Silicate-Sulfide Veins

Whether the sulfur is directly magmatic or incorporated into sediments from the sea and then concentrated, enormous quantities of metal sulfides have been found in veins, associated with quartz, feldspars, other silicates and carbonates (Fig. 23-6). Iron, copper, lead and zinc are abundant in this form, and silver, mercury, molybdenum and other metals are often mined from such sources. The metals seem to have a certain mobility in the aqueous fluid that permits their emplacement, for textures suggesting recrystallization and replacement of one mineral by another are as common in the sulfide ores as in silicate rocks. A characteristic feature is the development not of single well-defined veins but of zones, as if the ore had formed in a porous body of fragmented rock such as a shear zone. In fact, faults with recognizable offset structures can sometimes be found running into the ore bodies, to confirm the association of shearing with mineralization. A consequence is that groundwater affects the ores rather freely, and in the conversion from sulfide to sulfate, carbonate, oxide and different hydrated forms, there is further opportunity for separation of the metals and concentration of different ores at different levels (*secondary enrichment*).

Carbonate Veins

Except for those mentioned above, carbonate veins are commonly more or less pure calcite, very widespread and of only structural significance.

CHAPTER TWENTY-FOUR

METAMORPHISM AND MAGMATISM IN GEOSYNCLINES

The purpose of this chapter is to review the processes outlined in Chaps. 1 and 2 and to see to what extent they are still plausible in the light of the facts that have been presented.

Fundamental hypotheses are:

1. The earth has a surface layer of crystalline rocks at least 20 km thick under the continents and several kilometers thick under the oceans;

2. There are two separate levels of magma production, the upper generating granitic magma and the lower basaltic magma; magma from both sources intermittently rises into the top 25 km of the earth;

3. By sedimentary processes, material is carried from the surface of the continents into deposition basins at the continental margins, and so down into the crust;

4. The granitic components of the sediments are the main source for the granitic magma at (2); the basaltic components of the sediments possibly find their way into the basaltic magma at (2), but do not constitute its main source;

5. Beside being returned to the surface as magma, the sedimentary components also return in part as metamorphic rocks without melting.

Things we should like to know are:

At what rate does the sedimentary material sink?

At what rate does it lose its contained water while sinking?

How far does it sink before the first magma begins to form by partial melting?

What really happens to the basic solid residues from partial melting?

It is widely believed that the melting process is intermittent—that at any one locality, an episode of sediment subsidence is followed by an episode of metamorphism and magmatism and then by an episode of tranquillity—though of course as one region reaches tranquillity, another may be just beginning the cycle, or *orogeny,* with subsidence. Further questions then arise: what is it that initiates a cycle, and are peridotites an essential or an accidental adjunct to the process? An assessment of possible theories of orogeny requires data from stratigraphy, geophysics, geochemistry and structural geology and is therefore quite outside the scope of this book. Here we simply assume that a trough of sediments *does* sink, and then see what petrological consequences follow.

Among petrological consequences, the loss of water during sinking is preeminent. Curve F in Fig. 1-5 reminds us that where water is absent, there are negligible prospects for the formation of much granitic magma above the Mohorovicic discontinuity; and discussion of the quartz-water system ended with the conclusion that where water is present, its amount is a crucial factor controlling the amount of magma likely to form. We shall therefore seek to explain why, at the base of a trough of sinking sediments, an abnormal amount of water is available to facilitate melting.

A crucial observation is that in some belts of slightly metamorphosed sediments, *lawsonite* occurs in place of epidote-clinozoisite, and *jadeite* occurs in place of albite. Both these minerals point to the conjunction of high pressure with low temperature. The field relations suggest that the sediments involved have been part of a sediment trough or *geosyncline,* and accord with the expectation, based on physics, that where sediments are sinking particularly rapidly into the crust, the geothermal gradient will be low; instead of 500°C being reached at 25 km depth, we might find 300°C at 25 km and 500°C at 40 km (see Figs. 1-5, 18-2 and 24-1). More extreme deviations have been suggested, for example by Rittmann, whose calculations indicate 200°C at 40 km as possible: even this suggestion only involves a geothermal gradient 1°C/km smaller than the smallest actually measured (6°C/km).

The effect of uncommonly high pressure for a given temperature is to retard all the dehydration reactions that normally accompany burial. The

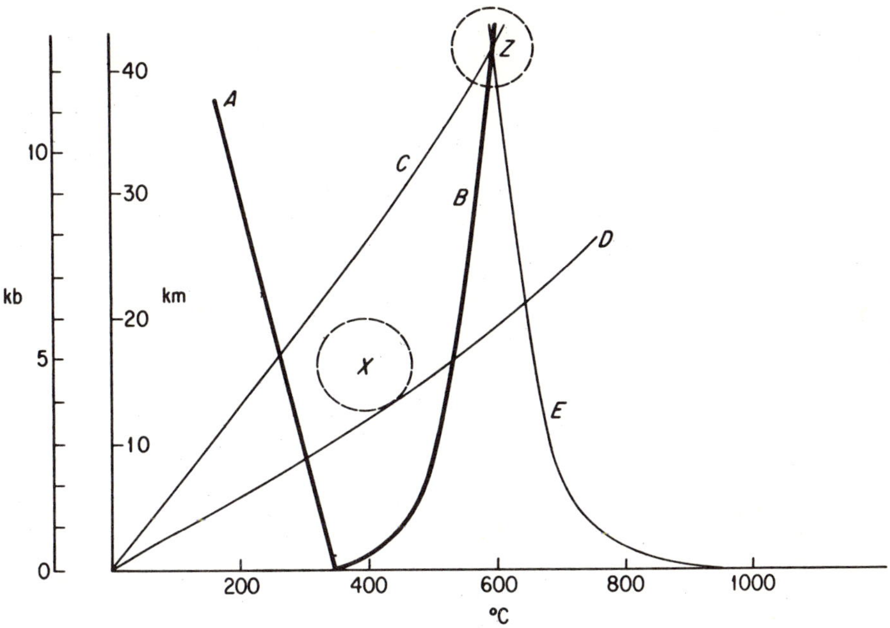

Fig. 24-1 Conditions for dehydration and melting in the crust. *A*, Reaction boundary cl ⟶ gt where water escapes; *B*, reaction boundary cl ⟶ gt where water is retained; *C–D*, range of geothermal conditions believed common (curve C tending toward geosynclinal conditions); *E*, onset of melting of granite in presence of water; *X*, conditions where reaction cl ⟶ gt normally occurs; *Z*, conditions where reaction cl ⟶ gt may occur in a geosyncline. (Compilation from Figs. 1-5 and 21-3.)

curve *B* in Fig. 24-1 could represent the reaction point for chlorite + quartz ⟶ garnet + water, as discussed in Chap. 21. This reaction runs at perhaps only 300°C in permeable rocks (curve *A*), but if the water cannot readily leave the reaction site and the pressure is high, chlorite may persist up to 600°C. This means that water from the dehydration of chlorite may pass directly into a hydrous magma (melting curve *E*). Using the equation

$$\underset{\text{MUSCOVITE}}{2\ KAl_3Si_3O_{10}(OH)_2} + \underset{\text{EXTREME Mg-CHLORITE}}{Mg_6Si_4O_{10}(OH)_8} + \underset{\text{QUARTZ}}{2\ SiO_2}$$

$$= \underset{\text{GARNET}}{2\ Mg_3Al_2Si_3O_{12}} + \underbrace{\underset{\text{POTASH FELDSPAR}}{2\ KAlSi_3O_8} + \underset{\text{WATER}}{6\ H_2O}}_{\text{MAGMA COMPONENTS}}$$

we calculate the following:

$$\left.\begin{array}{l} \text{100 g shale, 33 g muscovite} \\ \qquad\qquad\quad \text{23 g chlorite} \\ \qquad\qquad\quad \text{20 g albite} \\ \qquad\qquad\quad \text{24 g quartz} \end{array}\right\} = \left\{\begin{array}{l} \text{33 g solid residue (garnet)} \\ \text{67 g granitic magma with } 6\tfrac{2}{3}\% \\ \quad \text{water by weight} \end{array}\right.$$

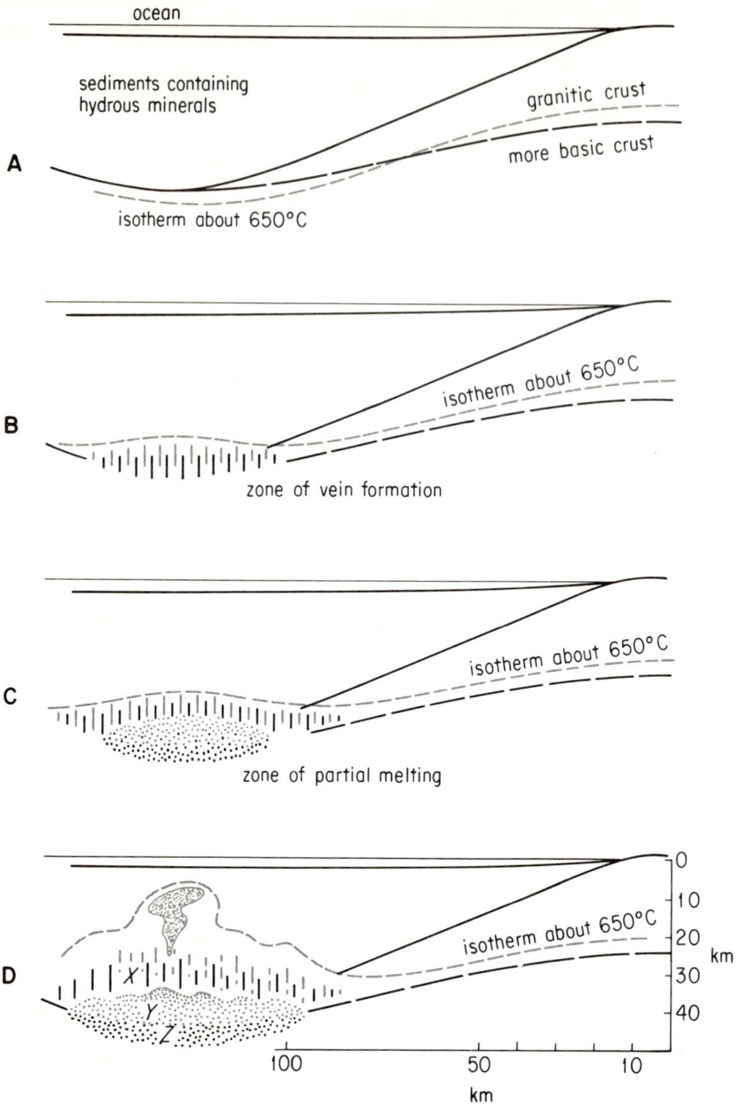

Fig. 24-2 The formation of a granite intrusion in a geosyncline.

This is of course only an approximate example of the kind of thing that could happen (in a more calcareous shale the basic solid residue could be amphibole or pyroxene), but it makes clear the point that a geosyncline provides exactly the conditions where the water of hydrous sediments is available for magmatic activity.

Whenever subsidence of cool sediments into the crust ceases, the isotherms will begin to return to their normal level. Events that accompany this stage are shown in Fig. 24-2. Diagram A shows the dimensions of the crustal downwarp envisioned, and a single isotherm which might be at 600°C or 700°C. In diagram B, melting has begun but the proportion of melt to solid is too small for the melt fraction to have mobility. In diagram C, a large enough body of granitic material has developed for the hydrostatic effects of Chap. 11 to begin operation. In diagram D, a minor amount of hydrous, low-melting magma has freed itself of stiffer, more basic contaminants and is being forced by gravity up to shallow levels in the crust.

The zone *X* that lies between sediments that have suffered only isochemical metamorphism, above, and sediments that have been extensively melted, below, is the zone where incursion of granitic components from below has been added to the normal processes of metamorphism. The extent and mechanism of the additions were discussed in Chap. 22, and the discussion of pegmatites in Chap. 23 is also closely relevant. The upshot is that although the rocks seem to contain some added granitic material, there is no agreement about its quantity or its transport. The most we can say with confidence is that the granitic components rise because of the low density of the phases they enter, and that the aqueous fluid that rises simultaneously plays a large part in their emplacement. In fact, departure of the last water from levels *Y* and *Z* marks the end of their chemical activity and the start of their existence as passive segments of the upper (acid) and lower (more basic) parts of the crust respectively.

This exceedingly cursory introduction does not do justice to the great variety and intriguing nature of geosyncline theories, but if the reader is encouraged to pursue further studies of the subject, the purpose will have been achieved.

CHAPTER TWENTY-FIVE

METHODS OF STUDY OF METAMORPHIC ROCKS

The study methods of metamorphic petrology are sufficiently similar to those of igneous petrology for reference to be made to Chap. 12.

Field Studies 1. **Observable Processes**

None of the processes of metamorphism is observable. However, an activity of metamorphic petrologists is comparable with that of an igneous petrologist who, to aid his study of a Paleozoic basalt, visits Iceland or Hawaii; the metamorphic petrologist can and should visit unmetamorphosed rocks that match his idea of the forebears of the metamorphosed rocks he studies. In this way, although he does not see the metamorphic process, he at least refreshes his mind about what starting materials are plausible.

Field Studies 2. **Products**

1. Size and shape of rock units. Geophysical methods are more restricted, because on metamorphism many contrasts between one rock unit and another, e.g., in density, are diminished. On the other hand, what goes on below the surface can be extrapolated from surface observations more reliably in metamorphic rocks, on account of their internal structures such as layering and folding.

Thus *mapping* plays a part in unravelling a metamorphic region that is of unparalleled importance and complexity.

2. Metamorphic effects. These include change of mineralogy, change of bulk composition, change of texture in hand specimen and change of structure (folding, fracturing, boudinage etc.). The observation that rock A is chilled against rock B shows that rock A is magmatic regardless of the locality at which the effect occurs; but there is more or less no observation in metamorphism whose significance is not dependent on its locality. Thus, metamorphic effects are recorded on isograd maps, structure maps etc., at the same time as the extent of rock units is recorded on formation maps. The study of structural effects alone is often an enormous undertaking (Turner and Weiss, 1963, Chaps. 3–6).

Laboratory Studies 1. **Geometrical**

1. Whole rock. Because the texture may include such features as porphyroblasts, reaction rims and pseudomorphs bearing on the mineralogical history of the rock and its unmetamorphosed parent, textural observations are of great importance. There is also a large field known as *petrofabric analysis*, in which the orientation of many individual grains is studied. For comparison purposes, a few rocks have been deformed artificially under conditions of known stress, so that to some extent it is possible to infer the stresses a rock has suffered from the orientation of its grains.

2. Individual minerals. Again, zircon and feldspar are the minerals whose internal geometry are of interest, feldspar because of the connection between twinning and history, and zircon because of the belief that it preserves, better than any other mineral, geometric traces of its premetamorphic state in spite of later metamorphism.

Laboratory Studies 2. **Compositional**

All the methods in Chap. 12 are applicable. The use of trace elements in determining the origin of problematic rocks, such as amphibolites, may increase in importance. In addition, two important applications of isotope analysis should be mentioned—chronometry and thermometry. These methods can be used to estimate the age and thermal history of magmatic rocks but the methods were not described in Chap. 12 because the importance of the results is more apparent after a review of the unsolved problems of metamorphism.

Chronologic methods depend on the fact that radioactive nuclei decay spontaneously at a rate independent of their physical or chemical state. For example, uranium 238 decays to lead 206; thus, in principle, if the amount of lead 206 produced and the amount of uranium 238 still remaining in a rock can be measured, and the decay rate is known, the time interval since the storing of lead 206 in the rock began can be calculated. The factors that complicate the application of this principle are numerous, and the reproducibility of the results inadequately known, but the data obtainable by such methods is invaluable. It is a particular convenience that some methods estimate the time since crystallization of the rock, while others estimate the time since the rock was last metamorphosed.

Thermometric isotope methods depend on the fact that if isotopes A and B substitute for each other in two phases Y and Z, the ratio

$$\frac{\frac{A}{B} \text{ in phase } Y}{\frac{A}{B} \text{ in phase } Z}$$

at equilibrium changes with temperature but is very insensitive to other variables. The same kind of relation holds for elements, such as Fe and Mg, but interference by third components is difficult to allow for between elements. Again, although the principle is simple, both measurement and interpretation of results present difficulties; but the value of such an independent check on the temperature of natural metamorphic reactions is apparent.

Laboratory Studies 3. **Physical**

Density and porosity of metamorphic rocks are of considerable importance in formulating theories. In addition, the deformation behavior under different kinds of stress is amenable to laboratory study. Unfortunately it is very hard to gain exactly the information required. Density and porosity of drill cores from the stable crust do not necessarily match those of the same rock in course of orogeny, and deformation behavior of a chemically inert rock is not the same as that of one in which reactions are proceeding. However, tests can be run at high temperatures and several kilobars confining pressure, and better and better approximations to metamorphic conditions are currently being made. Already, results set useful limits to what is plausible in metamorphic theory.

Laboratory Study of Materials other than Rocks

1. Chemical analogs. Experiments at controlled high temperatures and pressures have been and will continue to be as vital in studying metamorphic equilibria as magmatic; even where reaction is too sluggish for laboratory imitation, the measurement of thermodynamic quantities on artificial pure silicates gives a basis for calculating equilibria. Specific heat and heat of solution, measured in a calorimeter, are the quantities usually obtained.

2. Physical analogs. A variety of model experiments on the development of rock structures has been performed, many involving the hypothesis that, under slow deformation, a crystalline rock creeps like a very viscous fluid; thus putty, clay, pitch, syrup and wax have been used, as well as rubber and sand. Model experiments on change of texture are less easy; up to the present, ice seems to be the material that can most readily be recrystallized under controlled stresses, and sulfides recrystallize more readily than silicates (Roberts, 1965). Attempts to imitate the change of metamorphic texture by reaction are in their infancy.

3. Mathematical analogs. This is a rather sophisticated term for the rough, uncertain and fragmentary attempts that have been made at a quantitative description of metamorphism; but as emphasis passes from the question, "What changes occur in metamorphism?" to "How fast?" and "By what mechanism?"—matters which are for ever beyond reach of observation—it is to be expected that more comprehensive calculations will be made.

ABBREVIATIONS AND UNITS

ab	albite
act	actinolite
Al-cl	aluminous chlorite
an	anorthite
and	andalusite
anthoph	anthophyllite
aq	in aqueous solution
aug	augite
bi	biotite
Ca-gt	calcium garnet
Ca-py	calcium pyroxene
cl	chlorite
cl'oid	chloritoid
cord	cordierite
ct	calcite
cumm	cummingtonite
di	diopside
dol	dolomite
en	enstatite
ep	epidote
fo	forsterite
gt	garnet
hb	hornblende
he	hedenbergite
hy	hypersthene
kf	potash feldspar
ky	kyanite
Mg-cl	magnesian chlorite
musc	muscovite
ol	olivine
parag	paragonite
phlog	phlogopite
plag	plagioclase
py	pyroxene
pyroph	pyrophyllite
qz	quartz
serp	serpentine
sill	sillimanite
staur	staurolite
trem	tremolite
woll	wollastonite

1° latitude $\cong$ 110 km

km = 10^5 cm

m = 10^2 cm

mm = 10^{-1} cm

$\mu = 10^{-3}$ mm $= 10^{-4}$ cm $\cong 10$ on ϕ scale

Å $= 10^{-4}\ \mu = 10^{-7}$ mm $= 10^{-8}$ cm

n on ϕ scale $= \frac{1}{2}^n$ mm (e.g. $5\ \phi = \frac{1}{2}^5 = \frac{1}{32}$ mm)

Avogadro's Number $= 0.6 \times 10^{24}$

Mass of H atom $= 1.7 \times 10^{-24}$ g

1 year $= 3.2 \times 10^7$ sec

1 atmosphere $\cong$ 1 bar $= 10^6$ dyne/cm^2 $\cong$ 1 kg/cm^2

1 cal $= 41$ cc-atm $= 4.2$ joule $= 4.2 \times 10^7$ erg (dyne-cm)

PART FIVE

APPENDICES

APPENDIX ONE

DATA TABLES

TABLE A1-1 ELEMENTS

	Atomic Number	*Atomic Weight*	*Main Isotopes*	*Ionic Radius (Å)*
Aluminum	13	27		0.51
Calcium	20	40		0.99
Carbon	6	12	12, 14	0.16
Chlorine	17	35		1.81
Chromium	24	52		0.63 (+3), 0.52 (+6)
Fluorine	9	19		1.36
Iron	26	56		0.74 (+2), 0.64 (+3)
Lithium	3	7		0.68
Magnesium	12	24		0.66
Manganese	25	55		0.80 (+2), 0.60 (+4)
Oxygen	8	16	16, 18	1.40
Phosphorus	15	31		0.35
Potassium	19	39	39, 40	1.33
Silicon	14	28		0.42
Sodium	11	23		0.97
Sulfur	16	32	32, 34	1.84 (−2), 0.30 (+6)
Titanium	22	48		0.76 (+3), 0.68 (+4)
Argon	18	40	36, 40	
Lead	82	207	204, 206, 207, 208	
Rubidium	37	85	85, 87	
Strontium	38	88	87, 88	
Uranium	92	238	235, 238	

TABLE A1-2 MINERAL DENSITIES

Magnetite	5.0–5.2	Amphibole	3.0–3.5
Garnet	3.6–4.3	Mica	2.7–3.1
Kyanite	3.6	Feldspar	2.5–2.8
Olivine	3.2–3.6	Quartz	2.65
Pyroxene	3.2–3.6	Kaolinite	2.6

TABLE A1-3 MINERAL VARIETIES AND FORMULAS

Olivine Forsterite-fayalite $\binom{Mg}{Fe}_2SiO_4$

Monticellite $CaMgSiO_4$ has the same structure; larnite Ca_2SiO_4 does not; serpentine $Mg_6Si_4O_{10}(OH)_8$ approximates olivine + water, with a small imbalance of SiO_2.

Pyroxenes 1. Orthopyroxenes: enstatite-hypersthene $\binom{Mg}{Fe}SiO_3$
2. Clinopyroxenes:

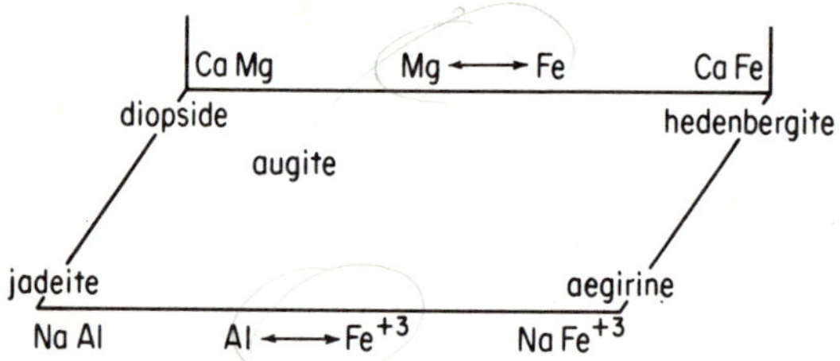

Uncommonly high Ti gives titanaugite; wollastonite $CaSiO_3$ does not have pyroxene structure.

Amphiboles 1. Ca-free: anthophyllite, cummingtonite $\binom{Mg}{Fe}_7Si_8O_{22}(OH)_2$
2. Ca-bearing }
3. Alkali }

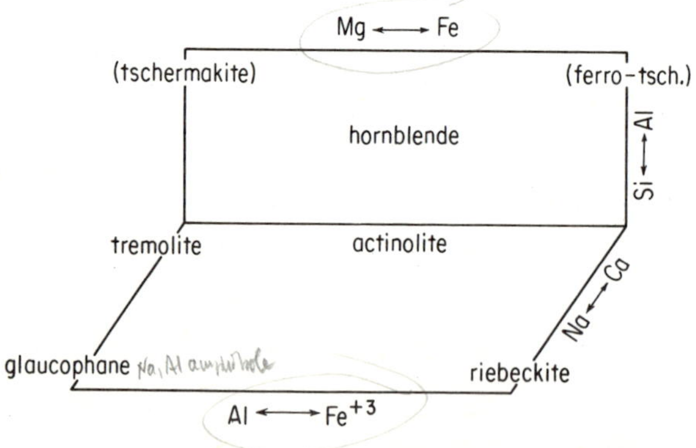

Sheet Minerals

Talc	$Mg_3Si_4O_{10}(OH)_2$	Pyrophyllite	$Al_2Si_4O_{10}(OH)_2$
Phlogopite Biotite	$K\left(\begin{smallmatrix}Mg\\Fe\end{smallmatrix}\right)_3Si_3AlO_{10}(OH)_2$	Muscovite Paragonite	$KAl_2 \cdot Si_3AlO_{10}(OH)_2$ $NaAl_2 \cdot Si_3AlO_{10}(OH)_2$
Serpentine Mg-chlorite	$Mg_6Si_4O_{10}(OH)_8$ or $3\ Mg(OH)_2 \cdot Mg_3Si_4O_{10}(OH)_2$	Kaolinite	$Al_4Si_4O_{10}(OH)_8$ or $2\ Al(OH)_3 \cdot Al_2Si_4O_{10}(OH)_2$
Ideal vermiculite	talc + water	Ideal montmorillonite	pyrophyllite + water

Natural chlorite: $3\ \left(\begin{smallmatrix}Mg\\Fe\end{smallmatrix}\right)(OH)_2 \cdot (Fe,Mg,Al)_3Si_3AlO_{10}(OH)_2$; natural vermiculite carries Al, Fe and Ca in addition to Mg and Si; natural montmorillonite carries Mg and Na or Ca in addition to Al and Si; illite is approximately a mixture of pyrophyllite and muscovite with extra water in the K-layers; glauconite is approximately a mixture of illite and "Fe-muscovite," $KFe_2 \cdot Si_3AlO_{10}(OH)_2$, with Na, Ca and Mg (see Fig. 13-10).

Framework Minerals

1. Feldspars and related minerals:

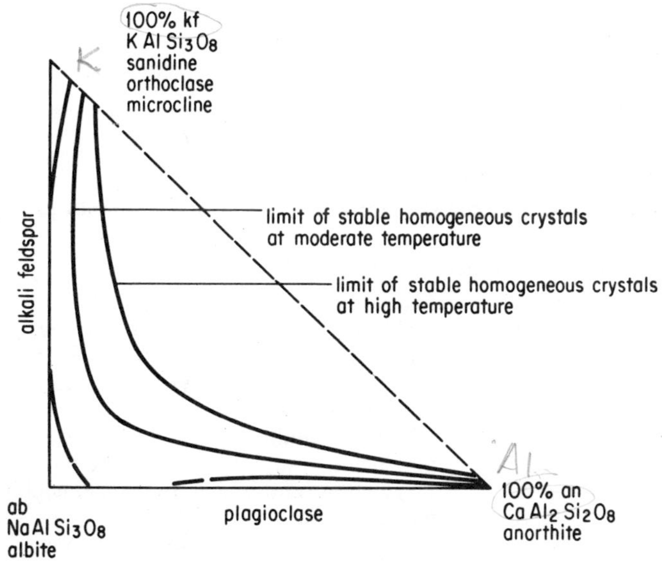

alkali feldspar = composition mainly ab + kf
plagioclase = composition mainly ab + an
perthite = intergrowth of alkali feldspars
antiperthite = kf inclusions in plagioclase host (An > 20)

albite = $An_0 - An_{10}$	labradorite = $An_{50} - An_{70}$
oligoclase = $An_{10} - An_{30}$	bytownite = $An_{70} - An_{90}$
andesine = $An_{30} - An_{50}$	anorthite = $An_{90} - An_{100}$

Scapolite ≡ plagioclase + NaCl, $CaCO_3$
Clinozoisite ≡ plagioclase + $Ca(OH)_2$ } not
Epidote ≡ clinozoisite with Fe^{+3} for Al } frameworks

	2. Feldspathoids ≡ feldspars reduced in SiO_2:	
	Nepheline	$NaAlSiO_4$-$KAlSiO_4$ solid solution
	Leucite	$KAlSi_2O_6$
	(jadeite	$NaAlSi_2O_6$, pyroxene structure)
Aluminous Minerals (other than above)	Andalusite, kyanite, sillimanite	Al_2SiO_5
	Mullite	$Al_6Si_2O_{13}$
	Staurolite	$Al_2SiO_5 \cdot \frac{1}{2}Fe(OH)_2$
	Chloritoid	$Al_2SiO_5 \cdot Fe(OH)_2$ $= Fe(OH)_2 \cdot FeAl_2(Si_2Al_2)O_{10}(OH)_2$ cf. chlorite
	Cordierite	$\binom{Mg}{Fe}_2Al_4Si_5O_{18}$
	Garnet, lime-free	$(Mg,Fe,Mn)_3Al_2Si_3O_{12}$
	Garnet, lime-	$Ca_3(Fe,Al)_2Si_3O_{12}$
	Tourmaline ≡ mixture of micas + Al_2O_3, B_2O_3	
Accessory Minerals	Pyrrhotite	FeS
	Pyrite	FeS_2
	Hematite	Fe_2O_3
	Magnetite	Fe_3O_4
	Ilmenite	$FeTiO_3$
	Sphene	$CaTiSiO_5$
	Apatite	$Ca_5(PO_4)_3(OH)$
	Zircon	$ZrSiO_4$

TABLE A1-4 THE DISTRIBUTION OF MAGMATIC COMPONENTS AMONG MINERALS

	Wt. % in Common Magmas	
SiO_2	40–75	free quartz; all silicates
Al_2O_3	12–20	feldspar, mica, amphibole (nepheline, pyroxene)
Fe_2O_3	< 4	amphibole, pyroxene, biotite (magnetite)
FeO	<10	olivine, amphibole, pyroxene, biotite (ilmenite, magnetite, pyrite)
MgO	< 8	olivine, amphibole, pyroxene, biotite
CaO	<11	plagioclase, amphibole, pyroxene (sphene)
Na_2O	< 8	plagioclase (amphibole, pyroxene, nepheline)
K_2O	< 8	feldspar, mica (leucite)
TiO_2	< 1	ilmenite, sphene, amphibole, pyroxene, biotite
MnO		amphibole, pyroxene, biotite
P_2O_5		apatite
H_2O		amphibole, mica, chlorite, epidote, clay minerals
CO_3^{-2}		calcite, scapolite
F^-		mica, apatite, fluorite
Cl^-		scapolite
S^{-2}		pyrite, pyrrhotite

Note: Parentheses enclose minerals that are rare or carry only small amounts of the component in question or occur in only small quantities in common rocks.

TABLE A1-5 PHYSICAL PROPERTIES OF ROCKS

Density	pumice	often < 1.0	
	limestone	1.9–2.7	
	sandstone	1.8–2.5	
	rhyolite glass	2.4	molten at 800°C 2.3–2.4
	rhyolite, granite	2.6	
	anorthosite	2.7	
	basalt glass	2.8	molten at 1200°C 2.6
	basalt, gabbro	2.9	
	peridotite	3.2–3.4	
Thermal Properties	heat capacity	0.21–0.24 cal/g at 200°C; 0.27–0.32 at 800°C	for common igneous rocks
	conductivity	5×10^{-3}–8×10^{-3} cal/cm-sec-°C	
	radioactive heat generation	6 cal/g-million years (granite)	
		2 (gabbro)	
		<1 (ultrabasic rocks)	
		probably > 10 for granitic rocks in early Precambrian time	

Data from Birch (1942), Barth (1962), Clark (1966).

APPENDIX TWO

IGNEOUS ROCK NOMENCLATURE

Terms were introduced in Fig. 6-1 for the limited purpose of summarizing what is known about rocks that have been studied up to the present. They conform to much of modern usage, but if the intention were to improve on modern usage, and to devise a system that can be used to assign names to single specimens in the future, some changes might be made. Streckeisen's thorough study of the matter resulted in the following proposals (Streckeisen, 1965); it has yet to be seen how widely these will be adopted by geologists.

To name a sample, its content of mafites* is first estimated as greater or less than 90 per cent (all percentages referred to in this appendix are by volume). If the mafites exceed 90 per cent, the rock is named from item 16 in the list below. If the mafites do not exceed 90 per cent,

> **1.** The quantity of quartz or feldspathoid† is estimated as a percentage of the *nonmafite total*, not as a percentage of the whole rock;

*mafites = mica, amphibole, pyroxene, olivine, iron minerals, melilite, garnet, epidote, apatite, zircon, sphene and primary carbonates.

†feldspathoids = leucite, nepheline, sodalite, nosean, haüyne, cancrinite, analcite etc.

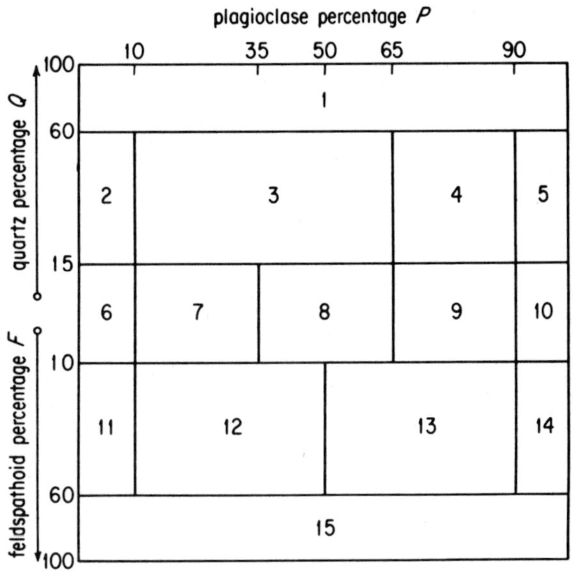

Fig. A2-1 The classification of igneous rocks after Streckeisen (1966).

2. The quantity of plagioclase in the range An_5 to An_{100} is estimated as a percentage of the *total feldspar*.

The rock can then be assigned to one of the fifteen numbered fields in Fig. A2-1, and named in accordance with the appropriate instruction in the list below. In the list, the percentage estimated at (1) above is referred to as Q, if quartz, and F, if feldspathoid; the percentage estimated at (2) is referred to as P; the percentage of mafites (in the whole rock) is referred to as M.

	Coarse Grained	*Fine Grained*
1. $Q > 90$	quartz rock	
$Q < 90$, $P < 65$	quartz granite	(no name provided)
$Q < 90$, $P > 65$	quartz granodiorite	
2. alkali granite		alkali rhyolite
3. granite		$P < 35$ rhyolite
(optional: $P < 35$	syenogranite)	$P > 35$ rhyodacite or quartz latite
($P > 35$	monzogranite)	
4. plag. An < 50	granodiorite	dacite
An > 50	granogabbro	
5. $M < 5$	quartz anorthosite	$M < 40$ quartz andesite
$M > 5$, An < 50	quartz diorite	$M > 40$ quartz basalt
An > 50	quartz gabbro	
6. alkali syenite		alkali trachyte
7. syenite		trachyte
8. monzonite		latite

9. plag. An < 50	monzodiorite	$M < 40$	latite-andesite
An > 50	monzogabbro	$M > 40$	latite-basalt
10. $M < 10$	anorthosite	$M < 40$	andesite
$M > 10$, An < 50	diorite	$M > 40$	basalt
An > 50	gabbro		
11. $M < 30$	foyaite		phonolite
M 30–60	malignite		
M 60–90	shonkinite		
12. plagifoyaite			tephritic phonolite
13. plag. An < 50	essexite		phonolitic tephrite
An > 50	essexite gabbro		
14. theralite			tephrite
15. feldspathoidite			extrusive
($F < 90$, $P < 50$	foyaitic f.		feldspathoidite etc.
$F < 90$, $P > 50$	theralitic f.		
$F > 90$)	f. *sensu stricto*)		
16. mafitite (peridotite, pyroxenite etc.)			extrusive mafitite (picrite etc.)

The most important departures in principle from the system in Fig. 6-1 are:

1. The estimation of Q and F as percentages of the nonmafite total rather than of the whole rock;
2. The extension of the term *granite* over the field of quartz monzonites;
3. The distinction of andesite from basalt by mafite content instead of plagioclase composition;
4. The application of the qualifier "alkali" on the basis of mineral proportions instead of oxide proportions (cf. Chap. 6).

All four proposals have merit and roots in tradition. Since uncertainty about the meaning that a name will have for one's readers seems likely to continue, it bears repeating that any form of words that expresses one's observations unambiguously is preferable to a name from a "system of nomenclature" that may imply more or less than is intended.

APPENDIX THREE

THE USE OF NORMS

A chemical analysis does not give a vivid mental image of the rock it represents. Further, if two analyses differ by one per cent in content of CaO, for example, it takes considerable experience and perspicacity to grasp at once what difference this will make to the mineralogy—will one rock carry more augite than the other, or more plagioclase, or will the main difference be in the plagioclase composition? To make the comparison of rock analyses more intelligible in mineral terms, the practice was developed of calculating *norms*. A norm is a list of mineral amounts that would occur in the rock if the oxides combined in certain simple ways instead of in the rather complicated ways actually found. For example, if the potassium in a rock entered only pure potash feldspar, $KAlSi_3O_8$, the amount of this mineral could be calculated from the K_2O weight percentage, whereas the actual amounts of alkali feldspar solid solution, biotite etc. cannot.

Since the purpose of calculating a norm is for *comparison*, it is essential that all norms be calculated in exactly the same way. If the amount of titanium, for example, is treated in one way by one operator and in another way by another, comparison of subsequent norms is useless. The two most popular standard procedures lead to the *CIPW norm* and the *Niggli molecular norm* (Johannsen, 1939; Niggli, 1954). Where the purpose is to give a rock description, one of these standard procedures should be followed in detail.

The purpose of this appendix is explanatory rather than descriptive. We wish to explore the connection between Figs. 7-1, 7-2 and 7-4 and to discover how a small change in the amount of one oxide may have a large effect on the resulting mineralogy. For these purposes, a simplified form of calculation is permissible. An example is set out in Table A3-1. The sequence of steps is as follows.

> **1.** Enter the analysis of interest in column (i). (In Table A3-1, the analysis entered is a fictitious basalt in the middle of the ranges given in Table 5-1, more or less as represented by the basaltic column of circles in Fig. 7-1. It contains no minor constituents or Fe_2O_3.)
>
> **2.** Enter the molecular weights of the oxides in column (ii).
>
> **3.** Enter 100 × (i)/(ii) in column (iii). The quantities in column (iii) are "molecular amounts"; they show the relative amounts of the different oxide molecules that are available in the rock; in theory, a sufficiently small sample of the rock would contain 87 molecules of SiO_2, 15 molecules of Al_2O_3 and so on. We then ask what minerals and what amounts can be made by combining these molecules in simple ways. The answer is built up by distributing the

TABLE A3-1 A SIMPLIFIED NORM CALCULATION: EXAMPLE 1

		i	*ii*	*iii*	*iv* kf	*v* ab	*vi* an	*vii* di–he	*viii*	*ix* ol	*x* en–hy	*xi*
a	SiO_2	52	60	87	12	18	20	16	21	6	15	
b	Al_2O_3	15	102	15	2	3	10					
c	FeO	12	72	17 } 35								
d	MgO	7	40	18				8	27	12	15	
e	CaO	10	56	18			10	8				
f	Na_2O	2	62	3		3						
g	K_2O	2	94	2	2							
h	mineral molecules				4	6	10	8		6	15	
						16 plag $Ab_{37}An_{63}$						
					×8	×8	×8	×6		×4	×3	
j	mineral volumes				32	48	80	48		24	45	(277)
k	volume percentages				12	17	29	17		9	16	
					58 feldspar			42 ferromagnesian				

amounts in column (iii) among the succeeding columns according to the following routine (steps 4 through 8).

4. Allot all the K_2O molecules to column (iv) to form potash feldspar and allot Al_2O_3 and SiO_2 molecules in the necessary proportions, 1 : 1 : 6. Draw a line across the remaining columns opposite K_2O to show that all K_2O has been allotted.

5. Allot all the Na_2O molecules with Al_2O_3 and SiO_2 in proportions 1 : 1 : 6 for albite, column (v). Draw a line opposite Na_2O.

6. Allot all remaining Al_2O_3 molecules to anorthite, column (vi), with CaO and SiO_2 in proportions 1 : 1 : 2. Draw a line opposite Al_2O_3; there will normally be CaO and SiO_2 remaining at this stage.

7. Allot all remaining CaO to diopside-hedenbergite, column (vii), with (FeO, MgO) and SiO_2 in proportions 1 : 1 : 2; once converted to molecular amounts, FeO and MgO need not be treated separately in the sequel. Draw a line opposite CaO.

8. Enter the remaining SiO_2 and (FeO, MgO) molecules in column (viii) and observe which is more abundant. If SiO_2 exceeds (FeO, MgO) by x, make x molecules of quartz and combine the remaining molecules to enstatite-hypersthene; if (FeO, MgO) exceeds SiO_2 by x, allot x molecules of SiO_2 and $2x$ molecules of (FeO, MgO) to olivine, and combine the remaining molecules to enstatite-hypersthene.

9. Note how many molecules of $KAlSi_3O_8$ the oxides in column (iv) generate and enter the number at the foot of column (iv); complete all columns similarly.

At this point, the oxides have been fully expressed as minerals and further procedures are variable according to the purpose in view. We may wish:

10A. To pigeonhole the rock in Fig. 6-1. No quartz or feldspathoid has appeared, so that the rock is in the middle row. The molecular ratio of kf to plag (4 : 16) closely approximates the volume ratio; that is, plagioclase constitutes 80 per cent of the total by volume, so the rock is an andesite or basalt. The molecular ratio of ab to an (6 : 10) indicates labradorite, so that the rock is a basalt, or gabbro if coarse-grained.

10B. To estimate mineral volume percentages. The volume of a mineral is approximately proportional to the volume of its contained oxygen*, so that proportional mineral amounts by volume

*One mole of oxygen when combined with the necessary ions to form

quartz gives	11.3 cc of mineral
feldspar	12.5
biotite	13.3
pyroxen	11.4
olivine	12.5

can be obtained by multiplying each number in row h by the number of oxygen atoms per molecule (8 in feldspar, 4 in olivine etc.) to give row j; these amounts are converted to percentages in row k. Agreement with Fig. 7-2 is as good as should be expected, remembering the imprecise nature of the indications Fig. 7-2 is intended to give (12 per cent of potash feldspar is rather much for such a basalt, but this is attributable to the rather high figure inserted for K_2O).

In Table A3-2, a slightly more siliceous analysis is treated by the same procedure. The rock turns out to be a diorite or andesite, with quartz instead of olivine, nearly sufficient to bring the rock into the quartz diorite class. Comparison with Fig. 7-2 is marred by the absence of any hydrous minerals in the calculation so far, but these can be introduced by using the conversions

a. kf + 3 en-hy + water ⟶ biotite + 3 qz

b. 2 di-he + 3 en-hy + qz + water ⟶ Al-free amphibole

c. 2 an + 3 en-hy + water ⟶ Al-amphibole + qz

Table A3-3 shows a few from the great range of possibilities. The line h_1 shows how the first-derived mineral molecular amounts are changed if re-

TABLE A3-2 A SIMPLIFIED NORM CALCULATION: EXAMPLE 2

		i	*ii*		*iii*	*iv* kf	*v* ab	*vi* an	*vii* di– –he	*viii*	*ix* qz	*x* en– –hy	*xi*
a	SiO_2	60	60		100	18	30	14	12	26	11	15	
b	Al_2O_3	15	102		15	3	5	7					
c	FeO	8	72	11	21				6	15		15	
d	MgO	4	40	10									
e	CaO	7	56		13			7	6				
f	Na_2O	3	62		5		5						
g	K_2O	3	94		3	3							
h	mineral molecules					6	10	7	6		11	15	
							17 plag $Ab_{61}An_{39}$						
						×8	×8	×8	×6		×2	×3	
j	mineral volumes					48	80	56	36		22	45	(287)
k	volume percentages					17	28	19	12		8	16	
						64 feldspar			28 ferromagnesian				

TABLE A3-3 THE CALCULATION OF POSSIBLE HYDROUS ASSEMBLAGES

	kf	ab	an	di– –he	qz	en– –hy			
From Table A3-2: h	6	10	7	6	11	15			
5 × (a) → h_1	1	10	7	6	26	—	+ 5 biotite		
j_1	8	80	56	36	52	—	60		(292)
k_1	3	27	19	12	18	—	21		
				biotite quartz-diorite					
3 × (b) → h	6	10	7	—	8	6	+ 3 Al-free am		
j_2	48	80	56	—	16	18	72		(290)
k_2	17	28	19	—	5	6	25		
				hypersthene hornblende diorite					
3 × (c) → h_3	6	10	1	6	14	6	+ 3 Al-am		
j_3	48	80	8	36	28	18	72		(290)
k_3	17	28	3	12	10	6	25		
				pyroxene hornblende monzonite					
3 × (b) → h_4	6	10	7	—	8	6	+ 3 Al-free am		
2 × (c) → h_5	6	10	3	—	10	—	+ 3 Al-free am + 2 Al-am		
j_5	48	80	24	—	20	—	120		(292)
k_5	16	27	8	—	7	—	41		
				hornblende diorite					
2 × (b) → h_6	6	10	7	2	9	9	+ 2 Al-free am		
3 × (a) → h_7	3	10	7	2	18	—	+ 2 Al-free am	+ 3 bi	
j_7	24	80	56	12	36	—	48	36	(292)
k_7	8	28	19	4	12	—	17	12	
				hornblende biotite quartz-diorite					

Note: The anhydrous assemblage from Table A3-2 is entered at line h. At the line 5 × (a), new numbers of molecules are entered that would be formed from those at h if reaction (a) on p. 322 were to proceed in quantities five times those shown, i.e., if 5 mol kf, 15 mol en-hy etc. were to react. The new numbers of molecules (variant h_1 on the anhydrous assemblage h) are converted to volumes and volume percentages in lines j_1 and k_1 as in Table A3-2.

Similarly, variant h_2 is derived from h by supposing that reaction (b) proceeds in quantities three times those shown. Variant h_5 is derived by supposing that in addition to reaction (b), reaction (c) also proceeds, in quantities twice those shown; variant h_7 also involves two reactions, (a) and (b).

action (a) proceeds until en-hy is exhausted; line h_2 shows how the amounts are changed if reaction (b) proceeds until di-he is exhausted; line h_5 shows how the amounts are changed if both kinds of pyroxene are converted wholly to amphibole, and so on. Each possibility for line h, the mineral molecular amounts, has been converted to mineral volume percentages; the rock names assigned and the points plotted in Fig. A3-1 show that a considerable variety of rocks can be derived from one set of oxide percentages. Thus, although the albite quantity hardly changes, the kf : an ratio runs from 1 : 6 to nearly 6 : 1 and the total feldspar runs from 48 to 64 per cent; among the dark minerals, hornblende may be the only one present or may

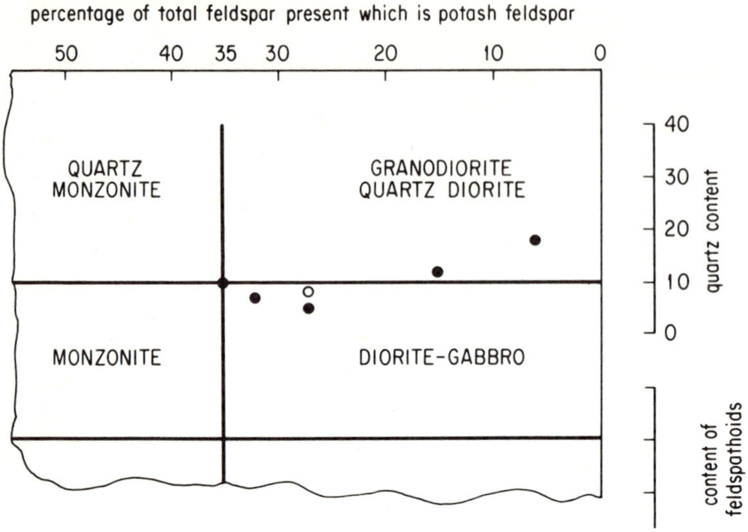

Fig. A3-1 Possible mineral proportions from one analysis. The scales and classification boundaries are taken from Fig. 6-1, top right corner; on these are plotted points corresponding to the volume percentages in Table A3-2 (open circle) and Table A3-3 (filled).

TABLE A3-4 A SIMPLIFIED NORM CALCULATION: EXAMPLE 3

		i	*ii*	*iii*	*iv*	*v*	*vi*	*vii*	*viii*	*ix*	*x*	*xi*	*xii*	*xiii*
					kf	(ab)	an	di–he		ol			ne	ab
a	SiO_2	57	60	95	30	(90)	4	4	−33	3			18	36
b	Al_2O_3	22	102	22	5	(15)	2						9	6
c	FeO	4	72	6 } 8				2	6	6				
d	MgO	1	40	2 }										
e	CaO	2	56	4			2	2						
f	Na_2O	9	62	15		(15)							9	6
g	K_2O	5	94	5	5									
h	mineral molecules				10		2	2		3			18	12
					×8		×8	×6		×4			×4	×8
j	mineral volumes				80		16	12		12			72	96 (288)
k	volume percentages				28		6	4		4			25	33

39 plag, $Ab_{86}An_{14}$ (columns vi + xiii)

Note: The conversion of all available Na to albite as in column (v) leads to a deficiency in SiO_2. The quantities entered in parentheses are therefore abandoned later in the calculation, and the quantities in columns (xii) and (xiii) are accepted instead.

be entirely prevented by pyroxene and biotite. Of course, not all the calculated possibilities are probable in nature: wholly aluminum-free amphibole in particular is not known from magmatic rocks. However, the calculations do show how the growth of one mineral can affect the proportions among the others. Whether such a magma develops mainly pyroxene or mainly hornblende or mainly biotite depends on the water vapor pressure and probably also on such matters as the abundances of titanium, fluorine and other minor components.

A final example, Table A3-4, shows how the calculation can be modified for silica-poor rocks. Proceeding as before, we find on attempting to fill column (viii) that the molecules of SiO_2 allotted are already 33 more than are present in column (iii). In this circumstance, clearly, the remaining (FeO, MgO) will make no mineral more siliceous than olivine; even the making of olivine to complete the (FeO, MgO) line increases the SiO_2 deficit

Fig. A3-2 **The CIPW normative classification of rocks. The figure illustrates the principles only; for details, see Johannsen (1939). Salic minerals are: quartz and feldspar; feldspathoids nepheline, leucite, kaliophilite; Al_2O_3, $ZrSiO_4$, NaCl, Na_2SO_4, Na_2CO_3.**

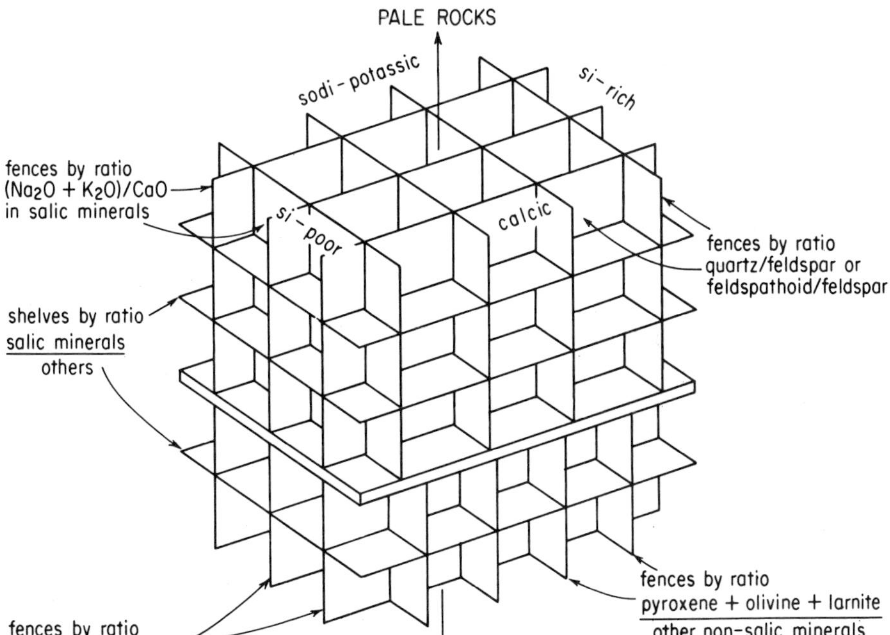

by 3, to 36. In most circumstances it is realistic to correct this deficiency by turning albite components into nepheline:

$$\begin{array}{lcl} 1\ Na_2O & & 1\ Na_2O \\ 1\ Al_2O_3 & \longrightarrow & 1\ Al_2O_3 + 4\ SiO_2 \\ 6\ SiO_2 & & 2\ SiO_2 \\ \text{ALBITE} & & \text{NEPHELINE} \end{array}$$

To make up a deficiency of 36 SiO_2, this change has to be made nine times; the resulting nepheline is entered in column (xii) and the albite that remains is entered in column (xiii).

NORMATIVE CLASSIFICATION

The classifications in Figs. 6-1 and A2-1 are based on volume percentages of actual minerals in the rock. An obvious alternative is to base a classification on the volume percentages or weight percentages of the ideal minerals in the norm. The authors of the CIPW norm suggested such a classification and this also is concisely set out by Johannsen (1939). Figure A3-2 merely shows the general nature of the system. It will be seen that, looking downwards from the top, the system resembles those in Figs. 6-1 and A2-1, but provision is also made for distinctions according to the abundance of nonsalic minerals (the salic minerals are listed under Fig. A3-2). As always, it is one's purpose that determines whether or not this is a useful classification.

APPENDIX FOUR

PERITECTIC RELATIONS AND BOWEN'S DISCONTINUOUS REACTION SERIES

At pressures higher than a certain minimum, both the systems potash feldspar–leucite and potash feldspar–quartz resemble the system quartz–albite in essential features. Since potash feldspar is chemically intermediate between quartz and leucite, phase diagrams for the two systems can be drawn contiguously on a single composition axis, as in Fig. A4-1A. The change in the diagram with lowering of pressure is shown in Fig. A4-1B and leads to the situation shown in Fig. A4-1C. Since X is the line where solid leucite begins to form, diagram C seems to indicate that on cooling pure molten potash feldspar, the first solid to form will be leucite. If this is so, we need to enquire what happens subsequently, because it is surely reasonable to expect that, *eventually,* pure solid potash feldspar should be obtained.

To resolve the uncertainty, it is convenient to consider the cooling of 100 g of a starting liquid not exactly equivalent to pure potash feldspar but equivalent to A_0 in Fig. A4-2 (which is the center of Fig. A4-1C enlarged). For purposes of arithmetic, some fictitious percentages have been inserted and key points lettered to emphasize the correspondence with Fig. 8-6A. At T_1, crystallization of leucite begins and the liquid changes composition along line X. At T_2, two phases are present, leucite and a liquid on X, the amounts being $100 \times AB/BC$ and $100 \times AC/BC$ respectively, i.e., 25 g

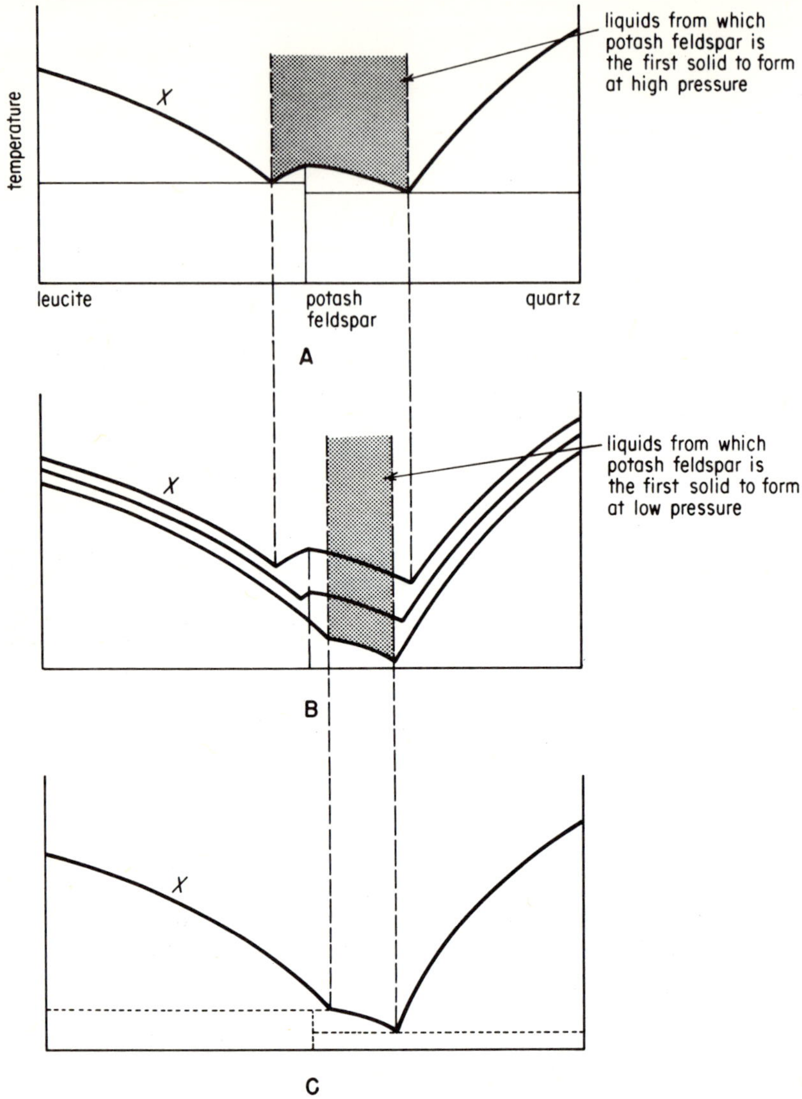

Fig. A4-1 Liquidus curves in the system leucite–quartz. (A) high pressure; (B) composite diagram illustrating the transition; (C) low pressure (Schairer and Bowen, 1935).

and 75 g (*check:* amount of quartz in liquid = 75 g × 40 per cent = 30 g, which does agree with the amount initially present at A_0). At T_3, infinitesimally above Z, the amounts have changed to: leucite 40 g, liquid Z 60 g (If the liquid still contains 30 g of quartz and the remainder is leucite, are the proportions 50 : 50 as is indicated for Z?). At T_4, infinitesimally below Z, the system has reached the point A_4 where the diagram indicates that leucite and potash feldspar are the two phases present. Applying the balance principle without further thought, we derive the quantities: 25 g leucite and

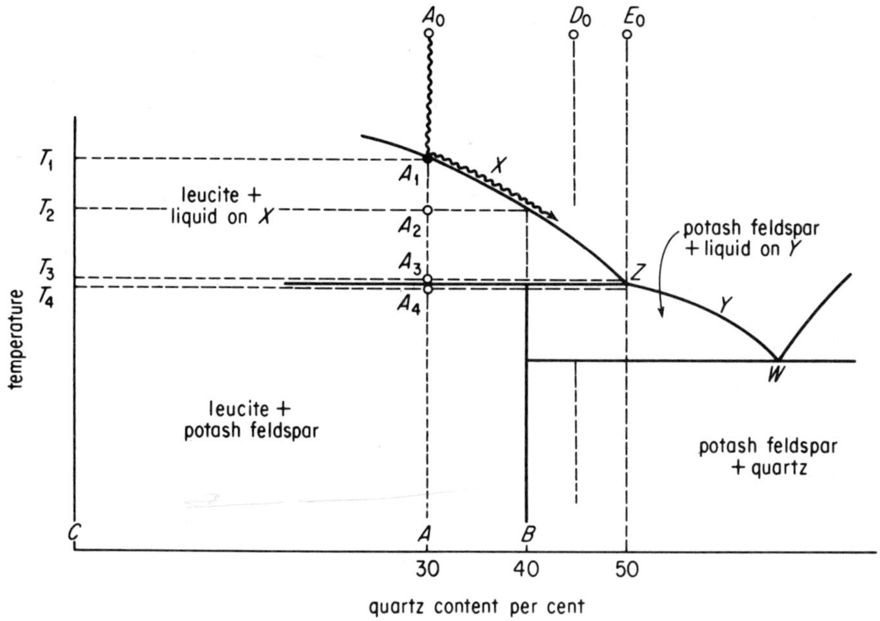

Fig. A4-2 Detail from Fig. A4-1C. The percentages 30, 40 and 50 are fictitious and for the purpose of the example–calculation only.

75 g potash feldspar. What, then, happened at the temperature of Z? Of the 40 g of leucite present at A_3, 25 g are still present at A_4, but the remaining 15 g, together with 60 g of liquid, have been replaced by 75 g of potash feldspar. Whereas at Z in Fig. 8-6A, consumption of the liquid was accompanied by accumulation of two solids, in Fig. A4-2, consumption of the liquid is accompanied by accumulation of one solid and *resorption* of the other.

To consolidate understanding of the system, the reader is invited to trace the progress of a starting liquid containing 45 g of quartz and 55 g of leucite, at D_0. Again, what happens *at* Z can be deduced from the quantities present at T_3 and the quantities present at T_4. It should appear that leucite and liquid are used up in the same proportions as before, viz., 1 g leucite to 4 g liquid, but in different quantities.

If the reaction at Z is written

$$\text{leucite} + \text{liquid} \longrightarrow \text{potash feldspar}$$

it is also interesting to ask:

1. What brings the reaction to a standstill in the sequence A_0, A_1, $A_2 \ldots$?

2. What brings the reaction to a standstill in the sequence D_0, etc.?

3. What happens to any liquid (of composition Z) that may still be present at temperature T_4 in some cooling sequences?

The answer to the third question is that its future is in no way different from that of a liquid that has reached T_4 after starting fresh at E_0—the presence of some ready-formed potash feldspar crystals has no effect on the liquid's progress down curve Y, and crystals of quartz will eventually form at the eutectic point W. (Points such as Z in Fig. A4-2 are not eutectic but *peritectic* points.) A feature of considerable interest arises here: even a liquid A_0 generates some liquid Z; in the sequence outlined above, this liquid was entirely consumed by reaction before the temperature reached T_4, but if, for any reason, reaction did not occur, the liquid could continue to cool down Y, with crystallization of potash feldspar, until crystallization of quartz began. Now reaction might be suppressed by separation of the early-formed crystals from the adjacent liquid; in this case, crystallization of *one* liquid could lead to early formation of leucite and late formation of quartz, in spite of the fact that these two minerals react with each other to form potash feldspar. This double result cannot be achieved as long as the relations between the minerals are as in Fig. A4-1A.

We turn now from the crystallization of potassic melts to the crystallization of basalt. The system forsterite–enstatite–quartz has, at least under common conditions, the same basic features as Fig. A4-1C, so that if rocks resemble in behavior this simple magnesian system, one basaltic liquid can lead to early formation of olivine and late formation of quartz; but if basaltic liquids differ from simple magnesian mixtures and behave more as in Fig. A4-1A, one liquid cannot give this double result. Does observation permit us to decide which pattern of behavior natural magmas follow?

It has been mentioned that in the Hawaiian Islands, for example, small quantities of quartz-bearing rocks accompany the large quantities of olivine-bearing rock. If we knew that all Hawaiian rocks were generated from one homogeneous initial magma pool, we could conclude that there was necessarily a peritectic relation as in Fig. A4-1C; but we do not know this and therefore two possibilities remain:

1. The parent magma of Hawaiian rocks is only slightly variable and the olivine and quartz derivatives are a consequence of a peritectic relation;

2. The parent magma of Hawaiian rocks, or the conditions of crystallization, are sufficiently variable to permit olivine-bearing rocks to form at one time and quartz-bearing rocks to form at another without a peritectic relation.

Another relevant observation is that some olivine phenocrysts in basalt are encrusted with pyroxene, to form a *corona*. There is no reason to expect this feature if pyroxene and olivine simply crystallize together eutectically, but it is exactly what would be expected if olivine is resorbed to permit pyroxene to grow. When pyroxene coronas are seen, therefore, it is correct to conclude that peritectic behavior has probably occurred. The absence of coronas is not an indicator one way or another.

Amphibole coronas on pyroxene are much more common but are less easy to interpret. In quickly cooled extrusive rocks, such coronas again indicate a peritectic relation, with pyroxene and magma being consumed during amphibole growth. However, in more slowly cooled, deep-seated rocks it is not so clear that the amphibole is a product of reaction between pyroxene and magma; it may also be a product of reaction between pyroxene and the late-stage aqueous fluid (stage 5 or stage 6 in Chap. 9); and in many field situations it may alternatively be a product of later metamorphism, entirely independent of magmatic processes.

Because of the coronas and because of the observations summarized in Fig. 7-2, Bowen summed up the possible peritectic relations in a *reaction series*, olivine–pyroxene–hornblende–biotite (Bowen, 1922). By stating that these four minerals constituted a series, he meant that any one could form by reaction from its predecessor and that a suitable liquid could generate all four in succession. However, he emphasized that liquids of other kinds could begin solidifying with the precipitation of pyroxene or hornblende and that these minerals were not necessarily preceded by olivine, any more than they are necessarily followed by biotite. Later experiments with both synthetic and natural mixtures have suggested, in addition, that the order of appearance can be changed—amphibole may appear after olivine but before pyroxene, or crystallization may start with amphibole and continue with olivine (Yoder and Tilley, 1962, pp. 448–53). Thus crystallization does not always proceed by the reactions implied in the reaction series, but as possibilities among other processes of petrogenesis, they rightly occupy a prominent position in current thinking.

APPENDIX FIVE

THE USE OF TRIANGULAR DIAGRAMS

TRIANGULAR COORDINATES

Those intending a serious study of petrology are advised to become familiar with plotting methods in triangular coordinates and to practice casting all the data in Chaps. 9 and 20 in this form.

1. A set of triangular coordinates is shown in Fig. A5-1A. If P, Q and R are three components, a combination of 20 per cent P, 30 per cent Q and 50 per cent R can be represented by a single point Y, and the composition represented by a single point Z can be read off as 70 per cent P, 20 per cent Q and 10 per cent R. The use of such coordinates is not confined to equilateral triangles; the same facts are just as well portrayed in diagram B.

2. The balance principle (Chap. 8) remains true in these diagrams. Thus in Fig. A5-2A, if a mixture A separates into b grams of B and c grams of C, then

$$b \times AB = c \times AC$$

and if the total weight is a grams, then

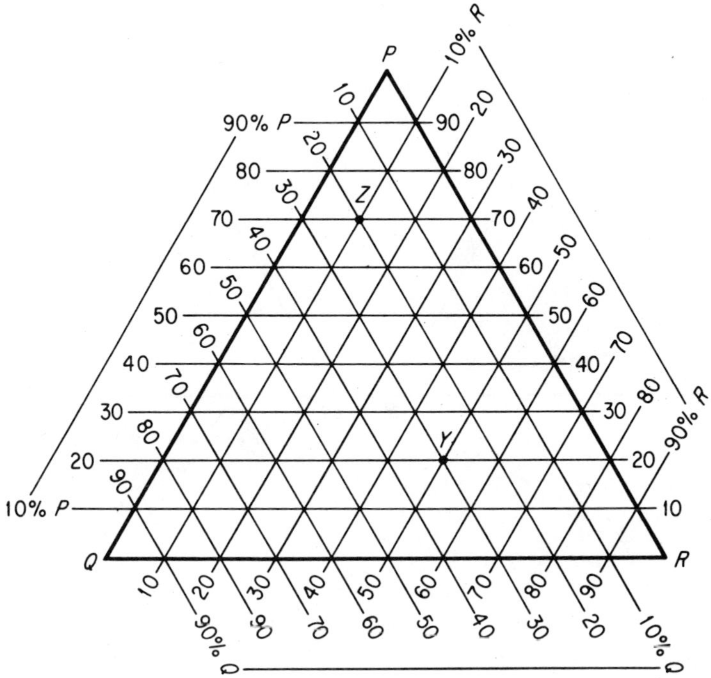

A

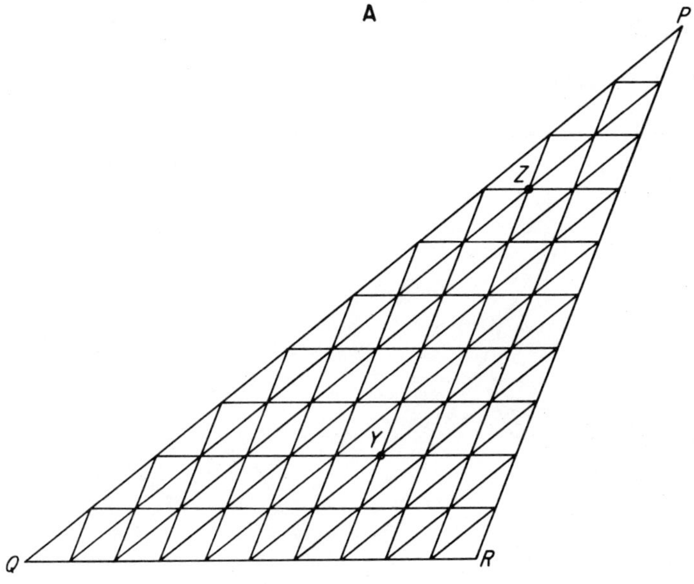

B

Fig. A5-1 Triangular coordinate systems.

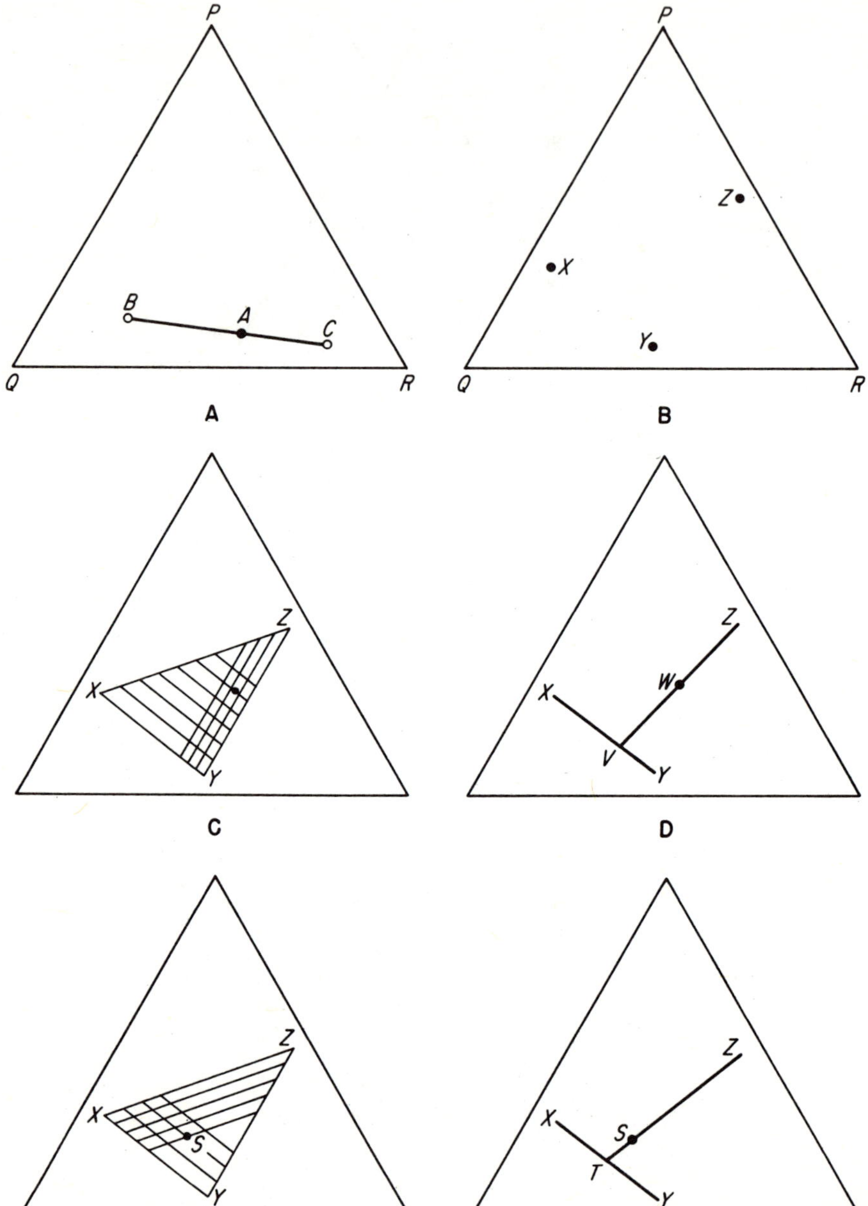

Fig. A5-2 Combination and partition of components.

$$b = a \times \frac{AC}{BC} \qquad c = a \times \frac{AB}{BC}$$

3. We may wish to show the bulk composition of a mixture of three phases such as 5 g of X, 10 g of Y and 15 g of Z, as in Fig. A5-2B. There are two ways to proceed:

a. convert to percentages (17 per cent X, 33 per cent Y, 50 per cent Z) and plot within XYZ as in diagram C;

b. perform two binary operations thus (diagram D): 5 g of X + 10 g of Y = 15 g of V, where V lies on XY and cuts off one-third of its length; 15 g of V + 15 g of Z = 30 g of W where W bisects VZ.

4. To find how much of phases X, Y and Z a given mixture S will generate, either of the above may be reversed. Diagram E shows the insertion of a grid that reads S = 40 per cent X, 40 per cent Y, 20 per cent Z; then quantity of $X = 0.40 \times$ quantity of S. Diagram F shows that s grams of S splits into $s \times \frac{4}{5}$ g of T and $s \times \frac{1}{5}$ g of Z, and then $s \times \frac{4}{5}$ g of T splits into $s \times \frac{2}{5}$ g of X and $s \times \frac{2}{5}$ g of Y.

5. The following rough rules apply:

a. When a mixture splits into three phases, it gives most of the phase nearest to the bulk composition and least of the phase furthest away;

b. The bulk composition of three phases lies closest to that of the phase of which most is present.

CRYSTALLIZATION AND MELTING IN THREE-COMPONENT SYSTEMS

The behavior portrayed in Fig. 9-2 is shown again in Fig. A5-3. Diagram A shows the binary systems, taking the components two at a time. Diagram B shows the temperature at which crystallization begins, for fluids of any composition within the system, and the subsequent behavior of one fluid. Diagram C shows the same behavior in the conventional triangular projection and diagram D shows how the liquidus temperatures, given by the height of the curved surface in diagram B, can be shown by contours in the projection.

Figure 9-4 is reduced in the same way in Fig. A5-4. In the ideal quartz–albite–potash feldspar system under 1 kb water pressure, the trough of low-melting compositions is somewhat displaced toward the albite side of the diagram, but the presence of a very small amount of anorthite in the albite will probably bring the trough to a more symmetrical position (Winkler, 1965).

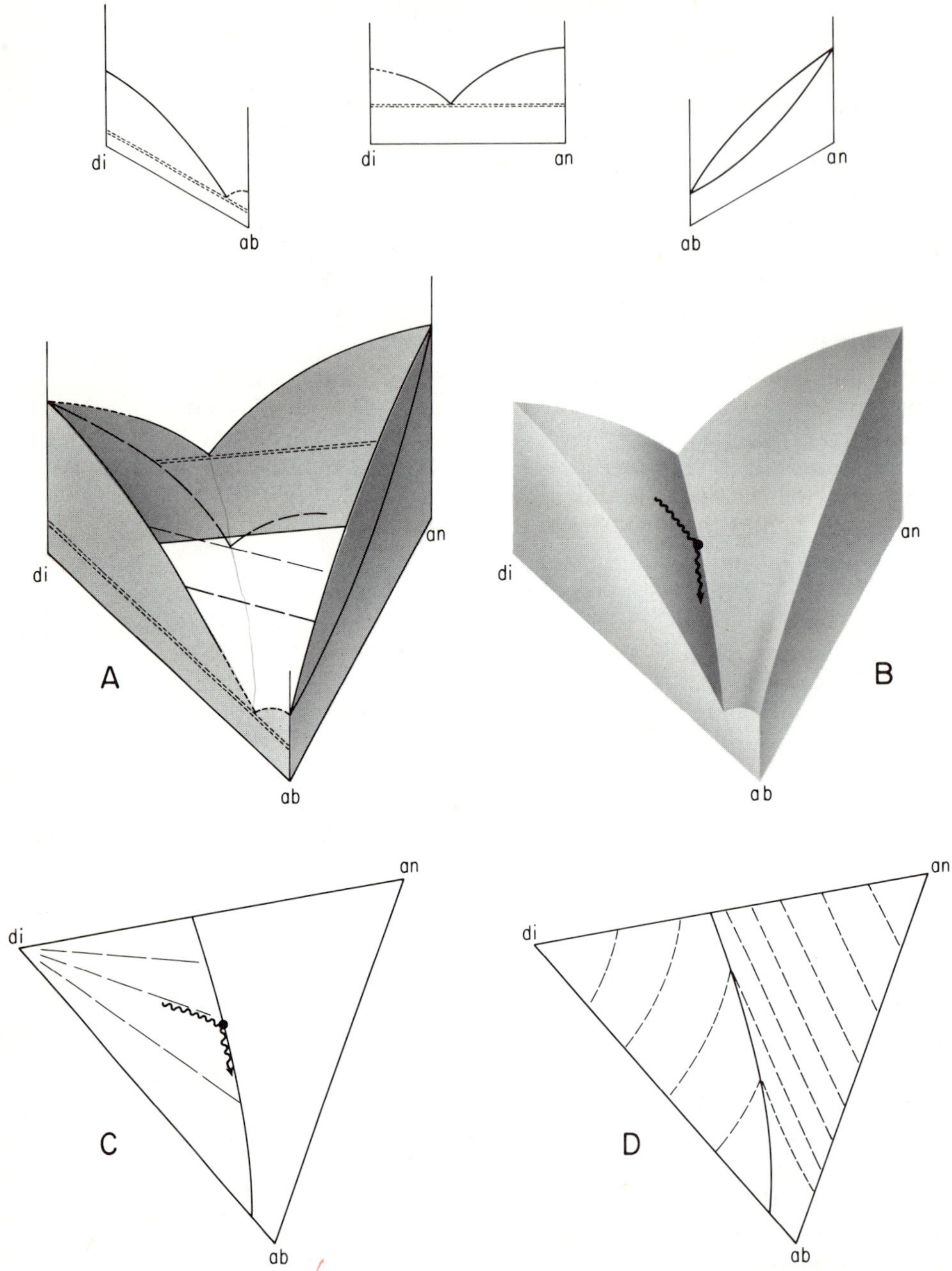

Fig. A5-3 The diopside–plagioclase system in triangular coordinates. As mentioned in Chap. 9, the solids that form from liquids of these three components are not always expressible in terms of these components alone; regions where departures from simple behavior are most marked are dotted (Schairer and Yoder, 1960; Bowen, 1915).

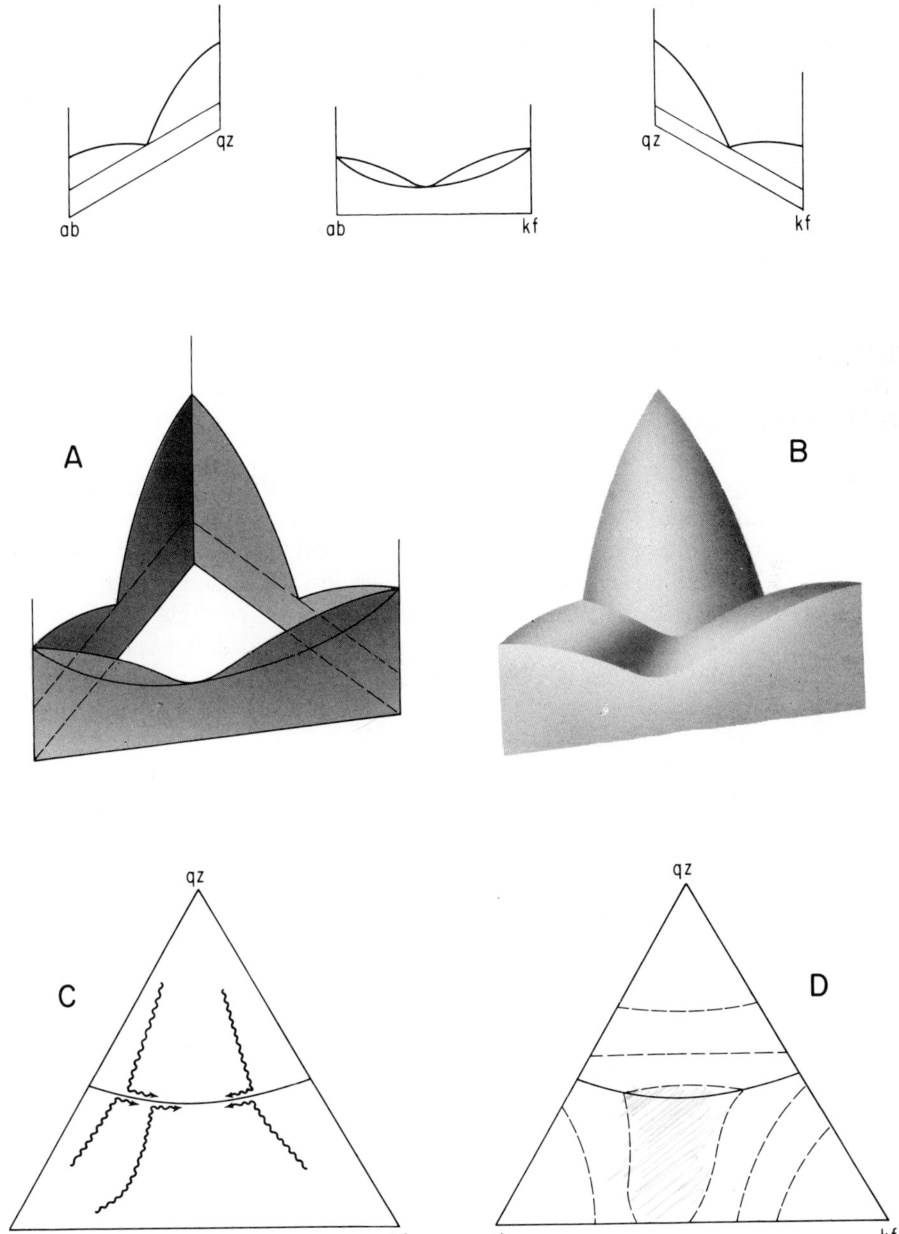

Fig. A5-4 The quartz-alkali feldspar system in triangular coordinates. (After Tuttle and Bowen, 1958.)

An elegant example of the power of these diagrams is the anorthite–nepheline–quartz system, Fig. A5-5. Because, under most circumstances, nepheline and quartz react to form albite, the system behaves as two independent systems, An–Ab–Ne and An–Ab–Qz. Both of these resemble the An–Ab–Di system in shape, though not, of course, in dimensions or temperatures, so that there is a ridge of high-melting compositions (plagioclases) down the middle of the main triangle. Because of this ridge, quite closely similar plagioclase-rich liquids may follow very divergent paths upon fractional crystallization, corresponding to the divergent trends in Fig. 7-4B.

THREE-COMPONENT DIAGRAMS IN METAMORPHISM

Reference to Fig. 20-2 shows that all the important minerals of metamorphism except quartz and plagioclase appear either on the base or on the right-hand face of the tetrahedron used. Consequently, three-component diagrams resembling these two faces are much used in discussions of metamorphism; they are known as *ACF* and *AKF* triangles respectively. It must be noticed, however, that the components that define these triangles are not simple oxides as in the tetrahedron (first defined in Fig. 13-6). Differences are as follows:

1. *AKF triangle:*
   a. The *A* component is *excess* Al_2O_3 that would remain after all

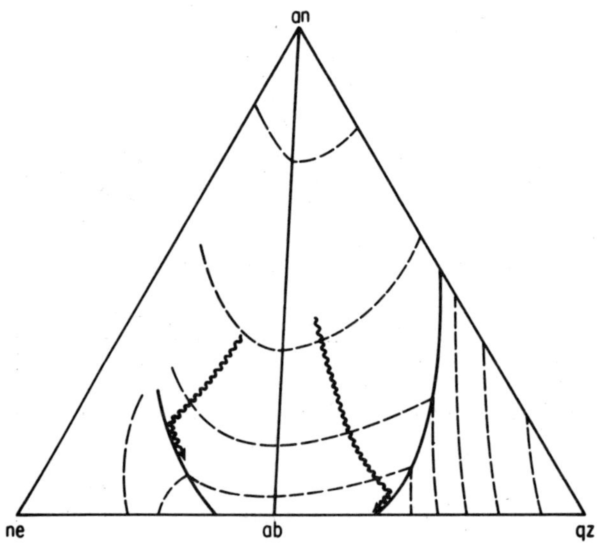

Fig. A5-5 The system quartz–anorthite–nepheline. (After Schairer, 1957.)

sodium, potassium and calcium in the rock has been turned to feldspar; in other words, amount of $A = Al_2O_3 - (Na_2O + K_2O + CaO)$, the subtraction being carried out in molecular amounts.*
b. The K-component is K_2O alone, without Na_2O.

EXAMPLE:

	wt. %	*mol. wt.*	*mol. amt.* (*wt. %/mol. wt.*)
SiO_2	79.7	60	...
Al_2O_3	7.25	102	071
Fe_2O_3	2.63	160	...
FeO	0.44	72	006
MgO	1.47	40	037
CaO	1.72	56	031
Na_2O	1.19	62	019
K_2O	1.00	94	011
H_2O	3.70	18	...
etc.	0.90	...	...

$A = 071 - (031 + 019 + 011) = 010$
$K = 011$
$F = 006 + 037 = 043$
In percentages, $A : K : F = 16 : 17 : 67$

This point is plotted at P in Fig. A5-6A, together with the more important minerals that appear in this triangle.

2. *ACF triangle:* This triangle is more used, the minerals are more variable and the rules defining the components more complex. The following are based on the original proposals of Eskola (1921).
a. If possible, estimate the ilmenite, sphene and magnetite by microscope examination. The results will come first in volume percentages; convert these to weight percentages i, s and m. Correct the weight percentages of Fe_2O_3, FeO and CaO by subtracting $(0.7\ m)$, $(0.5\ i + 0.3\ m)$ and $(0.3\ s)$ respectively.
b. Convert corrected weight percentages to molecular amounts and calculate A, C and F as follows:

$$A = Al_2O_3 + Fe_2O_3 - (Na_2O + K_2O)$$
$$C = CaO - (CO_2 + 3\ P_2O_5)$$
$$F = FeO + MgO + MnO$$

Although the corrections at (a) are usually small in comparison with the totals at (b), i, s and m need to be estimated with some precision; a quantitative method such as point counting is usually required.

*This procedure is taken from Turner and Verhoogen (1960, p. 514); for a slightly different procedure, see Winkler (1965, p. 44).

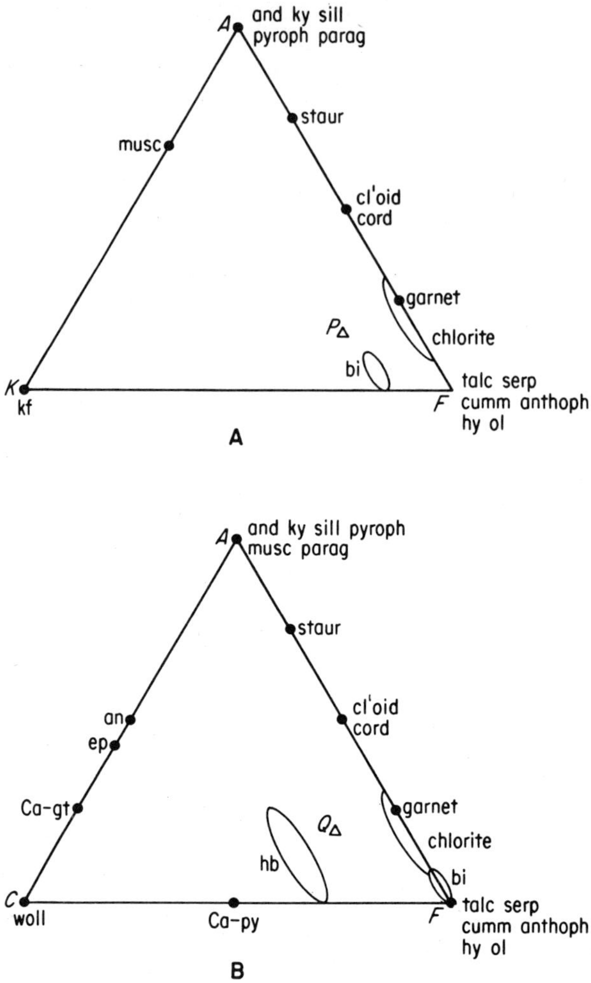

Fig. A5-6 *AKF* and *ACF* triangles.

EXAMPLE: $i = m = 1$ per cent; $s = 4$ per cent

	wt. %	*mol. wt.*	*mol. amt.*
SiO_2	59.1	60	. . .
Al_2O_3	15.4	102	150
Fe_2O_3	1.13 − 0.7 = 0.4	160	003
FeO	5.52 − 0.8 = 4.7	72	065
MgO	4.36	40	109
CaO	5.19 − 1.2 = 4.0	56	071
Na_2O	3.03	62	049
K_2O	4.44	94	047
MnO	0.03	71	0004
P_2O_5	0.31	145	002
CO_2	0.60	44	014
H_2O	0.53	18	. . .

$A = 150 + 3 - (49 + 47) = 57$
$C = 71 - (14 + 6) = 51$
$F = 109 + 65 = 174$
In percentages, $A : C : F = 20 : 18 : 62$

This point is plotted at Q in Fig. A5-6B, together with the more important minerals that appear in this triangle. The purpose of the somewhat elaborate definitions of A, C and F is to make the quantities match the aluminum, calcium and iron-magnesium that are available to enter the minerals shown. Thus if the analyzed rock carried biotite, hornblende and anorthite (in plagioclase), the position of Q would tell us how much of each to expect.

A more interesting question to consider, however, is what the rock could turn to upon metamorphism. By inspection of Q it appears that the rock could convert to a pyroxene-garnet-plagioclase rock (with quartz and potash feldspar not shown in the diagram). However, under metamorphism, there is no reason to suppose that the sphene, calcite etc. that were excluded before calculating the position of Q would remain inert; more probably they would enter reactions and could easily alter significantly the amount and nature of the pyroxene produced. For some purposes, therefore, a plot of the simple oxide sums ($Al_2O_3 + Fe_2O_3$), CaO and ($FeO + MgO + MnO$) would seem more appropriate, with no subtractions except Na_2O and K_2O to allow for the occurrence of feldspars. As elsewhere, the procedure to be followed must be subservient to the purpose in view.

Mineral Stability in Three-Component Systems

In addition to their use for showing rock compositions, triangular diagrams are also much used to show which minerals can coexist without reacting. As a simple example, the three components MgO, SiO_2 and H_2O will be discussed (Fig. A5-7; Bowen and Tuttle, 1949). The sequence (A) to (F) shows six assemblages stable at successively lower temperatures.

1. At high temperature, no hydrous solid is stable—the water present exists uncombined in contact with periclase and forsterite, or forsterite and enstatite, or enstatite and quartz; these possibilities are shown by the lines WP, WF, WE and WO.

2. On cooling, the first hydrous solid to form is brucite; as soon as brucite becomes stable, a composition in the triangle WPF will form either brucite, forsterite and excess water or brucite, forsterite and excess periclase, according to which component is exhausted first by the reaction

$$MgO + H_2O \rightleftharpoons Mg(OH)_2$$

$$P + W \rightleftharpoons B$$

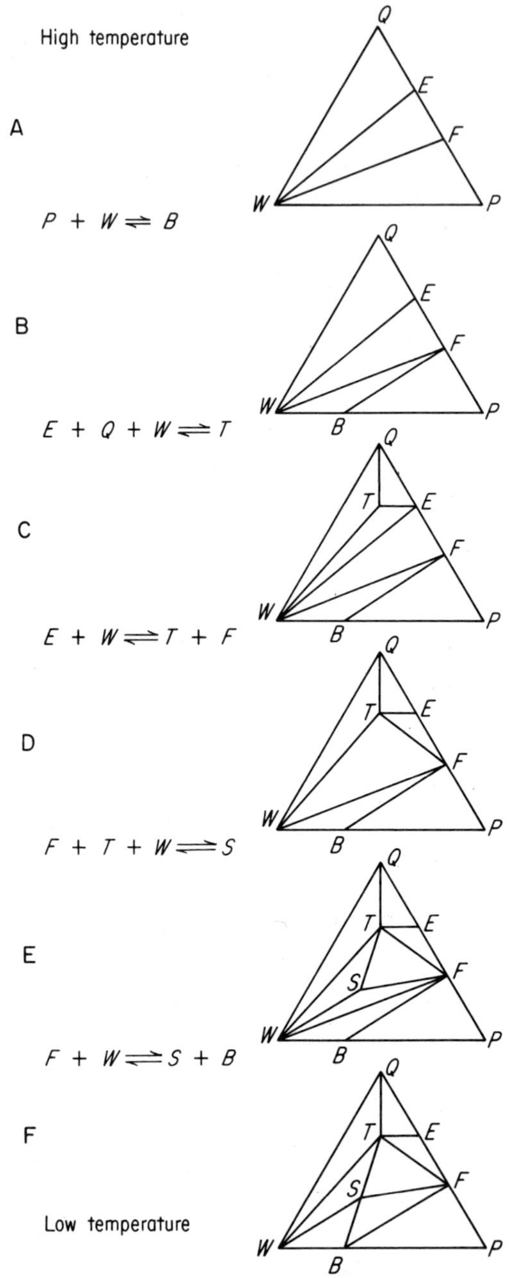

Fig. A5-7 Assemblages in the system periclase–quartz–water. *B*, brucite; *E*, enstatite; *F*, forsterite; *P*, periclase; Q, quartz; *T*, talc; *S*, serpentine; *W*, water (Bowen and Tuttle, 1949).

In fact, the appearance of the brucite point on the line *WP* symbolizes this reaction. Compositions in the triangle *WQF* show no change whatever as the temperature limit for brucite is passed.

3. The appearance of the talc point inside the triangle *WQE* symbolizes that the reaction

$$3E + Q + W \rightleftharpoons T$$

is the next to occur. If the composition is suitable, i.e., within the triangle *WQE*, the reaction will run until one reagent is exhausted and will generate one of the assemblages *W–T–Q*, *Q–T–E* or *E–T–W*. If the composition is unsuitable, i.e., within *WPE*, no change occurs as this reaction point is passed; an initial assemblage *W–E–F*, for example, would not yet have reacted at all.

Fig. A5-8 **The three types of reaction in a generalized three-component system.**

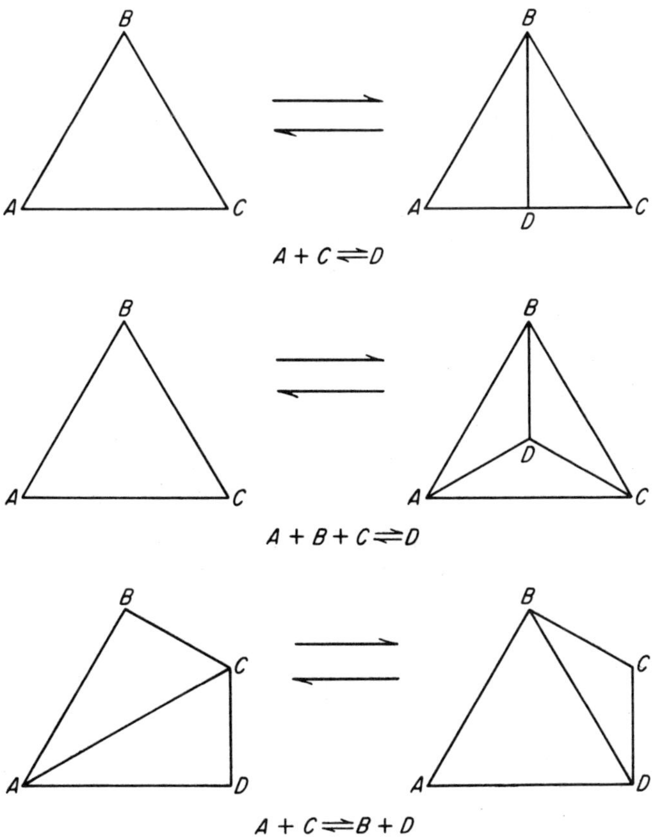

4. The next change with falling temperature is that the pair E–W ceases to be stable; the enstatite hydrates to talc thus:

$$5E + W \rightleftharpoons T + F$$

Below this point a composition in the quadrilateral $TEFW$ forms either the assemblage T–E–F (all water used up) or the assemblage T–W–F (all enstatite used up). This change is represented by the join WE giving way to the join TF.

5. The appearance of serpentine within WTF resembles the appearance of talc within WQE at (C).

6. The replacement of the join WF by the join SB resembles the replacement of the join WE by the join TF at (D).

The number of ways in which a three-component system can change by reaction is clearly very limited; in fact, the three types of change at (b), (c) and (d) exhaust the possibilities; these are shown in generalized form in Fig. A5-8. In four-component systems, such as Fig. 20-2, lines can still be drawn to show which minerals coexist, and every reaction in Tables 20-1 and 20-2 could be represented by a rearrangement of such lines. However, it has been emphasized that the order in which reactions occur in these tetrahedra is variable, so that the complete set of possible diagrams is voluminous. For further examples of the method by which a metamorphic change can be traced through a sequence of diagrams, reference must be made to a more detailed work (see, for example, Figs. 15 through 26 in Winkler, 1965).

APPENDIX SIX

REFERENCES AND FURTHER READING

Of the following two lists, the purpose of the references is simply to state the authority for various assertions in the text—the works mentioned are not necessarily recommended for reading in conjunction with this book. The suggestions for further reading are directed more to the purpose of guiding the reader who will, in many cases, profit by turning to a more advanced text before progressing into the journal literature. Titles have therefore been selected that provide:

1. More detailed information than is given here;
2. Lists of references arranged by topic that will guide the reader effectively into the field of his enquiry.

It is believed that these intermediate texts will be more useful than any list of original papers that could be included in a book at this introductory level.

REFERENCES

Allen, J. R. L., 1964, "Sedimentation in the Modern Delta of the River Niger, West Africa," in *Deltaic and Shallow Marine Deposits*, ed. L. M. J. U. van Straaten. Amsterdam: Elsevier Publishing Company, 26–34.

Anderson, G. M. and C. W. Burnham, 1965, "The Solubility of Quartz in Supercritical Water." Amer. J. Sci., **263,** 494–511.

Arrhenius, G., 1963, "Pelagic Sediments," in *The Sea,* v. 3, ed. M. N. Hill. New York: Interscience Publishers.

Baas Becking, L. G. M., I. R. Kaplan and D. Moore, 1960, "Limits of the Natural Environment in Terms of pH and Oxidation–Reduction Potentials." Jour. Geol., **68,** 243–84.

Barth, T. F. W., 1962, *Theoretical Petrology* (2nd ed.). New York: John Wiley & Sons, Inc.

Birch, F., ed., 1942, *Handbook of Physical Constants* (1st ed.). Geol. Soc. Amer. Spec. Pap. 36.

Bowen, N. L., 1913, "The Melting Phenomena of the Plagioclase Feldspars." Amer. J. Sci., **35,** 577–99.

———, 1915, "The Crystallization of Haplobasaltic, Haplodioritic and Related Magmas." Amer. J. Sci., **40,** 161–85.

———, 1922, "The Reaction Principle in Petrogenesis." Jour. Geol., **30,** 177–98.

Bowen, N. L. and O. F. Tuttle, 1949, "The System $MgO–SiO_2–H_2O$." Geol. Soc. Amer. Bull., **60,** 439–60.

———, 1950, "The System $NaAlSi_3O_8–KAlSi_3O_8–H_2O$." Jour. Geol., **58,** 489–511.

Bradley, W. F., R. E. Grim and W. C. Clark, 1937, "Study of the Behavior of Montmorillonite on Wetting." Zeit. Krist., **97,** 216–22.

Burnham, C. W., 1959, "Contact Metamorphism of Magnesian Limestones at Crestmore, California." Geol. Soc. Amer. Bull., **70,** 879–920.

Chilingar, G. V. and L. Knight, 1960, "Relationship between Pressure and Moisture Content of Kaolinite, Illite, and Montmorillonite Clays." Am. Assoc. Petroleum Geologists Bull., **44,** 101–6.

Christie, O. H. J., 1962, "Feldspar Structure and the Equilibrium between Plagioclase and Epidote—Discussion." Amer. J. Sci., **260,** 149–57.

Clark, S. P., ed., 1966, *Handbook of Physical Constants* (2nd ed.). Geol. Soc. Amer. Mem. 97.

√ Cotton, C. A., 1952, *Volcanoes as Landscape Forms.* New York: John Wiley & Sons, Inc.

Dawson, J. B., 1962, "The Geology of Oldoinyo Lengai." Bull. Volcanologique, **24,** 349–87.

Degens, E. T., 1965, *Geochemistry of Sediments.* Englewood Cliffs, N. J.: Prentice-Hall, Inc.

Doll, C. G., ed., 1961, *Centennial Geologic Map of Vermont.* Montpelier, Vt.: Vermont Development Department.

Eskola, P., 1921, "The Mineral Facies of Rocks." Norsk Geol. Tidsskr., **6,** 143–94.

Friedman, G. M., 1961, "Distinction between Dune, Beach, and River Sands from Their Textural Characteristics." J. Sed. Petr., **31,** 514–29.

Garrels, R. M. and C. L. Christ, 1965, *Solutions, Minerals and Equilibria.* New York: Harper & Row.

Goldberg, E. D. and M. Koide, 1962, "Geochronological Studies of Deep-sea Sediments by the Ionium-Thorium Method." Geochim. et Cosmochim. Acta, **26,** 417–50.

Goranson, R. W., 1938, "Silicate-Water Systems: Phase Equilibria in the $NaAlSi_3O_8$–H_2O and $KAlSi_3O_8$–H_2O Systems at High Temperatures and Pressures." Amer. J. Sci., **35–A,** 71–91.

Graton, L. C., 1945, "Conjectures Regarding Volcanic Heat." Amer. J. Sci., **243A,** 135–259.

Grim, R. E., 1953, *Clay Mineralogy.* New York: McGraw-Hill Book Company, Inc.

Holser, W. T. and G. C. Kennedy, 1958, "Properties of Water. Part IV. Pressure-Volume-Temperature Relations of Water in the Range 100–400°C and 100–1400 bars." Amer. J. Sci., **256,** 744–53.

———, 1959, "Properties of Water. Part V. Pressure-Volume-Temperature Relations of Water in the Range 400–1000°C and 100–1400 bars." Amer. J. Sci., **257,** 71–77.

Hubbert, M. K., 1937, "Theory of Scale Models as Applied to the Study of Geologic Structures." Geol. Soc. Amer. Bull., **48,** 1459–1520.

Jaeger, J. C., 1961, "The Cooling of Irregularly Shaped Igneous Bodies." Amer. J. Sci., **259,** 721–34.

Johannsen, A., 1939, *A Descriptive Petrography of the Igneous Rocks* (2nd ed.). Chicago: University of Chicago Press.

Jones, G. E., R. L. Starkey, H. W. Feely and J. L. Kulp, 1956, "Biological Origin of Native Sulfur in Salt Domes of Texas and Louisiana." Science, **123,** 1124–25.

Kennedy, G. C., G. J. Wasserburg, H. C. Heard and R. C. Newton, 1962, "The Upper Three-Phase Region in the System SiO_2–H_2O." Amer. J. Sci., **260,** 501–21.

Lane, E. W., *et al.*, "Report of the Subcommittee on Sediment Terminology." Amer. Geophys. Union Trans., **28,** 936–38.

Larsen, E. S., 1945, "Time Required for the Crystallization of the Great Batholith of Southern and Lower California." Amer. J. Sci., **243A,** 399–416. (See also Jaeger, 1961.)

McIntire, W. L., 1963, "Trace Element Partition Coefficients—a Review of Theory and Applications to Geology." Geochim. et Cosmochim. Acta, **27,** 1209–64.

Menard, H. W., 1964, *Marine Geology of the Pacific.* New York: McGraw-Hill Book Company, Inc.

Morey, G. W. and J. M. Hesselgesser, 1951, "The Solubility of Some Minerals in Superheated Steam at High Pressures." Econ. Geol., **46,** 821–35.

Murray, G. E., 1960, "Geologic Framework of Gulf Coastal Province of United States," in *Recent Sediments, Northwest Gulf of Mexico*, ed. F. P. Shepard, F. B. Phleger and T. H. van Andel. Tulsa: American Association of Petroleum Geologists, 5–33.

Naughton, J. J. and Y. Fujikawa, 1959, "Measurement of Intergranular Diffusion in a Silicate System: Iron in Forsterite." Nature, **184**, No. 4688, B.A. 54–56.

Niggli, P., 1954, *Rocks and Mineral Deposits,* trans. R. L. Parker. San Francisco: W. H. Freeman and Company.

Nockolds, S. R., 1954, "Average Chemical Compositions of Some Igneous Rocks." Geol. Soc. Amer. Bull., **65,** 1007–32.

Peacock, M. A., 1931, "Classification of Igneous Rock Series." Jour. Geol., **39,** 54–67.

Ramberg, H., 1952, *The Origin of Metamorphic and Metasomatic Rocks.* Chicago: University of Chicago Press.

———, 1963, "Experimental Study of Gravity Tectonics by Means of Centrifuged Models." Bull. Geol. Inst. Univ. Uppsala, **62,** 1–97.

Richardson, W. A. and G. Sneesby, 1922, "The Frequency-Distribution of Igneous Rocks." Miner. Mag., **19,** 306–13.

Rittmann, A., 1962, *Volcanoes and Their Activity,* trans. E. A. Vincent. New York: John Wiley & Sons, Inc.

Roberts, W. M. B., 1965, "Recrystallization and Mobilization of Sulfides at 2,000 Atmospheres and in the Temperature Range 50°–145°." Econ. Geol., **60,** 168–71.

Ross, C. S., 1955, "Provenience of Pyroclastic Materials." Geol. Soc. Amer. Bull., **66,** 427–34.

Runcorn, S. K., 1964, "A Growing Core and a Convecting Mantle," in *Isotopic and Cosmic Chemistry,* ed. H. Craig, S. L. Miller and G. J. Wasserburg. Amsterdam: North-Holland Publishing Company, 321–40.

Schairer, J. F., 1957, "Melting Relations of the Common Rock-forming Oxides." Am. Ceramic Soc. Jour., **49,** 215–35.

Schairer, J. F. and N. L. Bowen, 1935, "Preliminary Report on Equilibrium-Relations Between Feldspathoids, Alkali-Feldspars, and Silica." Amer. Geophys. Union Trans., 16th ann. mtg., 325–28.

Schairer, J. F. and H. S. Yoder, Jr., 1960, "The Nature of Residual Liquids from Crystallization, with Data on the System Nepheline–Diopside–Silica." Amer. J. Sci., **258A**, 273–83.

Scruton, P. C., 1960, "Delta Building and the Deltaic Sequence," in *Recent Sediments, Northwest Gulf of Mexico,* ed. F. P. Shepard, F. B. Phleger and T. H. van Andel. Tulsa: American Association of Petroleum Geologists, 82–102.

Shepard, F. P. and R. Young, 1961, "Distinguishing between Beach and Dune Sands." J. Sed. Petr., **31,** 196–214.

Shumway, G., 1960, "Sound Speed and Absorption Studies of Marine Sediments by a Resonance Method—Part III." Geophysics, **25,** 659–82.

Sippel, R. F., 1963, "Sodium Self-diffusion in Natural Minerals." Geochim. et Cosmochim. Acta, **27,** 107–20.

Smithson, S. B., 1963, "Granite Studies: I. A Gravity Investigation of Two Precambrian Granites in South Norway." Norges Geol. Undersøkelse, **214,** 53–140.

Streckeisen, A., 1965, "Die Klassifikation der Eruptivgesteine." Geol. Rdsch., **55,** 478–91.

Travis, R. B., 1955, "Classification of Rocks." Quarterly of the Colorado School of Mines, **50,** 1–98.

Turner, F. J. and J. Verhoogen, 1960, *Igneous and Metamorphic Petrology* (2nd ed.). New York: McGraw-Hill Book Company, Inc.

Turner, F. J. and L. E. Weiss, 1963, *Structural Analysis of Metamorphic Tectonites.* New York: McGraw-Hill Book Company, Inc.

Tuttle, O. F. and N. L. Bowen, 1958, *Origin of Granite in the Light of Experimental Studies.* Geol. Soc. Amer. Mem. 74.

Twenhofel, W. H., 1950, *Principles of Sedimentation* (2nd ed.). New York: McGraw-Hill Book Company, Inc.

Vanoni, V. A., 1946, "Transportation of Suspended Sediments by Water." Am. Soc. Civil Eng. Trans., **111,** 67–102.

Walton, M., 1960, "Molecular Diffusion Rates in Supercritical Water Vapor Estimated from Viscosity Data." Amer. J. Sci., **258,** 385–401.

Weill, D. F. and W. S. Fyfe, 1964, "The Solubility of Quartz in H_2O in the Range 1000–4000 bars and 400–550°C." Geochim. et Cosmochim. Acta, **28,** 1243–56.

Weller, J. M., 1959, "Compaction in Sediments." Am. Assoc. Petroleum Geologists Bull., **43,** 273–310.

Williams, G., 1964, "Some Aspects of the Eolian Saltation Load." Sedimentology, **3,** 257–87.

Winkler, H. G. F., 1965, *Petrogenesis of Metamorphic Rocks.* New York: Springer-Verlag New York Inc.

Yoder, H. S., Jr., 1958, "Effect of Water on the Melting of Silicates." Carnegie Inst. Wash. Year Book 57, 189–91.

Yoder, H. S., Jr. and C. E. Tilley, 1962, "Origin of Basalt Magmas: an Experimental Study of Natural and Synthetic Rock Systems." J. Petr., **3,** 342–532. Material from the Journal of Petrology is used by permission of the Clarendon Press, Oxford.

FURTHER READING

All Parts

Holmes, A., *Principles of Physical Geology* (2nd ed.). New York: The Ronald Press Company, 1965. Copious illustrations and facts about almost every topic mentioned; simply written, without elaborate theory and speculation.

Part I

Jacobs, J. A., R. D. Russell and J. T. Wilson, *Physics and Geology*. New York: McGraw-Hill Book Company, Inc., 1959. Chapters 2, 4 and 5 refer to topics from Part I.

Part II

Barth, T. F. W., *Theoretical Petrology* (2nd ed.). New York: John Wiley & Sons, Inc., 1962.

Rittmann, A., *Volcanoes and Their Activity*, trans. E. A. Vincent. New York: John Wiley & Sons, Inc., 1962. Extends from volcanoes themselves to all aspects of magmatic activity.

Turner, F. J. and J. Verhoogen, *Igneous and Metamorphic Petrology* (2nd ed.). New York: McGraw-Hill Book Company, Inc., 1960. Less recent than Barth (op. cit.) but more comprehensive in its coverage.

Levin, E. M., C. R. Robbins and H. F. McMurdie, *Phase Diagrams for Ceramists*. Columbus, Ohio: American Ceramic Society, Inc., 1964. Contains a concise explanatory section as well as a great number of diagrams of importance for petrology.

Readers should also be aware of the gradual publication (chapter by chapter) of U.S.G.S. Professional Paper 440 *Data of Geochemistry* (6th ed.), ed. M. Fleischer. This document, when complete, will summarize the chemical data with which petrology must deal.

Part III

Dunbar, C. O. and J. Rodgers, *Principles of Stratigraphy*. New York: John Wiley & Sons, Inc., 1957.

Krumbein, W. C. and L. L. Sloss, *Stratigraphy and Sedimentation* (2nd ed.). San Francisco: W. H. Freeman and Company, 1963.

Termier, H. and G. Termier, *Erosion and Sedimentation*, trans. D. W. Humphries and E. E. Humphries. London: D. van Nostrand Company, Ltd., 1963.

Weller, J. M., *Stratigraphic Principles and Practice*. New York: Harper & Row Publishers, Inc., 1960.

These four books all contain discussions of rock-forming processes and their influence on the nature of resulting rock.

Pettijohn, F. J., *Sedimentary Rocks* (2nd ed.). New York: Harper and Brothers, 1957.

Hatch, F. H. and Rastall, R. H., *Petrology of the Sedimentary Rocks* (4th ed.), rev. by J. T. Greensmith. London: Thomas Murby & Co., 1965.

These two books survey known rocks and proceed derivatively to discussion of possible processes.

Degens, E. T., *Geochemistry of Sediments*. Englewood Cliffs, N. J.: Prentice-Hall, Inc., 1965.

Carbonate Rocks, ed. G. V. Chilingar, R. W. Fairbridge and H. J. Bissell. Amsterdam: Elsevier Publishing Company, 1966.

Two valuable synopses of special aspects of sedimentary rock formation.

Part IV

Ramberg, H., *The Origin of Metamorphic and Metasomatic Rocks*. Chicago: University of Chicago Press, 1952. An exemplary attempt to reconcile what appears to have occurred in the field with what physical chemistry shows to be possible.

Winkler, H. G. F., *Petrogenesis of Metamorphic Rocks*. New York: Springer-Verlag New York Inc., 1965.

Also works by Barth and Turner and Verhoogen already listed under Part II.

INDEX

References to illustrations are in bold type.

Aa (*see* Lava, blocky)
Abbreviations, 307
ACF triangle, 339, **340**
Acid aqueous solution, 175 (*see also* pH)
Acid rock, 33
  definition, 53
Actinolite:
  formula, 312
  reactions, 242
Activity of solute, 173
Adamellite, definition, 54
Adirondack anorthosite body, **47**, 103
Adirondack mountains, charnockite, 65
Adsorption, 172, 186, 213
Aegirine:
  formula, 312
  occurrence, 61
Age determination (*see* Dating)
Agglomerate, **39**, 40
  distinction from cataclasite, 232
  recognition, 43
AKF triangle, 338, **340**
Albite, formula, 313 (*see also* Plagioclase *and* Feldspar)
Algae, 140
  and calcium in earth's crust, 16
  varieties, 134
Alkali feldspar, 313
  crystallization and melting, 94, **95**
    in hydrous conditions, 293
Alkali Lime Index (A.L.I.), 61, **62**
Alkali rocks:
  from basaltic magma, 338
  and carbonatite, 100
  composition trend, **62**
  definition, 53, 318

Alkali rocks (*cont.*):
genesis, 99, 338
and limestone assimilation, 99
mineralogy, 61
occurrence, 65
Alkali trend in magmatic rocks, 61
Allen, J. R. L., 152
Allochthon, 146
Alnö, Sweden, 65
Amphibole; varieties, 312 (*see also* Actinolite, Hornblende *etc.*)
Amphibolite:
definition, 230
origins, 246
(*see also* Gabbro, Hornblende)
Amygdule, **39**, 40
Analcite, formula, 175
Anatexis (*see* Melting)
Andalusite:
formula, 314
reactions, 243
Anderson, G. M., 107
Andesine, definition, 313
Andesite, definition, **54**, 318
Anhedral crystal, **39**, 41
Anhydrite, occurrence, 143
Anorthite:
formula, 313
reactions, 92, 235, 242, 272
(*see also* Plagioclase *and* Feldspar)
Anorthosite:
definition, 63, 229, 318
genesis, 103
from basaltic magma, 92
mineralogy, 63
occurrence, **47**, 65
Anthophyllite:
formula, 312
reactions, 242
Antiperthite, definition, 313
Apatite:
in alkali rocks, 61
formula, 314
Aplite, 292
Appalachian Mountains, 66
Apron, reef, 135, **136**
Aragonite, precipitation, 141, 190 (*see also* Calcite, Calcium *and* Carbonate solutions in water)
Arenite, definition, 165
Argillaceous rock, definition, 167
Argillite, definition, 167
Arkose:
definition, 167, **168**
Arkose (*cont.*):
in metamorphism, 239, **240**
Arrhenius, G., 137
Ash, volcanic, 13
definition, 40
(*see also* Tuff)
Assimilation, 99, 254
Atmosphere:
energy distribution, 19
unit of pressure, 7, 307
Atomic weights, 311
Augen, **233**, 240
formation, 233
Augite (pyroxene), metamorphic reactions, 242
Australia, Precambrian, 6
Authigenic crystals, definition, 253
Autochthon, 146
Autometamorphism:
definition, 226
of diorite, 246
progressive, 249
Avogadro's number, 307

Baas Becking, L.G.M., 185
Bahama Islands, 141
Baking, marginal, **39**, 42
Balance principle, 69, **71**, 82, 93
in metamorphism, 235
in paratectic behavior, 327
in ternary systems, 332, **334**
Bank of detritus, 131
Barth, T. F. W., 9, 16, 98, 268, 273, 315, 350
Bar (unit of pressure), 7, 307
Basalt, 6
definition, **54**, 318
melting conditions, **9**
reaction with water, 63
source of magma, 9, 297
Basalt flood, 35, **36**
Basaltic magma:
crystallization, 86–94, 330
to alkali rocks, 338
source, 9, 14, 297
Basement rocks, density and mineralogy, 6
Basic rocks, 33
definition, 53
Batholith:
cooling time, **112**, 113
dimensions, **47**, 48
extent of melting, **115**, 116
growth in geosyncline, **300**, 301

Batholith (*cont.*):
inhomogeneity, 98
shape, **117**, 118
Bauxite, 161
Beach deposits, 154, 203
imbrication, **148**, 149
Bedding in metamorphic rocks, 251, 282, **283**
Bed load, 198
Bicarbonates in aqueous solution, 176
Binary behavior, 69–85
departure from, 69, 78, **87**
Binary diagram, 69
with paratectic point, 327, **328**, **329**
Binary scale, **70**
Binary system, 67, 237
Biotite:
formation from clay, 182
formula, 313
in magmatic rocks, **58**, 61, 63 90, 331
metamorphic reactions, 242
reaction to hypersthene, 63, 295
Birch, F., 8, 315
Bissell, H. J., 350
Bomb, volcanic, 40
range of ejection, 120
Borers, agent for reef disintegration, 135
Bottomset bed, 152, **153**
Boudinage, 294
Boulder, definition, 165
Boulder clay, 156
Bowen, N. L., 9, 83, 88, 95, 96, 292, 328, 331, 336, 337, 341, 342
Bradley, W. F., 186
Breccia:
definition, 165
volcanic (*see* Agglomerate)
Brine, 67
evaporation, 69
evaporite deposit, 143
British Columbia batholith, **47**, 48
Brucite, reactions 237, 341
Burnham, C. W., 107, 273
Bushveld complex, 46
Bytownite, definition, 313

Calcite:
cement, 179
in lakes, 131
metamorphic reactions, 242
precipitation, 141, 190
stability in sediments, **185**
(*see also* Calcium)
Calcium:
carbonate solutions, 176
fixation by organisms, 16, 135, 137
precipitated by organisms, 141
Calcium silicate (*see* Calc-silicate)
Calc-silicate minerals:
at magma/limestone contact, 100
in metamorphism, 249
Calc-silicate rock, definition, 230
Calorie, 307
Canyon, submarine, 206, **207**
Carbon, escape from mantle, 99
Carbonate solutions in water, 176
Carbonatite:
and alkali rocks, 100
genesis, 98
mineralogy, 64
occurrence, 66
Carbon dioxide:
in aqueous solutions, 176
equilibrium with periclase, 238
influence on metamorphic reactions, 238, 259
product of metamorphism, 259
in seawater, 141
in soil, 177
Caspian Sea, U.S.S.R., 143
Cataclasite, **233**
definition, 231
recognition, 232
Cementation, 158
by calcite, 179
localized, **159**
of quartz sand, 158, **159**, **214**
of spheres, **179**
Channel, on ocean floor, 206, **207**
Charnockite:
definition, 63, 229
occurrence, 65
Chemical potential (*see* Free energy)
Chert:
definition, 167
diagenetic, 142
from organisms, 138
precipitation, 192
Chibina alkali intrusive, U.S.S.R., **47**, 65
Chile, nitrate, 143
Chilingar, G. V., 212, 350
Chilling, marginal, **39**, 42
Chimborazo, Ecuador, 37
Chlorite:
formula, 313
in orogeny, 299
reactions, **242**

Chlorite (*cont.*):
reaction to garnet, 259, **260**, 261, **262**, **264**, 299
in sediment, 159, 181
source of magmatic water, 299
Chloritoid:
formula, 314
reactions, 242
Christ, C. L., 176, 183, 184, 186
Christie, O. H. J., 262
Chromium, in ultrabasic rocks, 64
CIPW classification, 325
Clapeyron's Equation, 258
Clark, S. P., 315
Clark, W. C., 186
Classes, genetic and observational, 44
Classification:
magmatic rocks, 52, **54**, 316, **317**
normative, **325**, 326
metamorphic rocks, 227–31, **228**
procedure, 24, 227, 231
purpose, 52, 164
sedimentary rocks, 25, 164–70
veins, 286, **288**
Clay:
adsorption properties, 186, 213
china, 254
compaction, 211, **212**
definition, 165
detrital, 169
(*see also* Clay minerals)
Clay flake, dimensions, **188**
Clay minerals:
absorption of water, 186, **188**
adsorption of alkalis, 187
composition range, **234**
diagenesis, 97, 159, **160**
formation, 180
formulas, **160**, 313
in metamorphism, 182, 240
rate of formation, 192
reactions, 242, 243
(*see also* Kaolinite, Montmorillonite etc.)
Clinopyroxene (*see* Pyroxene)
Clinozoisite, composition, 313
Clinton ore, 142
Coal, 139
Coesite, 20
Colloidal particles, 153, 187, **188**
Colony, organic, 131, 136
Columbia River basalt, **36**
Compaction, 211, **212**, **213**, 214
differential, **138**, 139
Compaction (*cont.*):
quartz sand, **214**
Components, number of, 68, 236
Concentration gradient and diffusion:
in metamorphism, 277
in suspensions, 195
Concentration of solute:
effective, 173
effect on free energy, 172
effect on stability field, 185, 261
thermodynamic, 173
Concentration profile:
suspensions, **196**, **200**, **206**
across vein, 289
Concordant bodies:
definition, 45
dimensions, **46**
Conduction:
heat through crust, 8, 15
around hot intrusion, 109–14
and viscous flow, 209
Conductivity, 109
of earth's crust, 8
values for rocks, 315
Cone, talus, **136**
Conglomerate, definition, 165
Contamination of magma, 100, 254
Continental shelf, present-day sediments, 144
Continental slope:
definition, 145, **207**
turbidity currents, 205, **207**
Continents:
calcium depletion, 16
depth to M discontinuity, 4
elevation, 6
layered structure, **4**, 6, 301
mineralogy, 6
Convection in earth's crust, 15, **17**, **117**, 118
Convergence, metamorphic, 246
Cooling:
during crystallization:
quartz-albite mixture, 75, **76**
quartz-water mixture, **80**
of intrusions, 109–14
Coordinates:
tetrahedral, **150**
triangular, 332, **333**
Coral (*see* Reef)
Cordierite:
formula, 314
metamorphic reactions, 242

Corona, mineral, 330 (*see also* Reaction rim)
Cosmic dust, 130
Cotectic crystallization, 89, 335, **336**, 338
Cotectic curve, **88**, **336**
Cotton, C. A., 36
Critical phenomena, 79
Critical point, water, **106**
Crust of the earth, 3, **4**
  fractionation by orogeny, 12, 297, **300**, 301
Crystallization of synthetic melts (*see* Systems)
Crystallization sequence:
  in magma, 86–94, 330
    to alkali rocks, 338
  in synthetic mixtures (*see* Systems)
Cummingtonite:
  formula, 312
  reactions, 242
Cumulative curve, grain-size frequency analysis, **166**
Current:
  density, 132
  submarine, viscous drag, 209
  turbidity, 132, 205, **207**
Cycle:
  convective, 16, **17**
  of erosion and remelting, 11, **12**, 97, 297

Dacite, definition, **54**, 317
Dating:
  ionium-thorium, 216
  potassium-argon, 218
  uranium, 303
Dawson, J. B., 66, 99
Dead Sea, 142
Deccan basalt, India, **36**
Defects, crystal lattice, 274, **275**
Definitions, rules for erecting, 26
Deformation models, 305
Degens, E. T., 159, 188, 350
Degrees of freedom, 239
Dellenite, definition, 54
Delta, **136**, 151, **152**, **153**
  compaction, 214
  slumping, 147
Density:
  continental rocks, 6
  crustal rocks, 5
  distribution in crust, 13
  distribution in earth, 14
Density (*cont.*):
  and intrusion mechanisms, 16, **17**, 116, **117**, 301
  minerals, 312
  oceanic rocks, 6
  rock types, 315
  water, 105, **106**, 290
Density current, 132
Density inversion, 14, 16, 116, 301
Description and definition, 26
Description of rocks:
  procedure, 45, 55
  purpose, 54
Desert, North African, 154
Desert valley gravel, 144
Detritus, definition, 169
Diabase, definition, 54
Diagenesis, 157–61, 211–14
  absorption of alkali metals, 187
  of clay, 97, 181
  distinction from metamorphism, 223, **225**
  localized, **159**
  marine clay, 97
  physics, 211
  at shoreline, 158
Diameter of grain, 166
Diamond pipes, 14
Diatoms, 134, 138
Differentiation:
  metamorphic 251, **252**, **282**, **283**
  metasomatic, **284**
  (*see also* Fractionation)
Diffusion:
  crystal lattice, **275**
  and flow velocity, **208**, 209
  iron in olivine, 278
  of material in metamorphism, 272–85
    into low-pressure sites, 283
    mechanisms, 274, **275**
  potassium, regional, 279
  sodium in feldspar, 278
  of suspended particles, 195
  in vein formation, 287
Dike, 46, **47**
  cooling time, **112**, 113
  and veins, 108
Dilation veins, 287, **288**, **289**
Diopside:
  composition, 312
  crystallization and melting (*see* Systems)

Diorite:
definition, **54**, 318
in metamorphism, 245
Discontinuity, Mohorovicic, 4, 10
Discordant bodies:
definition, 45
dimensions, **47**
Dispersal of components in sedimentation, **150**, 151
Doll, C. G., 270
Dolomite:
formation, 159, 191
formula, 159
metamorphic reactions, 242
Domes, gravity-generated, 116, **117**, 301 (*see also* Load structure)
Drillholes:
deep, 3
for oil, 163
water pressure, 212, 291
Duluth lopolith, **46**, 47, 94
Dunbar, C. O., 350
Dune sand, 154
Dunite:
definition, 64
genesis (*see* Peridotite)

Earth, primitive state, 14
structure, **4**
Earthquakes:
and crustal structure, 3
energy, 15
Eclogite:
compared with basalt, 92
definition, 92, 229
in mantle, 101
Eh, 182–86, 261
Eh-pH diagram, **183**, **184**, **185**
Electrode, hydrogen, 182
Electrons, concentration in water, 182
Energy, 19
of disorder, 21, **22**
earthquakes, 15
free (*see* Free energy)
gravitational, 14, 116
influence on partitions, 201
internal, 21, **22**
potential, 14, 116
and preferred orientation, 281
and pressure, 19
and sorting of sediment, 200
of submarine flow, 208
surface, 281, 284
transfer through crust, 13

Energy (*cont.*):
units, 307
of volcanic eruptions, 34, 121
Enrichment of ore, 296
Enstatite:
formula, 312
incongruent melting, 330
metamorphic reactions, 243, 341
Entropy, 21
Environments and processes, sedimentary, **133**
Epidiorite, 230
Epidote:
composition, 313
equilibrium with plagioclase, 263
reactions, 242
veins, 294
Epidotite, definition, 230
Equilibrium constant, 174
Equilibrium curve (*see* Stability field)
Erg, 307
Esker, 155
Eskola, P., 268, 339
Essexite, definition, **54**, 318
Etna, Mount, 15, **32**, 37
Euhedral crystal, **39**, 41
Eutectic point, **72**, 75, 95
ternary, **96**, **337**
(*see also* Cotectic, Peritectic)
Evaporite deposits, 142
Excrement:
as sediment, 139
and glauconite, 160
Excretion, influence on sediments, 131
Extrusive rock, definition, 45 (*see also* Volcanic rocks)

F (*see* Free energy)
Facies:
metamorphic, 267, **268**
sedimentary, 269
Fairbridge, R. W., 350
Fanglomerate, 170
distinction from cataclasite, 232
Fayalite, formula, 312
Feldspar:
in magmatic rocks, **58**, 59–64 (*see also* Anorthosite, Basalt *etc.*)
metamorphic reactions, 242
powder in water, 187
solubility in water, 107
varieties, 313
weathering, 180
(*see also* Plagioclase *and* Potash feldspar)

Feldspathoidite, definition, 318
Feldspathoid minerals:
  occurrence (*see* Alkali rocks)
  varieties, 314, 316
Felty texture, **39**, 41
Ferric ions in solution, 186
Ferrous ions in solution, **183**, 184, **185**
Fleischer, M., 350
Flocculation, 153
Flood basalt, 35, **36**
Flood deposits, 147
Flow, laminar and turbulent, 194
Flow rocks:
  distinction from metamorphosed sediment, 251
  recognition, 42
  (*see also* Lava)
Flow texture:
  extrusive rock, **39**, 41
  intrusive rock, 48
Flow velocity, dissipation, **208**, 209
Fluid, interstitial between crystals, **275**
Fluid inclusions in crystals, 104
Fluidity (*see* Viscosity)
Fluorine:
  in granite aureole, 273
  in veins, 295
Foliation in schist, **240**
Foraminifera:
  and calcium in earth's crust, 16
  in deep sea ooze, 137
Foreset bed, 152, **153**
Forests, 134
Forsterite:
  formula, 312
  metamorphic reactions, 341
  and quartz, binary system, 330
  (*see also* Olivine)
Fossils, diagenesis, 158
Foyaite, definition, 318
Fractionation, 12
  and crystallization, **115**, 116
  during crystallization:
    basalt, 91
    plagioclase, 85
  of earth by gravity, 14
  of earth's crust, 12, 297, **300**, 301
  iron and magnesium, 254, 261
  and melting, **115**, 116, 301
    of sediments, 94–98
  by sedimentary processes, 97
    during transport, 149
Framework minerals, 313
Freedom, degrees of, 239
Free energy:
  change with pressure, **20**
  change with temperature, 189, **190**
  change with temperature and pressure, 256, **257**
  definition, 21
  and diffusion, 276
  of solute, 172
Freezing:
  brine, 67
  water in rock pores, 186
Frequency analysis, grain-size, 166
Frequency distribution, skewed, **202**
Friedman, G. M., 154, 203
Fugitive components (*see* Volatile constituents *and* Water)
Fujikawa, Y., 278
Fusion (*see* Melting)
Fyfe, W. S., 81, 107

Gabbro:
  definition, **54**, 318
  hornblende, 63, 92 (*see also* Amphibolite)
Ganges R., 5
Garnet:
  from chlorite and quartz, 259, **260**, 261, **262**, **264**, 299
  formula, 314
  from pyroxene and plagioclase, 235, 256, 272, **273**
  reactions, 242
  residue from crustal melting, 299
  stability and manganese content, 261, **262**
Garrels, R. M., 176, 183, 184, 186
Gas:
  dissolved in magma, 33, 121 (*see also* Volatile constituents *and* Water)
  in metamorphism (*see* Vapor phase)
Genetic and observational classes, 44
Geosyncline, 298, **300**
  pressures and temperatures, **225**
Geothermal gradient, 8, **9**
  in orogeny, 298, **299**
  Precambrian, 16
Geothermometers, 218, 304
Gibbs, J. Willard, 239
Gibbsite, occurrence, 141, 160
Glacial deposits, 155
Glass:
  by chilling magma, 33
  by metamorphism, 249

Glauconite:
composition, **160**, 313
formation, 159
Glaucophane, 312
Gneiss:
composition contrasts, origin, **283**
definition, **228**
with disturbed bedding, **283**
granitic, origin, 274
laminated, **283**
lenticular, 283
mineralogy, 247
ortho-and para-, 230
textures, **240**
Gneissosity, mechanism, 282
Goethite, in colloidal suspension, 189
Goldberg, E. D., 137, 216
Goranson, R. W., 81
Grade, metamorphic, recognition, 265
Grading of sediment, 147, 203, 205
Grain diameter, 166
Grain growth in metamorphism, 273, **280**, 281
Grain size:
frequency analysis, 166
frequency distribution, skewed by sorting, **202**
ranges, 165
Granite, 6
definition, **54**, 231, 316
in geosyncline, **300**
hypersthene, 63
melting behavior, **96**, 115
melting conditions, 9, 114
by metamorphism, 230, 251, 274, 283
source of magma, 9, 94, 297
Granite eutectic point, **96**, **337**
Granite gneiss, origin, 274
Granodiorite:
definition, **54**, 316
mineralogy, 60
occurrence, 65
Granofels:
definition, **228**
in dynamothermal metamorphism, 247
in thermal metamorphism, 249
Granogabbro, definition, 317
Granophyre, **49**, 284
Granularity by thermal metamorphism, **250**
Granulation by dynamic metamorphism, **233**
Graphic texture, **49**
Graphite, oxidation as heat source, 249
Graton, L. C., 121
Gravel:
definition, 165
desert valley, 144
glacial outwash, 155
river, 148
Gravity:
as cause of doming, 116, 301
influence on sedimentation, 144
and rock density, 5
Gravity anomaly:
and doming, 118
and isostasy, 5
Greenschist, occurrence, 246
Greensmith, J. T., 350
Greenstone, occurrence, 246
Greywacke, definition, 168
Grim, R. E., 186, 188
Groundwater:
acid, 177, 179
acid and alkaline, **185**
(*see also* Water)
Growth of grains (*see* Grain growth)
Guano, 139
Gypsum in evaporites, 143

Halite, in evaporite, 143 (*see also* Brine)
Hatch, F. H., 350
Hawaii, 7, 10, 32, 35, **36**, 330
Heat:
and density rearrangement, 13, 16
latent, of crystallization, 113
radioactive, quantities, 16, 315
Heat capacity of rocks, 315
Heat change:
during crystallization, 75, 81, 113
on reaction, 21
Heat transfer:
through crust:
by conduction, 8, 15
by magma movement, 15, 16, 115
at igneous contact, 109–14
Heavy mineral analysis, 218
Hedenbergite:
composition, 312
solid solution with diopside, 90
Hegau, West Germany, 100
Hematite:
in colloidal suspension, **188**, 189
formula, 314
occurrence in sediments, 142, **161**
stability in sediments, 183–85

Herefoss granite, Norway, **47**
Hesselgesser, J. M., 107
Himalaya Mountains, **4**
  gravity effect, 5
Holmes, A., 349
Holser, W. T., 106
Homogeneity, 228, 251
Hornblende:
  composition, 312
  metamorphic reactions, 242
  from plagioclase and pyroxene, 63, 272, **273**
Hornblende gabbro, 63, 92
Hornblendite, definition, 230
Hornfels:
  definition, 231
  genesis, 249
  spotted, 250
Hubbert, M. K., 125
Humic acid, **188**, 189
Hutton, James, 12
Hybridization of magma, 100
Hydrogen atom, mass, 307
Hydrogen electrode, 182
Hydrogen ion concentration (*see* pH)
Hydrous minerals:
  in diagenesis, 212
  in magmatic rocks, 59, 63
    norm calculation, 322
  in metamorphism, 245, 248, 258
  in orogeny, 299
Hypersthene:
  biotite reaction, 63, 295
  formula, 312
  hornblende reaction, 272, **273**
  in magmatic rocks, 63
  metamorphic reactions, 242
Hypersthene granite, 63
Hypersthene pegmatite, 295
Hypothesis-testing, 119

Iceland basalt flows, 35, **36**
Ice transport of sediment, 155
Idaho batholith, **47**
Igneous rocks, definition, 13, 24 (*see also* Magmatic rocks)
Ignimbrite:
  mechanism, 38
  texture, **39**, 41
Illite:
  composition, **160**, 313
  diagenesis, 159, **160**
  formation from feldspar, 180
Illite (*cont.*):
  indicator of paleosalinity, 218
  metamorphic reactions, 242
Ilmenite, formula, 314
Imbrication, **148**, 149, 203
Immiscibility, **80**, 81
Inclusions:
  fluid, in crystals, 104
  projecting, evidence of reaction, **265**
Incongruent melting, 327, **328**, **329**
Intensity of metamorphism:
  levels, **241**, 242, 243, **244**, 267
  recognition, 265
Interface between crystals, disordered, **275**
Intergrowth, vermicular, **265**
Intermediate rock, definition, 53
Interstitial fluid in recrystallization, **275**, **280**, 282
Intrusive rock:
  definition, 45
  recognition, 50
Inversions, density, 14, 16, 116, 301
Ionic radius, 311
Iron:
  concentrate at limestone contact, **284**, 285
  exchangeable with magnesium, 90, 236
  not exchangeable with magnesium, 254, 261
  in solution, 183–86
  transport processes, 142
    by colloids, **188**
Iron deposits, 142
Iron minerals, stability in sediments, 182–85 (*see also* Hematite *etc.*)
Isograd, 267, 269, **270**
Isostasy, 5, **6**
Isotopes:
  common, list, 311
  in dating, 303
  in geothermometry, 218, 304
  oxygen, in dolomite, 159

Jacobs, J. A., 350
Jadeite:
  composition, 312, 314
  occurrence, 298
Jadeitic pyroxene in eclogite, 92
Johannsen, A., 319, 325, 326
Jones, G. E., 140
Joule, 307

Kamchatka, U.S.S.R., 37
Kame, 155

Kaolinite:
diagenesis, 159, **160**
formation from feldspar, 180
formula, 313
at igneous margins, 254
Kaplan, I. R., 185
Karabugaz, Gulf, U.S.S.R., 143
Karroo region, S. Africa, 94
Katmai, Mount, 37
Kennedy, G. C., 81, 106
Kimberley, S. Africa, 14
Knight, L., 212
Koide, M., 137, 216
Krumbein, W. C., 350
Kuh-i-Namak, Persia, 118
Kyanite:
detrital, 266
formula, 314
reactions, 242

Labrador, Canada:
anorthosite, 65
Precambrian, 6
Labradorite, definition, 313
Laccolith, **46**, 48
Lag deposit (*see* Residue)
Lake Superior iron ore, 142
Laminar flow, 194
Landslide, 146
Lane, E. W., 165
Langesundfjord, Norway, 295
Lapilli, **39**, 40
Larnite:
formula, 312
occurrence, 100, 249
Larsen, E. S., 112, 113, 278
Lassen Peak, California, 37
Laterite, 161
Latite, definition, **54**, 316
Lava, 32–43
blocky, **39**, 40
pillow, **39**, 40
ropy, **39**, 40
(*see also* Flow rocks)
Lawsonite, occurrence, 298
Layered intrusions, 65, 66
Layering:
caused by metamorphism, 251, **252**, 282, **283**
of continents, 6, 301
Leaching, 287
Leucite:
crystallization and resorption, 327, **329**
Leucite (*cont.*):
formula, 314
occurrence, 61, 100 (*see also* Alkali rocks)
Leucite monzonite and syenite, definitions, **54**
Levin, E. M., 350
Lime garnet, occurrence, 100, 243
Lime mud, formation, 142
Limestone:
assimilation in magma, 100
and carbonatite, 98
cementation, 179
definition, 25
iron ore at igneous contact, 285
in metamorphism, 249
solution, 176
Lineation in schist, **240**
Liquidus line, 75, 83
Liquidus surface, **88**, **96**, **336**, **337**
Lithification, 158
localized, **159**
quartz sand, **214**
(*see also* Cementation, Compaction *and* Diagenesis)
Lithium, in veins, 295
Load structure in sediments, 214
Loess, 155
Lopolith, **46**, 48, 94
layered, 65–66
Lutite, definition, 165

McIntire, W. L., 246
McMurdie, H. F., 350
Mafite, definition, 316
Magdalena R., Colombia, 206
Magma:
basaltic (*see* Basaltic magma)
carbonate content, 99
compositions, 34
definition, 8
density, 315
in geosyncline, **300**
properties, 32
sources, 8, 14, 297
Magmatic rocks:
classification, 52, 53, **54**, 316, **317**
composition, 56–64, **58**, 314
composition range in tetrahedral diagram, **150**, **234**
composition trends, **62**
relative abundances, 56, **57**
weathering, 149
Magnesite, 238

Magnetite:
  formula, 314
  at limestone contacts, 285
  in metamorphism, 261
  in sediments, 183–86
Malignite, definition, 318
Manganese in garnet, 261, **262**
Mantle:
  composition, 101
  definition, 5
  escape of low density constituents, 14
Marble, definition, 230
Marl:
  definition, 167
  in metamorphism, 246
Mauna Loa, Hawaii, **32, 36**
Megacryst, definition, 41
Meionite, 172
Melilite, occurrence, 249
Melting:
  cause of density inversion, 16, **17**
  of crustal rocks, 8, **9**, 12, 94–96, 114–16, 297–301
  in geosynclines, 300
  incongruent, 327, **328, 329**
  influence of water, 9, 82, 115, 298
  of limestone, 98
  in metamorphism, **225**, **244**, 245, 249
  partial, 12, **115**, 116, 299–301
  of sediments to granite, 94–98
  of synthetic mixtures, (*see* Systems)
Menard, H. W., 32, 35, 148, 206
Merwinite, occurrence, 249
Metamorphic convergence, 246
Metamorphic differentiation 251, **252**, 282, **283**
Metamorphic rocks:
  definition, 13
  recognition, 250
  terminology, 227–31
Metamorphism, 12
  conditions and limits, 223, **225**
  contact, 285 (*see also* Metamorphism, thermal)
  dynamic, 232
  dynamothermal, 239–47
    reactions, 242
    textures, **240**, 246
  facies, 267, **268**
  general trends, 245
  influence on homogeneity, 251
  material movement, 272–85
  and melting, **244**, 245, 249, 297, **299**
Metamorphism (*cont.*):
  minerals generated, **234**, 236
    assemblages, **241**
  pressure and temperature conditions **225, 244**
  recognition of intensity, 265
  retrograde, 245, 266
  sequence of reactions, 240–49, 258
  thermal, 243, 247
    duration, 113, 248
    reactions, 243
    temperatures, 114
    textures, 249, **250**
  varieties, 224, **225**
Metasomatism:
  contact, 254, 285
  definition, 224
  extent, 272
  of greenstone, 246
  mechanisms, 274, **275**
  in orogeny, **300**, 301
  potassium, 274, 279
  selective, 284
Metastable minerals, 190
Methane, 14
Mexico, Gulf of, 153
Micro-, prefix in rock name, 54
Microcline, formula, 313 (*see also* Feldspar *and* Potash feldspar)
Micropegmatite, **49**
Migration (*see* Transport *and* Diffusion)
Mineral assemblages in metamorphism, **241**
  representation of stable pairs, 341, **342, 343**
  stability fields, **244**, 257, 267
Mineral composition diagrams:
  igneous, **150**
  metamorphic, **234**
  sedimentary, **150**
Mineral densities, 312
Mineral formulas, 312
Mineral molecular volumes, 321
Mineral proportions, magmatic rocks, **58**
Mineral varieties and formulas, 312
Mines, deep, 3
Mississippi R., 151
Modal analysis, 124
Mode, definition, 59
Models, 125, 305
  of dome structures, **117**, 118
Mohorovicic discontinuity, 4, 10
Mold, of organism, 158

Mole, definition, 173
Monticellite:
formula, 312
occurrence, 249
Montmorillonite:
composition, 313
diagenesis, 159, **160**
dimensions, **188**
formation from feldspar, 180
water absorption, 186, **188**
(*see also* Clay minerals)
Mont Pelée, 32, **36**
Monzodiorite, definition, 318
Monzogabbro, definition, 318
Monzogranite, definition, 316
Monzonite, definition, **54**, 317
Moore, D., 185
Moraine, 155
Morey, G. W., 107
Mound, detrital, 131
Mountains and isostasy, 5, **6**
Mount Etna, 15, **32**, 37
Mount Katmai, 37
Mud, definition, 165
Mudflow, 146
Mullite:
formula, 314
occurrence, 243, 249
Murray, G. E., 153
Muscovite:
formula, 313
iron-bearing, **160**
in magmatic rocks, **58**, 61
metamorphic reactions, 242
in sediments, 159, **160**
Mush, crystal-water, 102, 105
Mylonite, **233**
definition, 231
recognition, 233
Myrmekite, **265**, 284

Naughton, J. J., 278
Navajo Sandstone, 154
Nepheline:
formula, 314
occurrence, 61, 100 (*see also* Alkali rocks)
Nepheline monzonite and syenite, definitions, **54**
New Caledonia, ultrabasic rock, 66
New Red Sandstone, England, 154
New Zealand, ultrabasic rock, 66
Nickel, in ultrabasic rock, 64
Niger R., Nigeria, 152
Niggli, P., 319
Nitrate deposits, 143
Nockolds, S. R., 54
Norm, 59, 319
Normative analysis, 59, 319
Normative classification, **325**, 326
Nuée ardente, 38

Observational and genetic classes, 44
Oceanic crust, 7
density, 6
depth to M discontinuity, 4
Oceans, elevation, 6
Oil shale, 139
Oligoclase, definition, 313
Olivine:
crystallization and resorption, 330
formula, 312
in magmatic rocks, **58**, 59, 90–93, 330
metamorphic reactions, 242
(*see also* Peridotite)
Oolites, hematite, 142
Ooze:
calcareous, 137
rate of accumulation, 137, 156
siliceous, 138
Ophitic texture, **49**, 254
recrystallized, **265**
Order and disorder in sedimentary minerals, 191
Ore, stratigraphic control, **295**
Ore shoots, **295**, 296
Organic colloids, dimensions, **188**, 189
Organic molecules, dimensions, **188**
Organic precipitates, 141
Organic remains, 135
Organic structures, 134
Orientation, petrofabric analysis, 303
Orientation, preferred (*see* Preferred orientation)
Orinoco R., Venezuela, 152
Orogeny, 298
Orthoclase, formula, 313 (*see also* Feldspar *and* Potash feldspar)
Orthogneiss, 230
Orthopyroxene, varieties, 312 (*see also* Enstatite *and* Hypersthene)
Overgrowth:
hematite on fossils, 142
quartz on sand grains, 158
Oversaturated rock, definition, 53
Oxidation potential, 182–86, 259

Oxidation reactions in metamorphism, 260
Oxygen, volume in minerals, 321
Oxygen pressure and metamorphic reactions, 259

Pacific Ocean:
  diatom ooze, 138
  elevation, 6
  foraminiferal ooze, 137
  turbidite, 148
  volcanic deposits, 35
Pahoehoe (*see* Lava, ropy)
Paleosalinity, 218
Paragneiss, 230
Paragonite:
  albite reaction, 242
  formula, 313
Para-granite, 230
Parana basalt, Brazil, **36**
Partial pressure, 238, 260
Partition of component between coexisting phases, 201
  and geothermometry, 218, 304
Peacock, M.A., 61
Pebble, definition, 165
Pegmatite, definition, 49
Pegmatite veins, formation, 291
  anhydrous, 295
Pegmatitic stage in magmatic event, **292**, 293
Pelée, Mont, 32, **36**
Pelite, definition, 165, 230
Pellets:
  glauconite, 160
  limestone, 139
Percolation, 213
Periclase:
  reaction with water and CO_2, 237
  reaction with water and quartz, 341
Peridotite:
  genesis, 90, 101
  mineralogy, 64
  occurrence, 66
Peritectic behavior, 93, 327, **328**, **329**
Perthite, definition, 313
Petrofabric analysis, 303
Pettijohn, F. J., 350
pH:
  definition, 175
  effect of adsorption, 187
  and Eh, 183–86
  Eh-pH diagram, **183**, **184**, **185**
Phacolith, **46**, 48
Phase, definition, 68
Phase boundary (*see* Phase diagram *and* Stability field)
Phase diagram, 75
  published compilations, 350
  (*see also* Systems)
Phase rule, 233–39
  mineralogical, 236
Phenocryst, **39**, 41
Philippine Trench, **4**
Phi scale, 165, 307
Phlogopite:
  formula, 313
  reactions, 242
Phonolite, definition, **54**, 318
Photosynthesis, 141
Phyllite:
  definition, **228**
  spotted, 250
Plagifoyaite, definition, 318
Plagioclase:
  crystallization and melting, 82
    in hydrous conditions, **83**, **292**, 293
    with pyroxene present, **88**, 89, 335, **336**
  equilibrium with epidote, 263
  twinning, 123
  zoning, **84**
  (*see also* Feldspar)
Plants and soil chemistry, 141, 177
Plateau basalt (*see* Basalt flood)
Poikilitic texture, 48, **49**
Poikiloblastic texture, 229
Porosity, 211, **212**
Porphyritic texture, **39**, 41
Porphyroblast, 229, **240**, **250**
  and material movement, 273
Porphyroblastic texture, 229
Potash feldspar:
  biotite reaction, 63, 295
  incongruent melting, 327, **328**, **329**
  (*see also* Feldspar *and* Systems)
Potassium:
  adsorption on clay, 187
  cycle in erosion and remelting, 97
  metasomatism, 274, 279
Potential, oxidation, 182–86, 259–61
Precambrian shields, density and mineralogy, 6
Precambrian time:
  carbon distribution, 99
  geothermal gradient, 16

Precipitates, sedimentry, 140 (*see also* Aragonite, Quartz *etc.*)
Preferred orientation:
mechanisms, 279
varieties, **240**, 246
Pressure:
and change of state:
albite-anorthite system, 85
basaltic magma, 92
quartz-albite system, 77
quartz-water system, 81
and energy, 19
and mineral stability, 20
in metamorphism, 256
in sediments, 190
partial, 238, 260
for quartz-coesite transition, 20
and rock overburden, 7, 105, 291
units, 307
Processes, sedimentary, relation to environment, **133**
Psammite, definition, 165, 230
Psephite, definition, 165
Pseudomorph, 158, **265**
Pumice, **39**, 40
Pyrite:
formula, 314
stability in sediments, **185**, 186
Pyroclastic rocks, relation to other sediments, 163 (*see also* Volcanic rocks, Agglomerate *and* Tuff)
Pyrophyllite:
formation from clay, 159, **160**
formula, 313
metamorphic reactions, 242
Pyroxene:
in magmatic rocks, **58**, 59, 64, 87–93, 331
metamorphic reactions, 100, 235, 242
varieties, 312
(*see also* Aegirine, Enstatite *etc.*)
Pyroxenite, definition, 64, 229 (*see also* Ultrabasic rock)
Pyrrhotite, formula, 314

Quartz:
and coesite, 20
in magmatic rocks, **58**, 59–63
(*see also* Systems *and* Basalt, Granite *etc.*)
precipitated in sediments, 192
solubility in water, **107**, 192
transport in solution, 107, 277, 291
veins, 107, **108**, 192, 287, 291
Quartz diorite, definintion, **54**, 317
Quartzite, definition, 229
Quartz latite, definition, **54**, 317
Quartz monzonite, definition, **54**, 317

Radiation within earth, 15
Radioactivity:
and age determination, 303
and crustal melting, 114
heat generation values, 315
as heat source, 8, 16
Radiolaria, 138, 192
Rainwater, CO_2 content, 176
Ramberg, H., 117, 118, 262, 268, 351
Rankinite, occurrence, 249
Rastall, R. H., 350
Rate of accumulation, sediments, 137, 156
Rate of reaction and mineral abundance, 190, **191**
Reaction boundary (*see* Stability field)
Reaction rim, 265 (*see also* Corona)
Reactions:
anorthite, enstatite/garnet, quartz, 235, 256
basalt-amphibolite, 63
basalt-eclogite, 92
biotite-hypersthene, 63, 295
calcite-wollastonite, 100, 243, 259, 277
chlorite-garnet, 259, **260**, 261, **262**, **264**, 299
hypersthene, plagioclase/hornblende, garnet, 272, **273**
and material movement, 272
in metamorphism, 242, 243
Reaction series, 331
Reaction textures, **265**
Recognition:
intrusive magmatic rocks, 50
metamorphic intensity, 265
metamorphic rocks, 250
sedimentary rocks, 162, 251
volcanic rocks, 42–45, 163, 253
Recrystallization:
in diagenesis, 158, 213, **214**
under nonuniform stress, **280**, 281
(*see also* Texture)
Reef, 131, 134, **136**
Reef detritus, 135, **136**
Replacement:
of fossils, 158
limestone by hematite, 142
Replacement veins, 287, **289**

Residue:
from partial melting, 12, **115**, 298
sedimentary, 143
graded, 203
river gravel, **148**, 149
skewed size distribution, **202**
by surf action, 154
Resorption:
leucite, 329
olivine, 330
Rhyodacite, definition, **54**, 316
Rhyolite, definition, **54**, 316
Richardson, W. A., 57
Riebeckite:
composition, 312
occurrence, 61
Ring dike, **47**
Ripple marks on ocean floor, 164
Rittmann, A., 36, 298, 350
River deposits, 147, 148
Robbins, C. R., 350
Roberts, W. M. B., 284, 305
Rodgers, J., 350
Roots under mountains, 5
Ross, C. S., 36
Rounding, 209, **211**
by surf, 154
Roundness, 210, **211**
Rubble, definition, 165
Rudite, definition, 165
Runcorn, S. K., 15
Russell, R. D., 350

S (*see* Entropy)
Sahara Desert, N. Africa, 154
Salic minerals, varieties, 325
Salt (*see* Evaporite deposits)
Salt domes, 117
Sand:
beach, 154, 203
definition, 165
heavy mineral, 144, 165
windlain, 154, 203
Sandstone, definition, 25, 165
Sanidine:
formula, 313
in metamorphism, 243, 249
Saturated rock:
definition, 53
mineralogy, 59
Scapolite:
composition, 313
reaction to albite, 172
Scarp erosion, 145
Schairer, J. F., 87, 88, 328, 336, 338
Schist:
definition, **228**
formation, 244
mineralogy, 247
varieties, **240**
Schistosity, mechanism, 279, **280**
Science, methods of, 119
Scoria, definition, 40
Scouring, 201 (*see also* Winnowing)
Scruton, P. C., 151
Secondary enrichment of ore, 296
Sedimentary rocks:
classification, 25, 164
composition range in tetrahedral diagram, **150**, **234**
definition, 13, 25
diagenesis, 157, 211
forms and extents, 134–56
melting, 12, 94–98, 297–301
processes of formation, 130
recognition, 162, 251
textures, 134–56, 165
varieties, 131
Seismic waves, 3
and rock density, 6
Serpentine:
formula, 312
metamorphic reactions, 344
occurrence (*see* Peridotite)
Shale:
composition range, **150**
definition, 165
melting, 299
in metamorphism, 244
water content, 212
Shape of detrital grains, 203, 209, **211**
Shearing:
effect on rock homogeneity, 252
and ore veins, 296
Shear stresses in the crust, 7, 226
Sheet silicates, varieties, 313
Shelf, continental, sediments, 144
Shepard, F. P., 203
Shield volcano, 35, **36**
Shifta formation, Israel, 139
Shonkinite, definition, 318
Shumway, G., 212
Sicily, 15
Sierra Nevada batholith, **47**
Silica, 33
diffusion in metamorphism, 277

Silica *(cont.)*:
  noncrystalline, 192
  in organisms, 138
  precipitated in sediments, 192
  solubility, **107**, 192
  in solution, 181
  (*see also* Chert *and* Quartz)
Sill, **46**
Sillimanite:
  formula, 314
  in magma, 254
  reactions, 242, 243
Silt, definition, 165
Sippel, R. F., 278
Skarn:
  definition, 230
  iron-rich, 285
  (*see also* Calc-silicate minerals)
Skjaldbreit, Iceland, **36**
Slate:
  distinction from phyllite, 228
  formation, 212, 244
Slickensiding, 233
Slope, continental (*see* Continental slope)
Sloss, L.L., 350
Slump, 147
  submarine (*see* Turbidite *and* Turbidity current)
Sneesby, G., 57
Sodium:
  cycle in erosion and remelting, 97
  in diagenesis of clay, 97, 187
Solid solution:
  albite-potash feldspar, 95
  in metamorphism, 245, 261, **262**
  and mineral stability, 261, **262** (*see also* Solute free energy)
  plagioclase, 82
  pyroxenes, 90
Solidus, 83
Solubility in water:
  aragonite, 180
  calcite, 178
  feldspar, **107**
  quartz, **107**, 192
  silica, **107**, 192
Solute:
  free energy, 172
  stability and concentration, 172, 185, 264
  transport around intrusions, 107, 274–79
  in vapor phase (*see* Veins)
Solution (*see* Solid solution, Solubility *and* Solute)
Sorting of sediment:
  influence of shape, 203, 210
  by nonuniform flow, 200, **202**
    in rivers, **148**
  by selective trapping, 140
  by selective uptake (*see* Winnowing)
  (*see also* Grading)
Sphene, formula, 314
Sphericity, 210, **211**
Spherule, 41, 43
Spherulite, **39**, 41
Spicules:
  sponge, 138
  volcanic, **33**, 34
Sponges:
  in reefs, 134
  detritus, 138
Spotted hornfels and phyllite, 250
Spurrite, occurrence, 249
Stability field of mineral:
  in metamorphism, 256–64
    effect of solid solutions, 261–64, **262**
    effect of volatiles' escape, 258, **260**, 267
    variation with bulk chemistry, 266
  in sedimentary environment, 183–86
  (*see also* Phase diagram *and* Free energy)
Stability field of mineral assemblage, **244**, 257, 267
Staurolite:
  formula, 314
  reactions, 242
Steam:
  density, 105, **106**
  distinction from water, 79
  occurrence, 105, 291
  and vein formation, 290
  (*see also* Gas, Vapor Phase, Volatile constituents *and* Water)
Stillwater, Montana, 64, 66
Stock, **47**, 48
  veins emanating, 107, **108**
Stoichiometry, 78, 87
Streckeisen, A., 54, 316, 317
Stress, shear, in crust, 7, 226
Sudbury, Ontario, 64
Suite of rocks, 61
Sulfide:
  oxidation as heat source, 249
  production by organisms, 140
  stability in sedimentary environment, **184**, 186
Sulfide veins, 296
Sulfur deposits, 139

Sulfur-sulfide reactions, 186
Supercritical fluid, 79
Surf, 153
Surface energy, 281, 284
Surface phenomena, 172, 186, 213
Suspension, 194–200
  and graded deposits, 204
Swamps, 139
Sweden, Precambrian, 6
Syenite:
  definition, **54**, 317
  mineralogy, 61
  occurrence, 65
Syenogranite, definition, 316
System:
  binary, 67, 237
  condensed, 68
  definition, 68
  quasi-binary, 78, 87
  ternary, 335
  with vapor phase, 68, 237
Systems, individual:
  albite-anorthite, 82
  albite-anorthite-diopside, **88**, 89, 335, **336**
  albite-anorthite-water, **292**, 293
  albite-potash feldspar, 94
  anorthite-quartz, 77
  anorthite-quartz-nepheline, 338
  diopside-albite, 87
  diopside-anorthite, 89
  diopside-anorthite-albite, **88**, 89, 335, **336**
  diopside-quartz, 77
  forsterite-quartz, 330
  leucite-quartz, 327, **328**, **329**
  periclase-water, 237
  periclase-water-CO_2, 238
  periclase-water-quartz, 341, **342**
  quartz-albite, 71
  quartz-potash feldspar, 94
  quartz-potash feldspar-albite, 96, 335, **337**
  quartz-water, 79, 107
  water-NaCl, 67, 69

Talc:
  formula, 313
  metamorphic grade index, 266
  reactions, 242, 343
Talus, 145
Talus cone, **136**
Tanzania, 99
Temperature:
  crustal, 8, **9**, 16, 298
Temperature (*cont.*):
  geothermometry, 218, 304
  at igneous contact, 109–14
  lava, 32
  and mineral stability, 189, **190**, 256
Tephrite, definition, **54**, 318
Termier, G., 350
Termier, H., 350
Ternary diagrams, 332
  in metamorphism, 338
Ternary system, 335
Tetrahedral coordinate system, 150 (*see also* Mineral composition diagrams)
Texture:
  change of (*see* Recrystallization)
  due to metamorphism, 228
    dynamic, 232, **233**
    dynamothermal, **240**, 246
    thermal, 249, **250**
  due to reaction, **265** (*see also* Corona)
  intrusive magmatic rocks, 48, **49**
  sedimentary rocks, 134–56, 165
  synthetic models, 305
  volcanic rocks, 38–42, **39**
Theralite, definition, **54**, 318
Thulean basalt, **36**
Tie lines, 341, **342**, **343**
Till, tillite, 156, 170
Tilley, C. E., 9, 63, 83, 94, 103, 331
Tilleyite, occurrence, 249
Time:
  for cooling of intrusion, **112**, 113
  for growth of granite dome, 118
  for waning of transient flow, 209
Titanaugite, composition, 312
Tonalite, definition, 54
Topset bed, 152, **153**
Torridonian Sandstone, Scotland, 154
Tourmaline, composition, 314
Trace elements:
  in geothermometry, 304
  as guides to genesis, 246, 303
Trachyte, definition, **54**, 317
Trachytic texture, **39**, 41
Transport:
  in metamorphism, 272–79, **273**, **275** (*see also* Diffusion)
  of sediment:
    by ice, 155
    by water, 144–51, 194–211
    by wind, 154, 194–206
  in vein formation (*see* Veins)
Trapping of sediment by organisms, 131, 140, 204

Travis, R. B., 54
Tremolite, composition, 312
Trianglular coordinates, 332, **333**
Triangular diagrams:
  ACF, 339, **340**
  AKF, 338, **340**
Tridymite in metamorphism, 243, 249
Tschermakite, composition, 312
Tuff:
  extent and composition, 35, **36**
  by hot ash flow, 38
  recognition, 43, 253
  welded, **39**, 41
Turbidite:
  composition grading, 147, 151, 205
  definition, 132
  extent of deposit, 148, **207**
Turbidity current:
  behavior, 205, 206–9, **207**
  definition, 132
  deposits (*see* Turbidite)
Turbulence:
  distinction from laminar flow, 194
  effect on aqueous supension, 200
  effect on heat flow, 113
  (*see also* Suspension)
Turner, F. J., 268, 303, 339, 350
Tuttle, O. F., 9, 83, 95, 96, 292, 337, 341, 342
Twenhofel, W. H., 131, 143

U (*see* Energy, internal)
Ultrabasic rock:
  definition, 53
  mineralogy, 64
  occurrence, 66
Undersaturated rock, definition, 53
Units, listed, 307
Uptake of sediment, 200 (*see also* Winnowing)
Uranium in age determination, 303

Valley of Ten Thousand Smokes, Alaska, **36**
Vanoni, V. A., 206
Vapor phase:
  in binary system, 68
  and phase rule, 236, 237–39
  and stability fields, 258–61, 267
  (*see also* Gas, Steam, Volatile constituents *and* Water)
Vapor pressure, 238
Varved sediment, 156
Veins, 107–8, 286–96
  classification, 286, **288**
  mechanisms, 108, 286–90
  mineralogy, 294–96
  pegmatites, 291
  quartz, 192, 287, 291
    emanating from stock, 107, **108**
Velocity of seismic waves, 4
Velocity profile, submarine currents, **208**
Vent, volcanic, processes and dimensions, 120
Verhoogen, J., 268, 339, 350
Vermicular intergrowth, **265**
Vermiculite:
  composition, 313
  formation and diagenesis, 159, 181
Vermont isograds, **270**
Vesicle, **39**, 40
  formation, 81
Vesuvius, Italy, **32**, 37, 65, 100
Viscosity:
  magma, 32, 123
  rock, 118
  vein fluids, 294
Volatile constituents, 104–8
  composition, 34
  heat of unmixing, 113
  in metamorphism (*see* Vapor phase)
  volume, 121
  (*see also* Gas *and* Water)
Volcanic rocks:
  extents and compositions, 35, **36**
  metamorphism to granite, 274
  recognition, 42–45, 163, 253
  textures, 38–42, **39**
Volcanoes, 31–38
  cones, **36**, 37
  and crustal heat transfer, 15
  dimensions, **32, 36**
  shield, 35, **36**
  vent processes and dimensions, 120
Volume of mineral molecules, 321
Vredefort dome, Africa, 118

Walton, M., 278
Water:
  abundance in crust, 82, 102, 105, 213
  adsorbed on surfaces, 172, 187, **188**, 213
  boiling under pressure, 79, **106**, 290
  critical behavior, 79
  dehydration of sediments, 212, **213**, 298
  density, 105, **106**, 290
  Eh, pH values in nature, **185**

Water (*cont.*):
from hydrous minerals, 212, 298
influence on crystallization of magma, 63, 92, 295
influence on melting, 9, 82, 115, 298
influence on metamorphism (*see* Vapor phase)
interstitial between crystals, **275**
ionization, 174, 182, **185**
liberated in crystallization, 81, 90, 104, 113, 115, 121
percolation pattern, **213**
solute content (*see* Solubility)
transport medium (*see* Transport)
in volcanic eruption, 121
(*see also* Groundwater, Rainwater *and* Systems)
Wave action:
agent for reef detritus, 135
effect on beach sand, 153
Weathering, 12
distinction from metamorphism, 223, **225**
of feldspar, 180
of igneous rock, 149
Weill, D. F., 81, 107
Weiss, L. E., 303
Welded rock, **39**, 41
recognition, 43
Weller, J. M., 212, 350
Williams, G., 203, 206
Wilson, J. T., 350
Wind transport of sediment, 154, 194–206
Winkler, H. G. F., 96, 268, 335, 339, 344, 351
Winnowing, 144, 149, 154, 201, **202**, 210
Wollastonite:
calcite reaction, 100, 243, 259, 277
evidence of movement of silica, 277
formula, 312
occurrence, 100
stability range, 259

Xenocryst, 41
Xenolith, **49**, 50, **123**

Yellowstone Park, Wyoming, 104
Yoder, H. S., Jr., 9, 63, 83, 87, 88, 94, 103, 331, 336
Young, R., 203
Yukon Territory, tuff, 37

Zircon:
formula, 314
as guide to genesis, 123, 303
Zoning:
intrusive rock bodies, 49
plagioclase, **84**
xenoliths, **49**, 50